S0-CAH-414

Chemically Induced Magnetic Polarization

NATO ADVANCED STUDY INSTITUTES SERIES

Proceedings of the Advanced Study Institute Programme, which aims at the dissemination of advanced knowledge and the formation of contacts among scientists from different countries

The series is published by an international board of publishers in conjunction with NATO Scientific Affairs Division

A	Life Sciences	Plenum Publishing Corporation
B	Physics	London and New York
C	Mathematical and Physical Sciences	D.Reidel Publishing Company Dordrecht and Boston
D	Behavioral and Social Sciences	Sijthoff International Publishing Company Leiden
E	Applied Sciences	Noordhoff International Publishing Leiden

Series C – Mathematical and Physical Sciences

Volume 34 – Chemically Induced Magnetic Polarization

Chemically Induced Magnetic Polarization

Proceedings of the NATO Advanced Study Institute held at Sogesta, Urbino, Italy, April 17-30, 1977

edited by

L. T. MUUS, *Aarhus University, Denmark*

P. W. ATKINS, *Oxford University, U.K.*

K. A. McLAUCHLAN, *Oxford University, U.K.*

J. B. PEDERSEN, *Odense University, Denmark*

D. Reidel Publishing Company

Dordrecht-Holland/Boston-U.S.A.

Published in cooperation with NATO Scientific Affairs Division

Library of Congress Cataloging in Publication Data

Nato Advanced Study Institute, Urbino, 1977.
Chemically induced magnetic polarization.

(NATO advanced study institutes series: Series C, Mathematical and physical sciences; v. 34)
'Published in cooperation with NATO Scientific Affairs Division.'
1. Radicals (Chemistry)–Congresses. 2. Polarization (Nuclear physics)–Congresses. 3. Nuclear magnetic resonance–Congresses. I. Muus, L. T. II. Title. III. Series.
QD471.N29 1977 541'.378 77-10592
ISBN 90-277-0845-2

Published by D. Reidel Publishing Company
P.O. Box 17, Dordrecht, Holland

Sold and distributed in the U.S.A., Canada, and Mexico
by D. Reidel Publishing Company, Inc.
Lincoln Building, 160 Old Derby Street, Hingham, Mass. 02043, U.S.A.

Printed in The Netherlands

CONTENTS

PREFACE

Magnetic resonance has constantly been able to surprise with its ability to exhibit new phenomena. Just when it appears to be entering a quiet middle age it bursts into activity with some new manifestation of its versatility. This happened a few years ago, when observations on anomalous intensities were looked at more closely, and the pursuit of explanations and further evidence laid the foundations of the subjects treated in this volume.

In organizing the NATO Advanced Study Institute we attempted to bring together a number, but by no means all, of those who had contributed significantly to the subject, and to obtain from them a comprehensive and detailed exposition of the subject. We were particularly anxious to avoid a set of lectures that dealt solely with the theory of the subject, because much of the interest in chemically induced magnetic polarization is due to its usefulness in applications to chemical problems: it is a real chemical technique, not just an amusing diversion for theoreticians. We set about organizing the course with the idea of making it useful to people who wanted to use the technique (for, after all, in the case of nuclear polarization, CIDNP, the technique can be used in any laboratory with minor modification of standard equipment). In order to do that we started the course, as we start this volume, with a simple, straightforward introduction to the interpretation of CIDNP spectra, the radical pair theory, and some applications to the elucidation of reaction mechanisms. Then we have an introduction to the more involved techniques needed to look at electron polarizations, and a similar survey of the range of its applications.

At that point, in order to use the technique with confidence and to obtain the maximum amount of information, it is necessary to go more deeply into the underlying theory. At the ASI, and in this volume, we attempted to do this without losing sight of the reasons for doing it, and the chapters in this volume represent the mixture of theory and experiment often encountered in laboratories. The chapters also represent the various theories that are encountered: at this level of exposition, and in a subject where the interplay of quantum mechanical, statistical, and reactive processes is so intricate, there are several ways of looking at the processes involved. People new to the subject initially may find one way of talking about the subject more congenial than a more sophisticated version: both approaches will be found here.

This volume would not have appeared without the sustained effort of a large number of people. First there are the lecturers: they all worked very hard to prepare an interesting and authoritative set of lectures well before the ASI, and gave unstintingly of their time and expertise during the duration of the Institute. The editors of this book (who were also the organizers of ASI) were greatly assisted by the authorities at "Sogesta", where the Institute was held, and drew heavily on both their excellent facilities and their cooperation, for both the Institute itself and a portion of the editing of this book. We all would like to thank Mrs. Ruth Buch, who yet again devoted so much effort and efficiency to the organization of the ASI for several months before it took place. Finally, but by no means least, we would like to record our gratitude to the NATO Scientific Affairs Division for their generous financial support.

Aarhus, Odense, and Oxford.
May 1977.

L.T. Muus
P.W. Atkins
K.A. McLauchlan
J.B. Pedersen

LIST OF PARTICIPANTS

Professor Frank J. ADRIAN: The Johns Hopkins University, Applied Physics Laboratory, Johns Hopkins Road, Laurel, Maryland 20810, U.S.A.

Dr John AMMETER: Institute of Inorganic Chemistry, ETH, CH-8001 Zürich, Switzerland

Dr P. W. ATKINS: Physical Chemistry Laboratory, Oxford University, South Parks Road, Oxford OX1 3QZ, U.K.

Mr Paul BARBARA: Department of Chemistry, Brown University, Providence, Rhode Island 02912, U.S.A.

Dr Joachim BARGON: IBM Corporation, Research Division, Monterey and Cottle Roads, San José, California 95193, U.S.A.

Dr Fuat BAYRAKÇEKEN: Orta Dogu Teknik Universitesi, Teorik Kimya Bölümü, Ankara, Turkey

Dr Hans-Dieter BECKHAUS: Chem. Laboratorium der Universität, Albertstr. 21, D-7800 Freiburg, B.R.D.

Dr Reinhard BENN: Max-Planck-Institut für Kohlenforschung, Kaiser-Wilhelm Platz 1, 433-Mülheim a.d. Ruhr, B.R.D.

J.W.M. de BOER: Dept. of Physical Chemistry, Faculty of Science, University of Nijmegen, Toernooiveld, Nijmegen, Netherlands

Dr Marie BORZO: Institut de Chimie, Université de Liège, Liège, Belgium

Dr Robert E. BOTTO: Caltech, Pasadena, California 91109, U.S.A.

Dr Roselyne BRIERE: C.E.N.G., Laboratoire COP/DRF-B.P.85-Centre de TRI, 38041 Grenoble Cedex, France

Dr Charles BROWN: University Chemical Laboratory, University of Kent, Canterbury, Kent CT2-7NH, U.K.

Dr Marina BRUSTOLON: Physical Chemistry Institute, Padua, Italy

Mrs Ruth BUCH: Department of Chemistry, Aarhus University, 140 Langelandsgade, 8000 Aarhus C, Denmark

Dr Serge CAPLAIN: Univ. des Sciences et Techniques de Lille, Lab. de Chimie Organique Physique, B.P.36, 59650 Villeneuve d'Ascq, France

Professor G.L.CLOSS: Department of Chemistry, The University of Chicago, Chicago, Ill.60637, U.S.A.

Dr Odile CONVERT: Faculté des Sciences Paris VI, Lab. de Chimie Organique Structurale, 8 Rue Cuvier, 75230 Paris Cedex 05, France

Mr Karl-Michael DANGEL: Institut für Organische Chemie, Universität Tübingen, Auf der Morgenstelle 18-74 Tübingen 1, B.R.D.

Professor Nadine FEBVAY-GAROT: Faculté de Pharmacie, Laboratoire de Physique, Rue du Prof.Laguesse, 59045 Lille Cedex, France

Professor R.W. FESSENDEN: Radiation Laboratory, University of Notre Dame, Notre Dame, Ind. 46556, U.S.A.

Professor H. FISCHER: Physikalisch-Chemisches Institut der Universität, Rämistrasse 76, CH 8001 Zürich, Switzerland

Dr J. FOSSEY: CNRS, GR.12, 2 rue H.Dunant, 94 Thiais, France

Professor Jack H. FREED: Department of Chemistry, Cornell University, Ithaca, New York 14853, U.S.A.

Mr SØREN FRYDKJAER: Department of Chemistry, 140 Langelandsgade, DK-8000 Aarhus C, Denmark

W. Carl GOTTSCHALL: Department of Chemistry, University of Denver, University Park, Denver, Colorado 80210, U.S.A.

Dr Maurizio GUERRA: Lab. dei composti del carbonio contenenti etero-atomie e loro applicazioni, CNR, Via Tolara di Sotto 81/A, 40064 Ozzano Emilia(Bologna), Italy

Mr Richard A. HEARMON: Dpt. of Organic Chemistry, The University of Liverpool, The Robert Robinson Lab., P.O.Box 147, Liverpool, L69 3BX, U.K.

Dr Arnold J. HOFF: Department of Biophysics, State University of Leiden, Huygens Laboratory, Wassenaarseweg 78, Leiden, Netherlands

Mr Peter John HORE: Physical Chemistry Laboratory, Oxford University, South Parks Road, Oxford OX1 3QZ, U.K.

Mr Christopher JOSLIN: Physical Chemistry Laboratory, Oxford University, South Parks Road, Oxford OX1 3QZ, U.K.

Dr F.J.J. de KANTER: Gorlaeus Lab. of the State University of Leiden, Wassenaarseweg, P.O. Box 75, Leiden, Netherlands

Dr R. KAPTEIN: Physical Chemistry Laboratory, University of Groningen, Zernikelaan, Paddepool, Groningen, Netherlands

Mr Timothy Peter LAMBERT: Physical Chemistry Laboratory, Oxford University, South Parks Road, Oxford OX1 3QZ, U.K.

Dr Heinz LANGHALS: Chem. Laboratorium der Universität, Albertstr. 21, D-7800 Freiburg, B.R.D.

Professor R.G. LAWLER: Department of Chemistry, Brown University, Providence, Rhode Island 02912, U.S.A.

Mr Gary LEHR: Department of Chemistry, Brown University, Providence, Rhode Island 02912, U.S.A.

Dr K.A. McLAUCHLAN and Mrs McLAUCHLAN: Physical Chemistry Laboratory, Oxford University, South Parks Road, Oxford OX1 3QZ, U.K.

Dr Jean MARKO: Faculté de Pharmacie, Laboratoire de Physique, Rue du Prof. Laguesse, 59045 Lille Cedex, France

Mr Gareth MORRIS: Physical Chemistry Laboratory, Oxford University, South Parks Road, Oxford OX1 3QZ, England

Dr Karol A. MUSZKAT: Weizmann Institute of Science, Dept. of Structural Chemistry, Rehovot, Israel

Professor L.T. MUUS: Department of Chemistry, 140 Langelandsgade, DK-8000 Aarhus C, Denmark

Mr Jean-Yves NEDELEC: CNRS, GR 12, 2 Rue H. Dunant, 94 Thiais, France

Dr Luigi PASIMENI: Physical Chemistry Institute, Padua, Italy

Dr J. Boiden PEDERSEN: Department of Physics, Odense University, Niels Bohrs Alle, DK-5000 Odense, Denmark

Dr K. Sundarraya RAO: Department of Organic Chemistry, The University of Liverpool, The Robert Robinson Lab., P.O. Box 147, Liverpool, L69 3BX, U.K.

Dr Gûnther RIST: CIBA-GEIGY Ltd., K-127.686, CH-4002 Basel, Switzerland

Dr Heinz D. ROTH: Bell Laboratories, 600 Mountain Avenue, Murray Hill, New Jersey 07974, U.S.A.

Dr Wolfgang SCHWARZ: Technisch-Chemisches Laboratorium, ETH-Zentrum, CH-8092 Zürich, Switzerland

Dr Klaus G. SCHULTEN: Max-Planck Institut für biophysik Chemie, Postfach 968, D-3400 Göttingen-Nikolausberg, B.R.D.

Dr Jon SONGSTAD: Kjemisk Institut, Universitetet i Bergen, 5014 Bergen, Norway

Mr Larry L. STERNA: Department of Chemistry, University of California, Berkeley, California 94720, U.S.A.

Dr G. VERMEERSCH: Faculté de Pharmacie, Laboratoire de Physique, Rue du Prof. Laguesse, 59045 Lille-Cedex, France

Mr Rudolf WEBER: ETH-Zentrum, Technisch-Chemisches Laboratorium, CH-8092 Zürich, Switzerland

Mr H.-J. WERNER: MPI für biophysikal Chemie, Abt. Spektroskopie, Postfach 968, D-3400 Göttingen-Nikolausberg, B.R.D.

Dr S. King WONG: Department of Chemistry, The Univ. of Western Ontario, Chemistry Building, London, Ontario N6A 5B7, Canada

Mr Jon WITTMANN: Physical Chemistry Laboratory, Oxford University, South Parks Road, Oxford OX1 3QZ, U.K.

Mr Gary P. ZIENTARA: Box 427, Dept. of Chemistry, Cornell University, Ithaca, New York 14853, U.S.A.

CHAPTER I

INTRODUCTION TO CHEMICALLY INDUCED MAGNETIC POLARIZATION

R.Kaptein

Physical Chemistry Laboratory, University of Groningen,
Groningen, The Netherlands.

1. INTRODUCTION

Both esr and nmr lines with anomalous intensities have been observed in chemically reacting systems. The effects have been called chemically induced dynamic electron and nuclear polarization (CIDEP and CIDNP respectively). They arise from species (radicals or reaction products) that are formed with non-equilibrium spin state populations. CIDEP has first been found by Fessenden and Schuler in 1963 (1), but did not receive much attention at the time. Several years later in 1967 the CIDNP effect was discovered independently by Bargon and Fischer (2) and by Ward and Lawler (3). It took, however, until 1969 before a beginning was made in the understanding of both phenomena. The Radical Pair Mechanism (4,5) in one form or another, later complemented by the Triplet mechanism (6), seems to be capable of explaining virtually all CIMP effects and some related phenomena such as the magnetic field dependence of product yields (7) and of triplet yields in electron-transfer reactions (8) and the magnetic isotope effect (9).

In the Radical Pair mechanism (RPM) the polarization arises from local magnetic fields due to Zeeman and hyperfine interactions acting on the electron spins of the radicals that constitute a pair. In spite of being small local field differences cause a mixing of singlet and triplet electronic states of the pair. This results in spin-density preferences at the radical sites (CIDEP) and combined with spin-selective chemical reactions to nuclear spin state populations of the reaction products that may strongly deviate from Boltzmann equilibrium (CIDNP).

In the Triplet mechanism (TM) the polarization originates from the triplet sublevels of the photo-excited precursor and

L. T. Muus et al. (eds.), Chemically Induced Magnetic Polarization, 1-16. *All Rights Reserved.*

is transferred to the free radicals (CIDEP). In principle electron-nuclear cross-relaxation in the triplets or in the free radicals may further induce polarization in the nuclear spin system, but conditions are hard to satisfy. In the following sections these concepts will be further elaborated with some emphasis on the RPM.

2. OUTLINE OF THE RADICAL PAIR MECHANISM OF CIDNP

Let us consider the formation of a radical pair from a precursor B in a magnetic field and its subsequent reactions according to Scheme 1.

$$B \longrightarrow \overline{R_a\cdot + R_b\cdot} \longrightarrow R_a - R_b$$

precursor radical pair recombination

$$\overline{R_a\cdot + R_b\cdot} \searrow R_a\cdot + R_b\cdot \longrightarrow \text{escape products}$$

free radicals

Scheme 1.

In thermal reactions the precursor is usually a diamagnetic (singlet state) molecule. In photochemical reactions both singlet (S) and triplet (T) state precursors may occur. Alternatively, pair formation may occur by random encounters of free radicals (F pairs). When the pair is formed from a S precursor, it will initially also be in the singlet state (antiparallel spins). The radicals of the pair will separate by diffusion and the mutual exchange interaction J decreases. J will vanish eventually and S and T_o states (triplet state with M_S=0) are then degenerate. The isotropic magnetic interactions (hyperfine and Zeeman interactions) can now cause a coherent phase alteration of the electron spin states, which is equivalent to S-T_o mixing. This correlation of spin states is indicated by the bar in scheme 1. Random fluctuating magnetic fields would cause a loss of spin correlation but fortunately spin-spin relaxation times (T_2) of free fadicals are usually long (10^{-6} - 10^{-5} sec) compared to the time scale of S-T mixing and geminate recombination.

In the region where J is still appreciable S-T mixing leads to differences in spin density at the radical sites a and b and hence the radicals come out carrying CIDEP. This will be further discussed in section 5. S-T mixing will also affect the reactivity of the pair, because recombination is possible only from a S state. This latter condition, which is of fundamental importance for CIDNP is generally obeyed. The only exceptions are some electron-transfer reactions where radical pair collapse is favoured in the triplet state.

Nuclear polarization now arises because the S-T mixing is nuclear spin state dependent. Thus, certain nuclear levels will be preferentially populated in the recombination product and hence will necessarily be depleted in the escaping free radicals and the products thereof. This process which results in opposite polarizations in recombination and escape products is often denoted as "spin-selection" and is valid for CIDNP in high magnetic fields.

2.1. Vector model

It is instructive to consider a simple vector representation of the electron spins of a radical pair in a magnetic field (H_o). In Figure 1 the projections of the electron spin vectors are shown in the xy-plane, perpendicular to the direction of H_o. For the S state the spins are antiparallel and for the T_o state they are parallel. The spin vectors will precess about the direction of H_o and hence their projections rotate in the xy-plane. The crucial point is now that the vectors S_a and S_b may precess with different rates according to the effective magnetic field they experience. Thus, they alternate between S and T_o orientations and this describes mixing of S and T_o states.

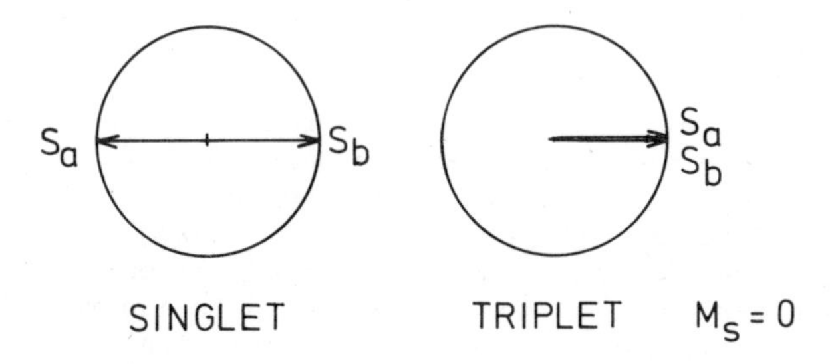

Fig.1 Projections of electron spin vectors S_a and S_b of a radical pair in the xy plane perpendicular to the magnetic field direction for a singlet and triplet (T_o) state.

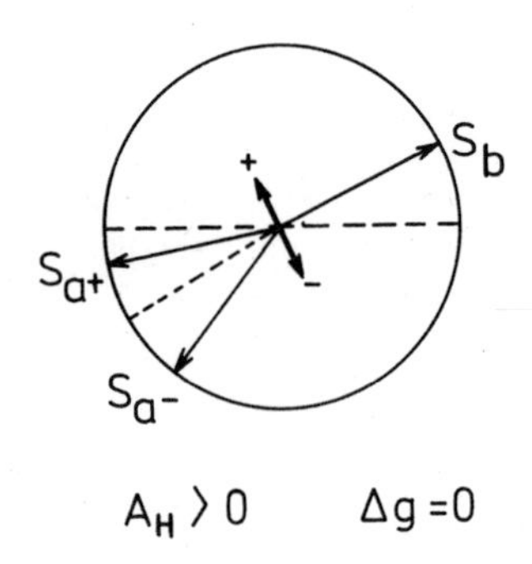

Fig.2 Precession of electron spin vectors of a radical pair $R_aH\cdot + R_b\cdot$ with $\Delta g = 0$. + and - denote the nuclear states. Heavy arrows indicate triplet component.

2.2 Magnetic isotope effects

Let us first consider a radical pair $R_aH\cdot + R_b\cdot$ with equal electronic g-factors (Δg=o) in which one of the radicals carries a proton. In addition to the external field the electron spin of radical <u>a</u> experiences a hyperfine field that is either parallel or antiparallel to the direction of H_o, depending on the proton spin state ($M_I=+\frac{1}{2}$ or $-\frac{1}{2}$). As is shown in Figure 2 the precessing rates are different for the $S_{a\pm}$ and S_b vectors, resulting in an admixture of the T_o state if the pair is formed from a S precursor. Compared to a radical pair $R_aD\cdot + R_b\cdot$ where the deuteron has a much smaller magnetic moment, the recombination rate will be decreased in the protonated case, because more triplet character is mixed in. This is the origin of the magnetic isotope effects (9,10).Note that the amount of S-T_o mixing is the same for the + and - pairs (magnitude of the heavy arrows in Figure 2), so that no nuclear polarization is generated in the recombination reaction. Apparently this is due to our assumption of Δg=o.

2.3 Net nuclear polarization

The case of a one proton pair $R_aH\cdot + R_b\cdot$ with different g-factors is depicted in Figure 3. The corresponding esr spectrum of $R_aH\cdot$ is a doublet with the centre shifted with respect to the single line of $R_b\cdot$. Precession of the spin vectors now leads to a larger triplet component for the pair with + nuclear state than for the - state. Hence the + pair will have a reduced recombination

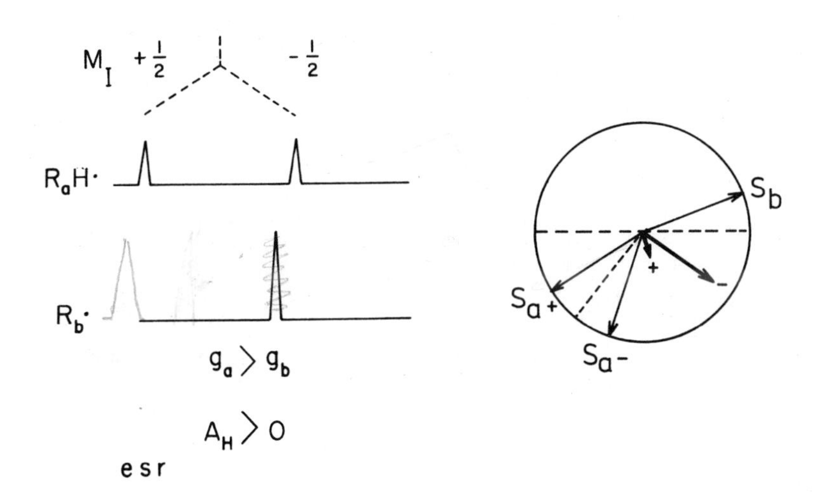

Fig.3 Esr spectra and electron spin vectors for a radical pair $R_aH\cdot + R_b\cdot$ with $\Delta g \neq 0$. Triplet components are different for the nuclear states + and - .

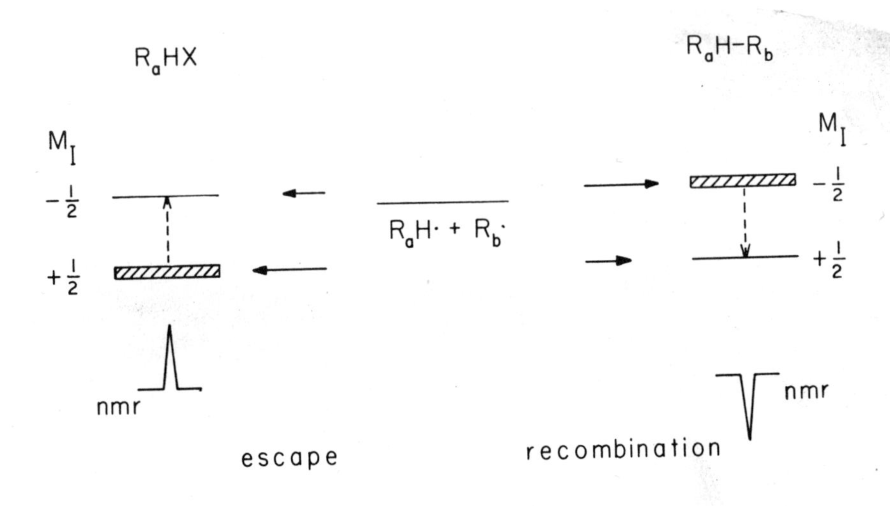

Fig.4 Populations of nuclear spin levels of the recombination and escape products of a radical pair $R_aH\cdot + R_b\cdot$ with $\Delta g \neq 0$. gb > ga

probability. As is shown in Figure 4 this results in emission (E) in the recombination product and enhanced absorption (A) in the escape product. Net CIDNP effects are thus seen to be related to g-factor differences in the radical pair.

2.4 Multiplet effects

The simultaneous occurrence of both emission and enhanced absorption within the multiplet of a nucleus coupled to other nuclei is called a multiplet effect. It is denoted by E/A when emission occurs at low field and absorption at high field in the multiplet and A/E for the reverse order. This effect also follows naturally from the radical pair mechanism. The simplest example is the two proton radical pair $R_aH_1H_2^{\cdot} + R_b^{\cdot}$ with $\Delta g=0$. The esr spectra of this pair are shown in Figure 5. The lines have been assigned to the four nuclear states assuming positive hyperfine coupling constants ($A_1, A_2 > 0$). Larger separations of the outer lines (++ and --) with respect to the single line of $R_b^{\cdot}$ arise from larger differences in effective fields acting on the radicals with ++ and -- nuclear states than those with +- and -+ states. Thus, precession rate differences are larger in the ++ and -- pairs meaning larger $S\text{-}T_o$ mixing and reduced recombination probabilities in the case of singlet born pairs. This leads to the population diagrams of Figure 6, where a positive nuclear spin-spin coupling constant J_{12} has been assumed. A/E multiplets are predicted for the recombination product $R_aH_1H_2\text{-}R_b$ and E/A for the escape product. It should be noted that no nuclear spins are flipped in this spin-sorting process so that polarization should be exactly opposite in the two types of products. Differences in magnitude may arise, however, because of relaxation in the free radicals and T_1 differences in the products.

In the case of T precursors S-T mixing for a given nuclear state enhances the recombination probability instead of reducing it, so that polarizations from T-pairs are opposite to those from S-pairs. F-pairs behave qualitatively like triplet born pairs. This can be understood by noting that during the first encounter of a random encounter pair the singlets (for which there is 25% probability) will partially recombine leaving relatively more triplets in solution.

3. CIDNP RULES

3.1 Net and multiplet effect rules

Two sign rules have been formulated for high field CIDNP (11). They are useful for the qualitative analysis of CIDNP spectra and serve to summarize the predictions of the Radical Pair

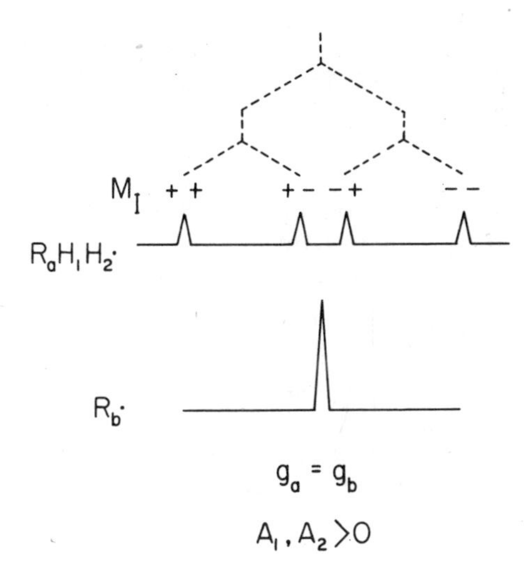

Fig.5 Esr spectra of a radical pair $R_aH_1H_2\cdot$ + $R_b\cdot$ with $\Delta g = 0$. For the assignments of the lines to nuclear states positive hyperfine coupling constants have been assumed.

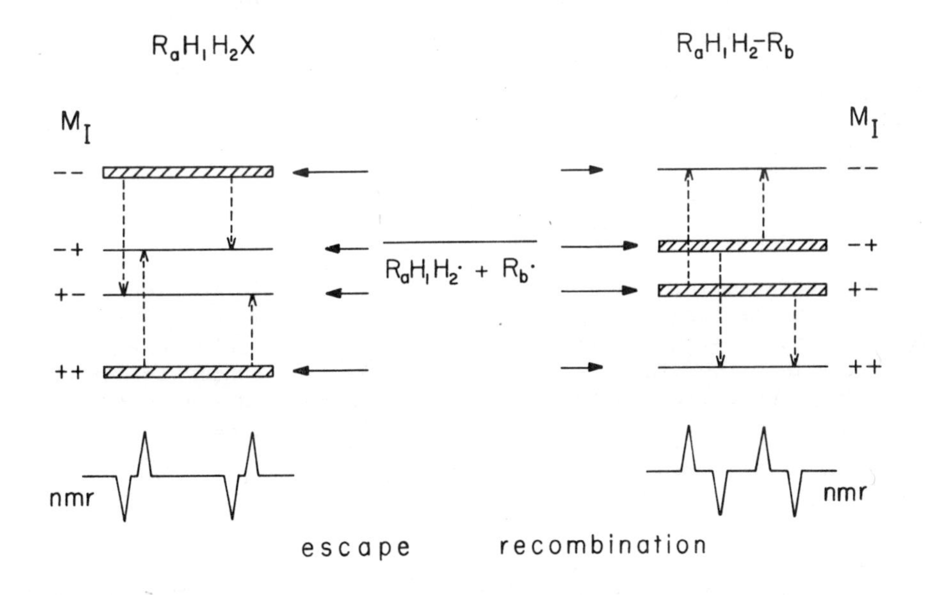

Fig.6 Populations of nuclear spin levels of the products of a radical pair $R_aH_1H_2\cdot$ + $R_b\cdot$ The E/A and A/E nmr spectra follow from the energy level diagram assuming a positive nuclear spin-spin coupling constant.

theory. Thus, the sign of the net polarization of nucleus i is given by a product of four signs

$$\Gamma_n(i) = \mu\ \varepsilon\ \Delta g\ A_i \begin{cases} + & A \\ - & E \end{cases} \qquad (1)$$

Similarly the "phase" of the multiplet effect of two (groups of) coupled nuclei i and j is given by

$$\Gamma_m(i,j) = \mu\ \varepsilon\ A_i A_j J_{ij}\ \sigma_{ij} \begin{cases} + & E/A \\ - & A/E \end{cases} \qquad (2)$$

These expressions contain the signs of the hyperfine coupling constants A_i and A_j, the sign of $\Delta g = g_a - g_b$, where g_a is the g-factor of radical a that carries nucleus i, and the sign of the nuclear spin-spin coupling constant J_{ij}. The sign conventions for the other parameters are as follows

$$\mu \begin{cases} + & \text{for a T precursor and for F-pairs} \\ - & \text{for a S precursor} \end{cases}$$

$$\varepsilon \begin{cases} + & \text{for recombination products} \\ - & \text{for escape products} \end{cases}$$

$$\sigma_{ij} \begin{cases} + & \text{when nuclei i and j reside in the same radical} \\ - & \text{when i and j reside in different radicals} \end{cases}$$

Rules (1) and (2) can be used to obtain information on the precursor multiplicity and on the absolute signs of magnetic parameters from CIDNP.

3.2 Examples from peroxide decompositions

Acetyl peroxide: net effects

The 60 Mc ^{1}H CIDNP spectrum taken during thermal decomposition of acetyl peroxide in hexachloroacetone (HCA) is shown in Figure 7. Apart from the parent compound (AP) all lines show net polarization. A CIDNP analysis must always start with the peak assignment, which is a simple matter in this case. Then a reaction scheme is set up including the radical pairs responsible for the observed polarizations. In our case a plausible scheme is the following

$$(CH_3COO)_2$$

$$\downarrow$$

$$\overline{2CH_3CO_2^{\cdot}} \longrightarrow CO_2 + \overline{CH_3^{\cdot} + CH_3CO_2^{\cdot}} \longrightarrow 2CO_2 + \overline{2CH_3^{\cdot}} \longrightarrow CH_3Cl, CH_4$$

$$\overline{CH_3^{\cdot} + CH_3CO_2^{\cdot}} \searrow CH_3COOCH_3 \qquad \overline{2CH_3^{\cdot}} \searrow C_2H_6$$

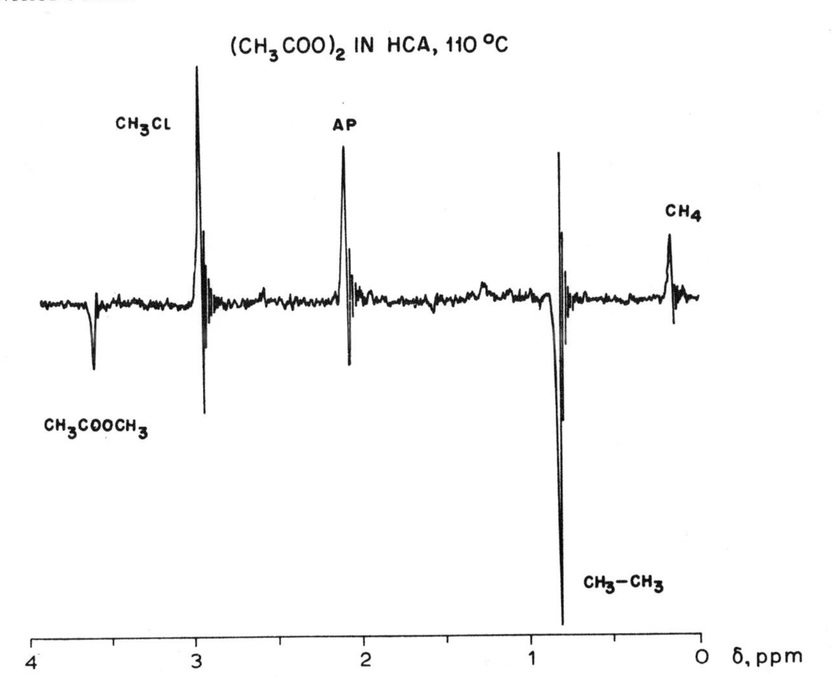

Fig.7 60 Mc ^{1}H CIDNP spectrum of the thermal decomposition of acetyl peroxide in hexachloroacetone.

The net effects must arise from the methyl-acetoxy radical pair ($\Delta g \neq 0$). For example, rule (1) for the emission line of methyl acetate would give

$$\Gamma_n(OCH_3) = -+-- = -(E)$$

because $\mu = -$ (S precursor), $\varepsilon = +$ (coupling product, $\Delta g = g(\text{methyl}) - g(\text{acetoxy}) < 0$ (the presence of the oxygen atoms causes g(acetoxy) to be larger than g(methyl), and $A(CH_3) < 0$ for the methyl radical. The spectrum of Figure 7 will be further discussed in the chapter on pair substitution effects.

Propionyl peroxide: multiplet effects.

Figure 8 shows the 60 Mc spectrum obtained during the decomposition of propionyl peroxide. In spite of the chemical analogy with the previous case the polarization is now a pure multiplet effect, which must originate from a pair with $\Delta g = 0$. Thus the propionyloxy radical does not live longer than 10^{-10} sec or, more likely, the decomposition is a concerted process.

$$(CH_3CH_2COO)_2 \longrightarrow 2CO_2 + \overline{2CH_3CH_2\cdot} \xrightarrow{RCl} CH_3CH_2Cl$$
$$\searrow C_4H_{10}$$

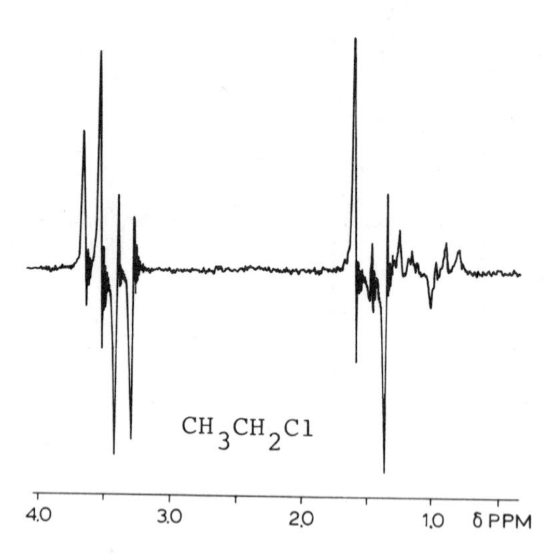

Fig.8 60 Mc ^{1}H CIDNP spectrum of the thermal decomposition of propionyl peroxide in hexachloroacetone.

The multiplet effect rule (2) applied to ethylchloride would give

$$\Gamma_n(CH_2CH_3) = ---+++ = - \quad (A/E)$$

since $\mu = -$ (S precursor), $\varepsilon = -$ (escape product), $A(CH_2) < 0$ and $A(CH_3) > 0$ in the ethyl radical, and $\sigma = +$ (the coupled protons come from the same radical).

5. BASIC RADICAL PAIR THEORY

In this section a basic account of the radical pair theory will be given. A simple high field version of the theory will describe how the combined effects of the time-evolution of the spin states and the diffusive motion of the radical pair leads to nuclear spin dependent recombination probabilities (CIDNP) and non-equilibrium spin densities in the escaping radicals (CIDEP).

5.1 Time-evolution of spin states

The starting point is the Schrödinger equation, which governs the time dependence of the wave functions $\psi(t)$,

$$i\,\frac{\partial\psi(t)}{\partial t} = \mathcal{H}\,\psi(t) \tag{3}$$

with the spin Hamiltonian $\mathcal{H}$ of the radical pair including Zeeman, exchange and hyperfine interactions (only isotopic interactions need to be included for fast tumbling radicals)

$$\mathcal{H} = (g_a S_{1z} + g_b S_{2z})\beta\hbar^{-1}H_o - J(\tfrac{1}{2} + 2 S_1S_2) + \sum_j^a A_jI_jS_1 + \sum_k^b A_kI_kS_2 \quad (4)$$

J is the exchange integral. It is convenient to expand the wave function in a coupled basis set of singlet $|S\rangle = \tfrac{1}{2}(|\alpha\beta\rangle - |\beta\alpha\rangle)$ and triplet $|T_o\rangle = \tfrac{1}{2}(|\alpha\beta\rangle + |\beta\alpha\rangle)$ states combined with the nuclear spin states $|\chi_n\rangle$. The remaining triplet states $|T_+\rangle = |\alpha\alpha\rangle$ and $|T_-\rangle = |\beta\beta\rangle$ do not mix at high field and will not be considered.

$$\psi(t) = C_{Sn}(t)|S\chi_n\rangle + C_{Tn}(t)|T_o\chi_n\rangle \quad (5)$$

S and T_o states are not stationary states because they are coupled by the Zeeman and hyperfine terms of $\mathcal{H}$. A S-T_o mixing coefficient can be defined as

$$Q_n = \langle S\chi_n|\mathcal{H}|T_o\chi_n\rangle = \tfrac{1}{2}\Delta g\beta\hbar^{-1}H_o + \sum_j^a A_jM_j - \sum_k^a A_kM_k \quad (6)$$

where the sums have to be evaluated with the nuclear spin quantum numbers M_j pertaining to the state $|\chi_n\rangle$.

Eq.(3) can now be solved with (5) for various initial conditions given by $C_S(0)$ and $C_T(0)$. For instance, for a singlet state precursor $C_S(0) = 1$ and $C_T(0) = 0$.

5.2 CIDEP

Electron spin polarization originates at interradical separations where the exchange integral is still appreciable. The simplest model is that in which J, which is very large at bonding distances, drops to a low but constant value at time t = 0 for an average duration τ and then vanishes. The solution of eq.(3) then becomes

$$C_{Sn}(t) = C_S(0)\left[\cos\omega_n t - \frac{iJ}{\omega_n}\sin\omega_n t\right] - \frac{iQ_n}{\omega_n}C_T(0)\sin\omega_n t \quad (7)$$

$$C_{Tn}(t) = C_T(0)\left[\cos\omega_n t + \frac{iJ}{\omega_n}\sin\omega_n t\right] - \frac{iQ_n}{\omega_n}C_S(0)\sin\omega_n t \quad (8)$$

where $\omega_n = (Q_n^2 + J^2)^{\frac{1}{2}}$. The electron spin density at radical a can be written (12)

$$\rho_{an}(t) = -\rho_{bn}(t) = \langle\Psi^*(t)|S_{1z} - S_{2z}|\Psi(t)\rangle =$$

$$= C_{Tn}(t)\,C_{Sn}^*(t) + C_{Tn}^*(t)\,C_{Sn}(t) \quad (9)$$

Since for electrons $<S_z> = +\frac{1}{2}$ in the state with higher energy, a positive ρ_a means esr emission for radical a. Evaluating expression (9) with (7) and (8) and keeping the main term yields

$$\rho_{an}(t) = 2\frac{Q_n J}{\omega_n^2}\left[|c_S(0)|^2 - |c_T(0)|^2\right]\sin^2\omega_n t \quad (10)$$

For an exponential distribution of life-times with average τ the total polarization of radical a is

$$\rho_{an} = \frac{1}{\tau}\int_0^\infty \rho_{an}(t)e^{-t/\tau}dt = 4Q_n J\left[|c_S(0)|^2 - |c_T(0)|^2\right]\frac{\tau^2}{1+4\omega_n^2\tau^2} \quad (11)$$

For realistic values of $\tau(10^{-11} - 10^{-10}$ sec) in non-viscous solutions eq.(11) predicts small polarizations. When the effect of re-encounters is included (12) larger polarizations can be accounted for. Nevertheless, the qualitative features of radical pair CIDEP follow from eq.(11) with (6) and rules can be formulated similarly to the CIDNP rules (1) and (2). Net electron polarization arises also only in the case of inequivalent radicals. The sign of the net effect for radical a is given by

$$\Gamma_n^{el} = \mu J \Delta g \begin{cases} + & A \\ - & E \end{cases} \quad (12)$$

where J stands for the sign of the exchange integral, which is usually negative (singlet state lower than triplet state). The multiplet effect now denotes positive and negative lines in the esr spectrum belonging to different hyperfine components and is given by

$$\Gamma_m^{el}(i) = \mu J \begin{cases} + & A/E \\ - & E/A \end{cases} \quad (13)$$

Thus, for the famous H-atom E/A polarization found by Fessenden (1) rule (13) becomes

$$\Gamma_m^{el}(H) = + - = - (E/A)$$

and requires that either the H-atom is formed from an excited T state or that the polarization is generated in random encounters with other radicals. A quantitative analysis of the H-atom CIDEP supported the latter possibility (13).

5.3 CIDNP

Nuclear polarization arises from the spin-selective chemical reactions of the radical pair. The recombination probability can be assumed to be proportional to the S character, $|c_{Sn}(t)|^2$, and will depend upon the extent of $S\text{-}T_0$ mixing. This process is not confined to regions of finite J. In fact it takes place predominantly at larger separations where J vanishes. Thus, we can set J = O and from eq.(7) it follows then that

$$|c_{Sn}(t)|^2 = |c_S(0)|^2 + \left[|c_T(0)|^2 - |c_S(0)|^2\right]\sin^2 Q_n t \tag{14}$$

or for a singlet precursor

$$|c^S_{Sn}(t)|^2 = \cos^2 Q_n t \tag{15}$$

and for a triplet precursor

$$|c^T_{Sn}(t)|^2 = \sin^2 Q_n t \tag{16}$$

The recombination probability further decreases in time because of the diffusive motion. It is appropriate to take for the re-encounter probability a time distribution function valid for three-dimensional random walk diffusion (14)

$$f(t) = m\, t^{-3/2}\, e^{-\pi m^2/p^2 t} \tag{17}$$

where $m \approx \tau_D^{1/2}$ (τ_D is the average time between diffusive steps) and $p = \int_0^\infty f(t)\,dt$ is the total re-encounter probability. Typically $m = 10^{-6}$ sec$^{1/2}$ and $p \approx 0.5$. The total recombination probability for the pair with nuclear state χ_n is then

$$P_n = \lambda \int_0^\infty |c_{Sn}(t)|^2\, f(t)\,dt \tag{18}$$

where the steric factor λ is the chance of reaction during a singlet encounter. Evaluating eq.(18) for a S precursor yields

$$P^S_n = \lambda\,[\,p - m\,(\pi Q_n)^{1/2}\,] \tag{19}$$

and for a triplet precursor

$$P^T_n = \frac{1}{3}\,\lambda m (\pi Q_n)^{1/2} \tag{20}$$

The factor $\frac{1}{3}$ arises because only one of the T states is involved. Expressions (19) and (20) have a square root dependence on Q_n, which is typical for the diffusion model.

5.4 Time scales of various processes important in CIMP

In Table 1 a summary is given of the various processes that play a role in CIDNP and CIDEP. The time scales pertain to small radicals and molecules in non-viscous solvents. It is useful to distinguish primary and secondary recombination of geminate pairs (14). The former takes place within the solvent cage and is too fast for CIDNP to develop. Therefore, secondary recom-

TABLE 1

Time scales of processes relevant to CIMP

Process	Time scale (sec)
Molecular tumbling	$10^{-12} - 10^{-11}$
Primary recombination	$10^{-11} - 10^{-10}$
Secondary recombination	$10^{-10} - 10^{-7}$
S - T mixing	$10^{-9} - 10^{-8}$
Electron T_1 in triplets	$10^{-10} - 10^{-8}$
Electron T_1 in radicals	$10^{-6} - 10^{-4}$
Electron T_2 in radicals	$10^{-6} - 10^{-5}$
Nuclear T_1 in radicals	$10^{-5} - 10^{-3}$
Nuclear T_1 in products	1 - 30

bination after diffusive excursions is responsible for the CIDNP effects, although a major part of the chemical yield may be due to primary recombination (cage effect). Electron polarization decays with the electronic T_1, whereas nuclear polarization in the escaped radicals decays somewhat slower with the nuclear T_1 in the radicals. Finally, it is the relatively long T_1 of nuclei in diamagnetic products that often makes CIDNP observable at all. This slow decay of polarization in the products makes possible high steady-state levels of polarization in the usual case where it is continuously generated.

6. THE TRIPLET MECHANISM OF CIDEP

In the triplet mechanism (6,15,16) the electron polarization is generated in the photo-excited triplet precursor. It arises from spin-selective intersystem crossing from excited S to T states governed by spin-orbit coupling rules. Since it depends on orientation, it vanishes when the triplet molecule is rotating very rapidly. Population differences of the triplet sub-levels may be transferred to the radical pair spin states resulting in CIDEP provided that the pair producing reaction is faster than relaxation in the triplet, which takes place on a time scale of $10^{-10} - 10^{-8}$ sec. The TM predicts equal signs of polarization for both radicals of the pair, whereas in the RPM the radicals are oppositely polarized. Another test for the TM has recently been devised by Adrian (16). He predicted a dependence on the direction of polarization of the exciting light, which has indeed been experimentally found (17).

The conditions for the TM seem to be somewhat more restrictive than for the RPM. Only fragmentations and electron-transfer reactions may be fast enough to satisfy the fast reaction condition.

By an Overhauser type cross-relaxation process the initial electron polarization may also be transferred to the nuclear spin system (18). A few examples of this effect have recently been found (cf. chapters by Adrian and Roth in this volume).

7. EXPERIMENTAL ASPECTS

7.1 CIDEP

Because of the short electron spin T_1 observation of CIDEP is usually not as easy as that of CIDNP. When the radicals are continuously generated deviation from equilibrium intensities is not large and therefore time resolved experiments have been developed. Radical generation is accomplished either photolytically (6,19) or by pulse-radiolysis with high energy electron beams (1). Time resolution on the microsecond time scale requires a spectrometer with fast time response somewhat at the expense of sensitivity. No such loss in sensitivity occurs when time resolution is limited to the 100-200 μsec region. Rotating sectors can then be used to modulate the light source (6), but quantitative results are more difficult to obtain. An interesting development is the use of polarized light in order to discriminate between the RPM and the TM (16,17).

7.2 CIDNP

The observation of CIDNP does not necessarily require spectrometer modifications. Thus, most of the original CIDNP work (2,3) has been carried out on simple nmr spectrometers such as the Varian A-60. Thermal reactions have usually been studied by heating the sample in the probe (or mixing reactants) and rapidly scanning the region of interest.

Photo-CIDNP studies can also be performed with unmodified spectrometers (20), but often sidewise light irradiation is desired, which necessitates probe modification. Photo-CIDNP lends itself better to quantitative studies, because the start and stop of the reaction can be controlled with a shutter.

Because of the relatively high reaction rates the time available for these experiments is limited and therefore, the advent of pulse Fourier transform (FT) techniques has been very important for CIDNP studies. Data collection times are reduced by at least an order of magnitude. When using the FT technique care should be taken to work at small flip angles ($\alpha < 30^o$), because at $\alpha = 90^o$ all homonuclear multiplet effects are lost (21). The FT method is ideally suited for the determination of relaxation times and CIDNP enhancement factors. Especially the various combinations of rf and light pulses seem promising in this respect and it may be expected that these techniques will be further explored in the near future.

REFERENCES

1. R.W.Fessenden and R.H.Schuler, J.Chem.Phys., 39, 2147 (1963).
2. J.Bargon, H.Fischer and U.Johnsen, Z.Naturforsch., 20a, 1551 (1967).
3. H.R.Ward and R.G.Lawler, J.Amer.Chem.Soc., 89, 5518 (1967).
4(a) G.L.Closs, J.Amer.Chem.Soc., 91, 4552 (1969).
(b) G.L.Closs and A.D.Trifunac, J.Amer.Chem.Soc., 92, 2183 (1970).
5(a) R.Kaptein and L.J.Oosterhoff, Chem.Phys.Letters, 4, 195 (1969);
(b) R.Kaptein and L.J.Oosterhoff, Chem.Phys.Letters, 4, 214 (1969).
6. S.K.Wong, D.A.Hutchinson and J.K.S.Wan, J.Chem.Phys., 58, 985 (1973).
7. R.Z.Sagdeev, K.M.Salikhov, T.V.Leshina, M.A.Kamkha, S.M. Shein and Yu.N.Molin, Zh.Eksp.Teor.Fiz., 16, 599 (1972).
8(a) K.Schulten, H.Staerk, A.Weller and B.Nickel, Z.Phys.Chem., 101, 371 (1976).
(b) M.E.Michel-Beyerle, R.Haberkorn, W.Bube, E.Steffens, H.Schröder, H.J.Neusser, E.W.Schlag and H.Seidlitz, Chem.Phys., 17, 139 (1976).
9. A.L.Buchachenko, E.M.Galimov, V.V.Ershov, G.A.Nikiforov and A.D.Perskin, Dokl.Ak.Nauk.SSSR, 228, 379 (1976).
10. R.G.Lawler and G.T.Evans, Ind.Chim.Belge, 36, 1097 (1971).
11. R.Kaptein, Chem.Commun., 732 (1971).
12. F.J.Adrian, J.Chem.Phys., 54, 3918 (1971).
13. N.C.Verma and R.W.Fessenden, J.Chem.Phys., 58, 2501 (1973).
14. R.M.Noyes, J.Amer.Chem.Soc., 78, 5486 (1956).
15. P.W.Atkins and G.T.Evans, Chem.Phys.Letters, 25, 108 (1974).
16. F.J.Adrian, J.Chem.Phys., 61, 4875 (1974).
17. A.J.Dobbs and K.A.McLauchlan, Chem.Phys.Letters, 30, 257 (1975).
18. H.M.Vyas and J.K.S.Wan, Chem.Phys.Letters, 34, 470 (1975).
19. P.W.Atkins, K.A.McLauchlan and P.W.Percival, Mol.Phys., 25, 281 (1973).
20. M.Tomkievicz and M.P.Klein, Rev.Sci.Instr., 43, 1206 (1972).
21. S.Schäublin, A.Höhener and R.R.Ernst, J.Magn.Res., 13, 196 (1974).

CHAPTER II

CIDNP EXHIBITED BY THERMALLY DECOMPOSING DIACYL PEROXIDES*

R. G. Lawler

Department of Chemistry, Brown University
Providence, R. I. 02912 USA

Diacyl organic peroxides, RCO_2O_2CR', are in several respects ideal compounds for study by CIDNP. They are easily prepared from readily available reagents and may be decomposed thermally at temperatures sufficiently high to permit their isolation and purification, but low enough to be accessible with routinely available nmr variable temperature accessories. Equally important, however, is the well defined radical pair initially formed from the decomposition.

$$RCO_2O_2CR' \xrightarrow{\text{heat}} [RCO_2^{\cdot}\ {}^{\cdot}O_2CR'] \xrightarrow{\text{fast}} R\cdot\ \cdot R'$$

The decarboxylation of the first formed acyloxy radicals seems to be nearly instantaneous except when R = CH_3 or C_6H_5 [1]. One is thus left with the geminate radical pair, $R\cdot\cdot R'$, formed in the singlet electron spin state. A wide variety of structural features may thus be designed into a radical pair either for use in studying the properties of the radicals themselves or, if desired, the production of enhancement in specific reaction products with especially interesting nmr spectra. One may if he wishes also allow the first formed radicals to react with suitable scavengers to produce new radicals which undergo diffusive encounters with randomly formed electron spin states [2].

$$R\cdot\cdot R' \rightarrow R\cdot + R'\cdot \xrightarrow{SX} RX + R'X + 2S\cdot \rightarrow S\cdot\cdot S$$

* This work was supported in part by grants from the National Science Foundation.

L. T. Muus et al. (eds.), Chemically Induced Magnetic Polarization, 17-28.

This lecture will present a portfolio of CIDNP spectra obtained during the decomposition of various symmetrical and unsymmetrical diacyl peroxides. The emphasis will be on illustrating the simple qualitative rules for relative intensities which interrelate the signs of magnetic resonance parameters and the life story of radicals and radical pairs [3].

The symbols employed in applying and discussing the simple rules are those now in common use [3].

Net Effect: $\Gamma_{ne} = \mu \epsilon A_i \Delta g$, $+ = A$

Multiplet Effect: $\Gamma_{me} = \mu \epsilon A_i A_j J_{ij} \sigma_{ij}$, $+ = EA$

Implicit in the use of these relations, of course, is the assumption that the radical-radical reaction proceeds preferentially via the <u>singlet</u> electron spin state.

1. SIGNS OF MAGNETIC RESONANCE PARAMETERS

Since extensive use must be made of the absolute signs of hyperfine splittings and nuclear spin spin splittings and the relative magnitudes of electron g-factors, a summary of these quantities for some commonly occurring structural and nuclear spin types is given below.

1.1 g-Factors

$g > 2.0025$: radicals with spin on heteroatoms ($RCO_2^{\bullet}$, $R_2\dot{C}X$)

$g \cong 2.0025$: π hydrocarbon radicals ($R_3C\bullet$)

$g < 2.0025$: σ radicals ($R\dot{C}O$, $Ph\bullet$)

1.2 Hyperfine splittings, A_i (starred atom)

$A_H < 0$: α -alkyl ($R_2\dot{C}H^*$)

$A_H > 0$: β -alkyl ($R_2CH^*CR_2^{\bullet}$)
π radicals with negative spin density ($R_2C = CH^*CR_2^{\bullet}$)

$A_C > 0$: α - or γ - alkyl ($R\text{-}C^*\text{-}C\text{-}C^*R_2^{\bullet}$)

$A_C < 0$: β -alkyl ($R\text{-}C^*\text{-}CR_2^{\bullet}$)

1.3 Spin-spin splittings, J_{ij} (starred atoms)

J > 0: ^1J ($-C^*-H^*$)

^3J (H^*C-CH^* , H^*C-CC^*)

J < 0: ^2J (H^*CH^* , H^*CC^* [variable])

^4J ($H^*C=C-C-H^*$)

2. CIDNP FROM SINGLET-BORN PAIRS (μ = -)

Each of the spectra presented below will be described by a scheme showing the reaction pathways leading to the principal products exhibiting CIDNP. The sign of the enhancement using the simple rules will be presented.

2.1 Pure net effect, ^1H

Figure 1A shows the CIDNP spectrum obtained during the decomposition of acetyltrichloroacetyl peroxide in a carbon tetrachloride solution of I_2 at 50°.

$$CH_3CO_2O_2CCl_3$$

$$\downarrow$$

$$2.7\,\delta,(E)\ CH_3^*\!-\!CCl_3 \leftarrow CH_3\cdot\ \ \cdot CCl_3 \xrightarrow{I_2} ICH_3^*\,(A),\ 2.2\,\delta$$

$$\epsilon = + \qquad A_i = -,\ \Delta g = - \qquad \epsilon = -$$

$$\Gamma_{ne} = -+-- = E \qquad\qquad \Gamma_{ne} = ---- = A$$

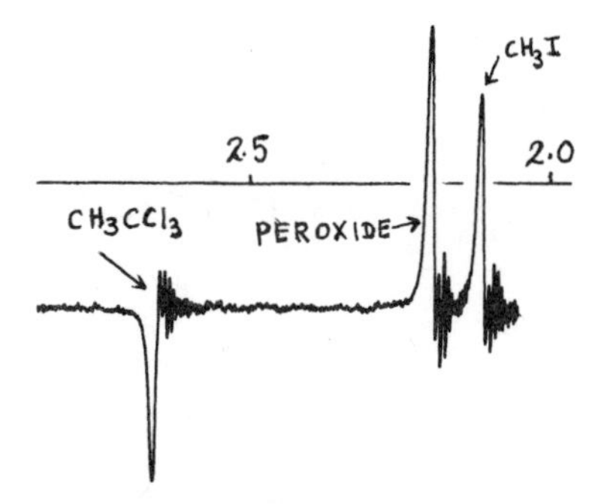

Figure 1A. Decomposition of $CH_3CO_2O_2CCCl_3$ in CCl_4 at 50° with added I_2.

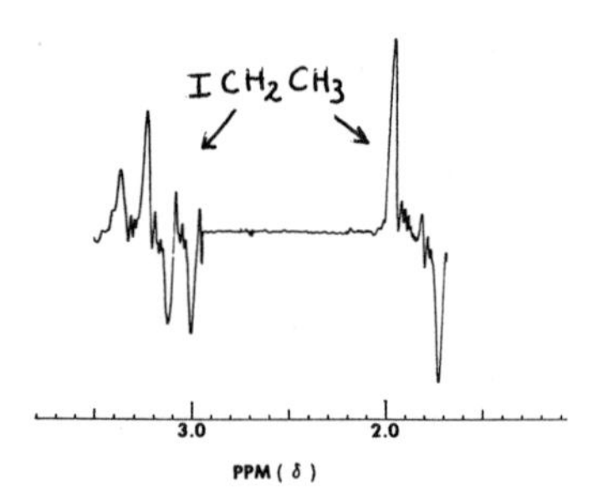

Figure 1B. Ethyl iodide formed from decomposition of propionyl peroxide in ODCB saturated with I_2 at 120°.

2.2 Pure multiplet effect, ^1H

Figure 1B shows the spectrum of ethyl iodide obtained as a scavenging product during the decomposition of dipropionyl peroxide in ortho-dichlorobenzene (ODCB) saturated with I_2 at 120°.

$$(CH_3CH_2CO_2)_2 \longrightarrow 2CH_3CH_2^{\bullet} \xrightarrow{I_2} \overset{*}{CH_3}-\overset{*}{CH_2}-I \qquad (AE)$$

$$A_\alpha = - \qquad \epsilon = -$$

$$A_\beta = + \qquad J = + \qquad \Gamma_{me} = ---+++ = AE$$

$$\sigma = +$$

The radical coupling product, butane, exhibits a multiplet effect too complicated to fit the simple rules and the disproportionation products ethane and ethylene both have single line nmr spectra which cannot exhibit a multiplet effect.

2.3 Mixed net and multiplet effects, ^1H

Figure 2 shows the spectrum of propionyltrichloroacetyl peroxide decomposing in carbon tetrachloride at 50°.

$$CH_3CH_2CO_2O_2CCl_3$$

$$\downarrow$$

$$\underset{CH_3CH_2CCl_3}{2.8\delta(E + [AE])} \longleftarrow CH_3CH_2^{\bullet} \; {}^{\bullet}CCl_3 \xrightarrow{CCl_4} \underset{CH_3CH_2Cl}{(A + [EA]), 3.5\,\delta}$$

$CH_3CH_2CCl_3$	$CH_3CH_2^{\bullet}\ {}^{\bullet}CCl_3$	CH_3CH_2Cl
$\epsilon = +$	$A_\alpha = -$	$\epsilon = -$
$J = +$	$A_\beta = +$	$J = +$
$\sigma = +$	$\Delta g = -$	$\sigma = +$

$$\Gamma_{ne} = -+-- = E \qquad\qquad \Gamma_{ne} = ---- = A$$

$$\Gamma_{me} = -+-+++ = EA \qquad\qquad \Gamma_{me} = ---+++ = AE$$

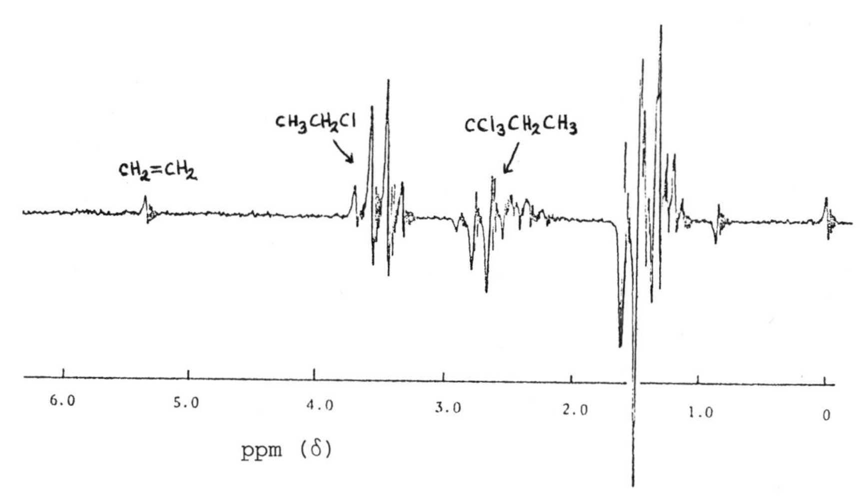

Figure 2. Decomposition of $CH_3CH_2CO_2O_2CCl_3$ in CCl_4 at 50°.

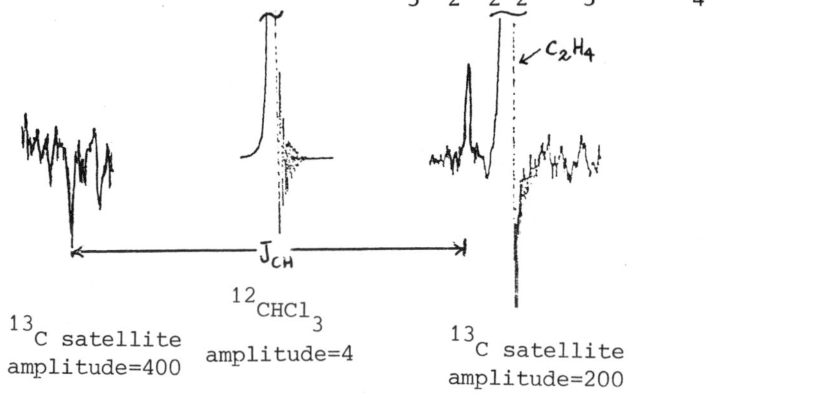

Figure 3. CIDNP from $HCCl_3$ and its carbon-13 satellites formed during the decomposition of $CH_3CH_2CO_2O_2CCl_3$ in CCl_4 at 72°.

Note that in this case the simple rules appear to predict the wrong type of multiplet effect for both types of product! The difficulty lies in the fact that the net effect is much larger than the multiplet effect in this case and small variations in intensity due to intensity borrowing make the low field lines of both multiplets of lower absolute intensity than they should be.

Figure 2 also exhibits an effect due to a difference in the magnitude of A_α and A_β in the ethyl radical: ethylene, at 5.3 δ produced as a disproportionation product of the radical pair, exhibits a weak enhanced absorption as expected if the CIDNP were dominated by the positive hyperfine splitting in the position of the ethyl radical.

2.4 Mixed net and multiplet effects, $^{1}H-^{13}C$

Figure 3 shows an expanded view of the region near 7δ during the reaction of propionyltrichloroacetyl peroxide in carbon tetrachloride at 72°. The lines observed are due to $H^{12}CCl_3$ and $H^{13}CCl_3$ formed in the disproportionation reaction of the radical pair.

$$CH_3CH_2\cdot \; {}^{13}\cdot CCl_3 \longrightarrow CH_2 = CH_2 + H^{*}\!-\!{}^{13}CCl_3 \quad (A + EA)$$

$A_\beta = +$ $\quad A_C = +$

$\Delta g = -$

$\epsilon = +$

$J = +$

$\sigma = -$

$$\Gamma_{ne} = -++- = A$$

$$\Gamma_{me} = -++++- = EA$$

In this case the multiplet effect in the spectrum of the proton from the β-ethyl position dominates because the magnitude of A_C is much larger than A_β. At the time this spectrum was obtained the magnitude of A_C in the trichloromethyl radical was unknown.

3. CIDNP FROM DIFFUSIVELY FORMED PAIRS (μ = +)

Diffusively formed pairs generated from diacyl peroxides require rather complicated chemistry for their production. The two examples shown here are fairly typical of the chemical complexity which must be introduced to explain the observed CIDNP spectra.

3.1 Pure net effect

In Figure 4 is shown the nmr spectrum obtained when di-(trichloroacetyl) peroxide is decomposed in a carbon tetrachloride solution containing tetramethylethylene (TME) at room temperature.

$$(CCl_3CO_2)_2 \longrightarrow 2CCl_3\cdot \xrightarrow{TME} \underset{\mathbf{1}}{Cl_3CC(CH_3)_2\dot{C}(CH_3)_2}$$

$$\mathbf{1} + \cdot CCl_3 \xrightarrow[\text{encounter}]{\text{diffusive}} \overline{CCl_3\cdot \quad \cdot C(CH_3)_2-C(CH_3)_2-CCl_3} \qquad A_\beta = + \quad \Delta g = -$$

$$\overline{CCl_3\cdot \; \mathbf{1}} \longrightarrow H^{*}CCl_3 + C^{*}H_2 = C(CH_3^{*})C(CH_3)_2CCl_3$$

$$\epsilon = +$$

$$\Gamma_{ne} = ++-+ = E$$

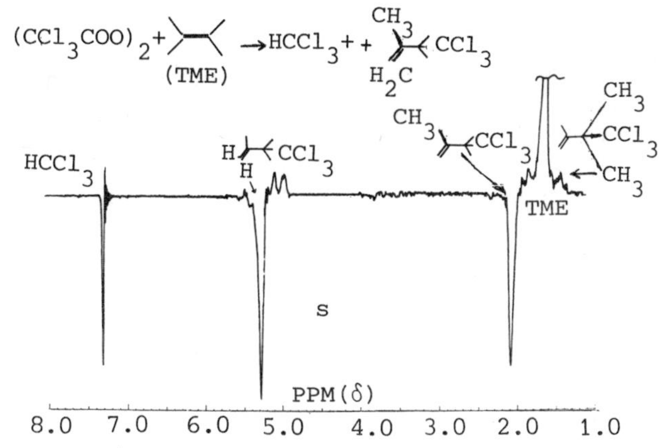

Figure 4. CIDNP spectrum taken during thermolysis of trichloroacetyl peroxide at room temperature in CCl_4 containing TME. The spectrum is the composite of three scans.

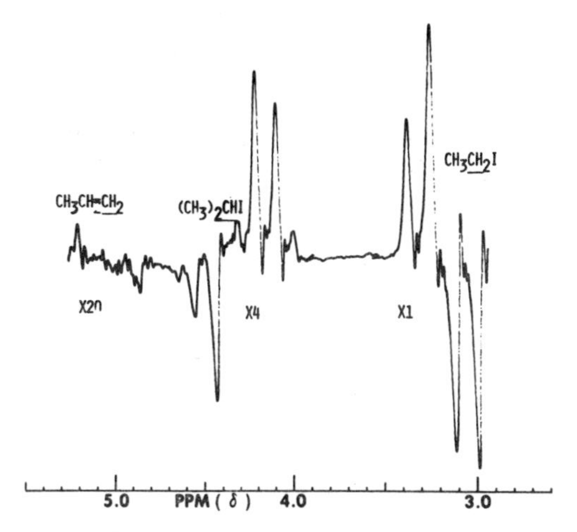

Figure 5. Spectrum recorded during the decomposition of a solution of propionyl peroxide and isopropyl iodide in ODCB at 100°.

In this case the other two possible diffusive encounter pairs, [$2CCl_3^{\bullet}$] and [1 1] may well be formed. However, the former cannot produce ^{1}H CIDNP (or indeed even ^{13}C CIDNP!), while the latter would produce a multiplet effect, probably much weaker than the net effect observed.

3.2 Pure multiplet effect.

Figure 5 shows the CIDNP spectrum obtained during the decomposition of dipropionyl peroxide in an ODCB solution of isopropyl iodide at 100°. The AE multiplet of ethyl iodide, previously discussed in Section 2.2, is seen at 3.2 δ . Additional multiplet effects are seen, however, in propene and the isopropyl iodide itself. These arise from diffusive encounters of two isopropyl radicals produced when the initially formed ethyl radicals react with isopropyl iodide.

$$(CH_3CH_2CO_2)_2 \rightarrow 2CH_3CH_2^{\bullet} \xrightarrow{CH_3CHICH_3} CH_3\dot{C}HCH_3$$

$$2CH_3\dot{C}HCH_3$$

$$\text{diffusive} \downarrow \text{encounter}$$

$$(AE)\ CH_3CH = CH_2 \longleftarrow CH_3\dot{C}HCH_3 \quad CH_3\dot{C}HCH_3 \xrightarrow{CH_3CHICH_3} CH_3CHICH_3\ (EA)$$

$\epsilon = +$	$A_1 = +$	$\epsilon = -$
$J = +$	$A_2 = -$	$J = +$
$\sigma = +$		$\sigma = +$

$\Gamma_{me} = +++-++ = AE$ $\qquad\qquad \Gamma_{me} = +-+-++ = EA$

The multiplet effect for propene is not strictly predictable by the simple rules since the two terminal vinyl protons are not magnetically equivalent and are also coupled to the methyl group. However, they are coupled strongly to the other vinyl proton and this leads to a spectrum which is approximately a doublet.

4. COMPARISON OF ^{1}H and ^{13}C CIDNP SPECTRA

It is now possible fairly routinely to obtain natural abundance carbon-13 spectra of samples exhibiting CIDNP. We present here two examples comparing such spectra to their proton resonance counterparts.

4.1 Pure multiplet effect

Figure 6A shows the ^{1}H spectrum of lauroyl peroxide, $(C_{11}H_{23}CO_2)_2$ decomposing at 100° in ODCB containing isopropyl iodide. Figure 6B shows the proton coupled ^{13}C spectrum obtained from the same peroxide in hexachloroacetone (HCA). The scheme leading to the observed products is the same in both cases.

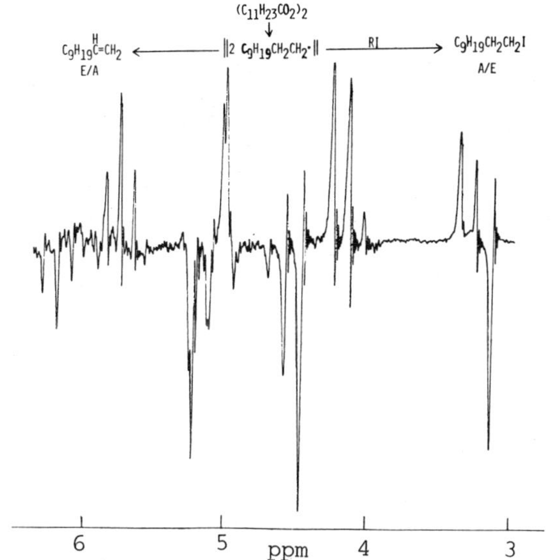

Figure 6A. Proton spectrum recorded during thermolysis of lauroyl peroxide in ODCB and isopropyl isodide.

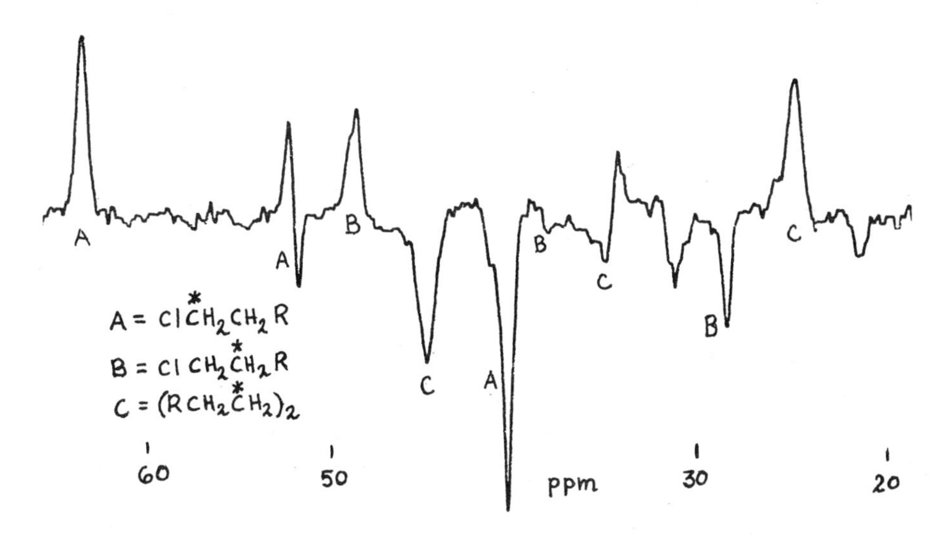

Figure 6B. Carbon-13 spectrum recorded during thermolysis of lauroyl peroxide in HCA. Broadband proton decoupling was not employed.

$$(C_{11}H_{23}CO_2)_2$$

$$\downarrow$$

$$\text{(EA) } C_9H_{19}CH = CH_2 \leftarrow C_9H_{19}CH_2CH_2^{\bullet} \; {}^{\bullet}C_{11}H_{23} \xrightarrow{RX} C_9H_{19}CH_2CH_2\text{-}X \text{ (AE)}$$

$\epsilon = +$ | $A_{C\alpha} = +,\ A_{H\alpha} = -$ | $\epsilon = -$

$J = +$ | $A_{C\beta} = -,\ A_{H\beta} = +$ | $J = +$

$\sigma = +$ | | $\sigma = +$

$\Gamma_{me}^{HH} = -{+}{+}-{+}{+} = EA$ | $\Gamma_{me}^{HH} = --{+}-{+}{+} = AE$

$\text{(EA) } (C_9H_{19}CH_2\overset{*}{C}\overset{*}{H_2})_2$

$C_9H_{19}CH_2\overset{**}{CH_2}\text{-}X$ (AE), 52 δ

$C_9H_{19}\overset{**}{CH_2}CH_2\text{-}X$ (AE), 38 δ

$\Gamma_{me}^{CH} = -{+}{+}-{+}{+} = EA$ | $\Gamma_{me}^{CH} = --{+}-{+}{+} = AE$

$\Gamma_{me}^{CH} = ---{+}{+}{+} = AE$

The principal combination product in this example is the dimer. It is interesting to note, however, that although its ^{13}C spectrum exhibits the expected multiplet effect, the ^1H spectrum of this product is <u>unpolarized</u> because the methylene groups in the product have the same chemical shift. A further note of interest is the "multiplet effect" exhibited by the center line of the triplet of the ^{13}C spectrum of $C_9H_{19}CH_2CH_2Cl$ at 52 δ . This arises from the coupling of the proton at the β position to the α carbon of the undecyl radical. Since the ^2J is probably negative for this pair of nuclei, one would predict $\Gamma_{me}^{C\alpha,H\beta} = --{+}{+}-{+}$ =AE, as observed.

4.2 ^1H CIDNP absent

Figure 7 shows the ^{13}C nmr spectrum obtained during decomposition of di-(trimethylpropionyl) peroxide in HCA. The remarkable aspect of this peroxide is that it produces no ^1H CIDNP in <u>any</u> of the products net effects are forbidden because the radicals are identical and multiplet effects cannot be observed because the reaction products give only nmr singlets. The latter restriction is removed, however, in the proton coupled ^{13}C spectrum which clearly exhibits multiplet effects for the coupling and scavenging products.

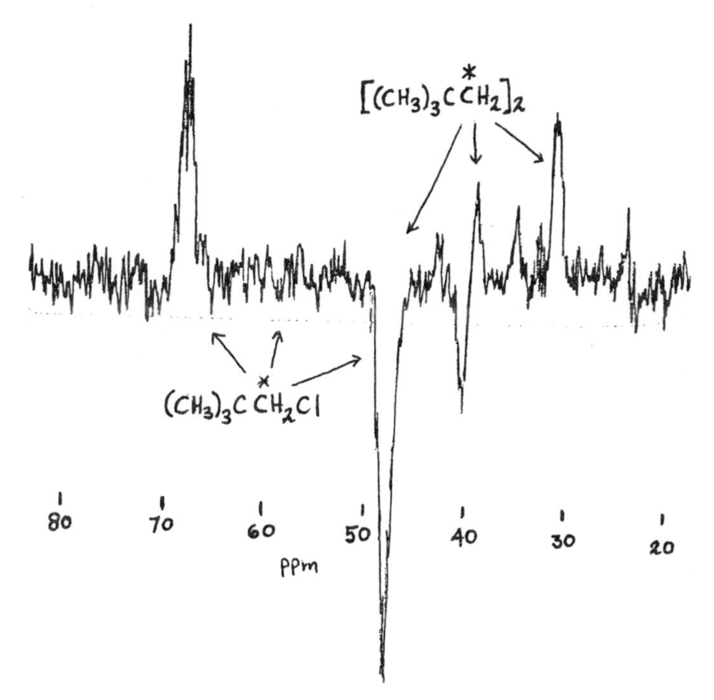

Figure 7. Carbon-13 spectrum recording during thermolysis of di-(trimethylpropionyl) proxide in HCA. Broadband proton decoupling was not employed.

$$(Me_3CCH_2CO_2)_2$$

$$\downarrow$$

$$40\delta\,(EA)\,Me_3CC^{*}H_2C^{*}H_2CMe_3 \longleftarrow 2Me_3CCH_2^{\bullet} \xrightarrow{HCA} Me_3CC^{*}H_2Cl\ (AE),\ 58\delta$$

$\epsilon = +$ $\qquad A_C = +$ $\qquad \epsilon = -$

$J = +$ $\qquad A_H = -$ $\qquad J = +$

$\sigma = +$ $\qquad \sigma = +$

$\Gamma_{me} = -{+}{+}{-}{+}{+} = EA$ $\qquad \Gamma_{me} = -{-}{+}{-}{+}{+} = AE$

Note that the center line of the coupling product multiplet at 40δ exhibits the same type of multiplet effect as discussed in the previous section. In this case the coupling arises from the methylene protons on the other radical fragment (the center line in the scavenging product at 58 δ is unpolarized). For that carbon-proton pair $^2J = -$, as before, but now $\sigma = -$, since the two nuclei occupied different radicals in the pair. Thus $\Gamma_{me} = -{+}{+}{-}{-}{-} = EA$, as observed.

CONCLUSION

Diacyl peroxides continue to be excellent vehicles for illustrating the basic principles of CIDNP. This lecture has attempted to describe the variety of CIDNP spectra which may arise from the simple chemistry associated with these radical precursors and display examples from both ^{1}H and ^{13}C nmr.

REFERENCES

1. R. Kaptein, H. Fischer, in "Chemically Induced Magnetic Polarization", A. R. Lepley and G. L. Closs, Eds., Wiley, 1973; Chaps. 4 and 5.

2. M. Lehnig and H. Fischer, Z. Naturforsch., 25a, 1963 (1970).

3. R. Kaptein, Chem. Commun., 732 (1971).

CHAPTER III

TIME-DEPENDENCE OF CIDNP INTENSITIES*

R. G. Lawler

Department of Chemistry, Brown University
Providence, R. I. 02912 USA

By its very nature the phenomenon of CIDNP is a time-dependent or "dynamic" effect. It was appreciated, even before the origin of the effect was correctly understood, that the observed nmr intensities reflect the competition between "pumping" of the nuclear spin states by a chemical reaction and relaxation of the populations to their equilibrium values [1]. It is our purpose here to illustrate and discuss some of the time dependent effects which have been observed in CIDNP, with special emphasis given to some of the distortions due to relaxation which can complicate the comparison between theoretical and experimentally observed spectra. In so doing we shall restrict the discussion as follows.

a) We will discuss only CIDNP spectra obtained from samples reacting in the same, high, magnetic field in which they are observed. This avoids special problems arising from the time-dependent field experienced by nuclei when a sample is allowed to react in one field and is then rapidly transferred to the spectrometer field [2].

b) We will ignore special effects which may occur in the very early stages of chemical reaction or radio frequency excitation of a sample exhibiting CIDNP. These include oscillations in pulsed nmr arising from homonuclear spin-spin coupling [3], non-linear effects arising from saturation in pulsed [4] or rapid scan [5] CIDNP spectra, and transient nutations which occur in CW nmr experiments when the chemical reaction rate is modulated [6].

* This work was supported in part by a grant from the National Science Foundation.

L. T. Muus et al. (eds.), Chemically Induced Magnetic Polarization, 29-37.

c) Although the effects we will discuss do not depend in general on specific mechanisms for chemical pumping, we will concentrate on the net and multiplet effect types of CIDNP arising from the radical pair mechanism.

1. CIDNP FROM NMR SINGLETS

The simplest possible CIDNP spectrum consists of an enhanced line arising from the single nmr transition for one spin one-half nucleus. The time dependence of such a line may be described by a simple modification of the Bloch Equations for the nuclear magnetization [1,7]. We discuss this simple model below.

1.1 Modified Bloch Equations

The intensity of a simple nmr line observed under non-saturating conditions is proportional to the magnetization of the nuclei, M^z, in the direction of the applied magnetic field. It is assumed that the sample is undergoing the dynamic changes shown in the scheme below.

$$R^o \xrightarrow{r} P^* \xrightarrow{T_1^{-1}} P^o$$

R and P designate the reactant and product molecules, respectively. A star is used to signify CIDNP and a superscript zero denotes species with equilibrium nuclear spin state populations. T_1 is the nuclear spin-lattice relaxation time and r is the rate of reaction in sec^{-1}. Note, however, that r will be time-dependent for any kinetic scheme except a first-order reaction.

We may thus write

$$\dot{M}_p^z = (V_p+1)\, \dot{M}_p^{z,o} - T_{1p}^{-1}\, (M_p^z - M_p^{z,o}) \tag{1.1}$$

$$\dot{M}_p^{z,o} = r\, M_R^{z,o} \tag{1.2}$$

where V_p is the CIDNP enhancement factor, (Verstarkungsfaktor) defined as the extra contribution to the intensity of P made by the reaction over what would be expected for a sample with nuclear spins at equilibrium.

1.2 Zero- and First-order Reactions

Equations (1.1) and (1.2) are easily solved for the case where $rM_R^{z,o}$ is either constant or decreases exponentially with time.

i) Zero-order reaction

$$M_p^z(t)-M_p^z(t^o)=V_p r T_{1p} M_R^z(o^o)[1-\exp(-t/T_{1p})] \tag{1.3}$$

$$M_p^z(t^o)=rM_R^z(o^o)t$$

ii) First-order reaction

$$M_p^z(t)-M_p^z(t^o)=V_p kT_{1p}(1-kT_{1p})^{-1}M_R^z(o^o)[\exp(-kt)-\exp(-t/T_{1p})] \tag{1.4}$$

$$M_p^z(t^o)=M_R^z(o^o)[1-\exp(-kt)]$$

The time dependence of M_p^z from Eqns. (1.3) and (1.4) is shown schematically in Figure 1 for a hypothetical CIDNP emission line.

In Figure 2 is shown the experimentally observed CIDNP spectrum obtained at several different times during the thermal decomposition of the mixed peroxide shown below.

$$ArCO_2O_2CCH_2CMe_3 \xrightarrow{R-Cl} ClC\overset{*}{H}_2CMe_3$$

The strong enhanced absorption peak near 3 ppm marked "C" is due to the methylene protons in neopentylchloride. For comparison, the weaker peak marked "P" arises from the nine unpolarized t-butyl protons in the peroxide molecule. Equation (1.1) is also a good description of the behavior of single lines, such as the ones in Figure 2, arising from a group of magnetically equivalent nuclei.

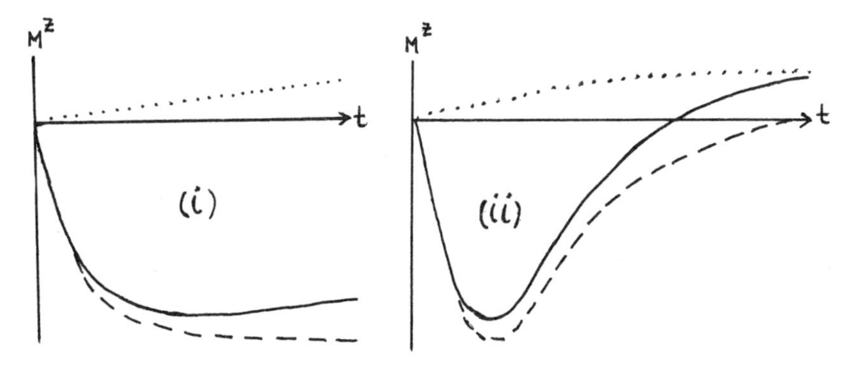

Figure 1. Schematic representation of the time dependence of the magnetization for zero-order (i) and first-order (ii) reactions producing CIDNP. Dotted (...), dashed (---) and solid(——) lines represent equilibrium, chemically pumped, and observed net contributions to the magnetization, respectively.

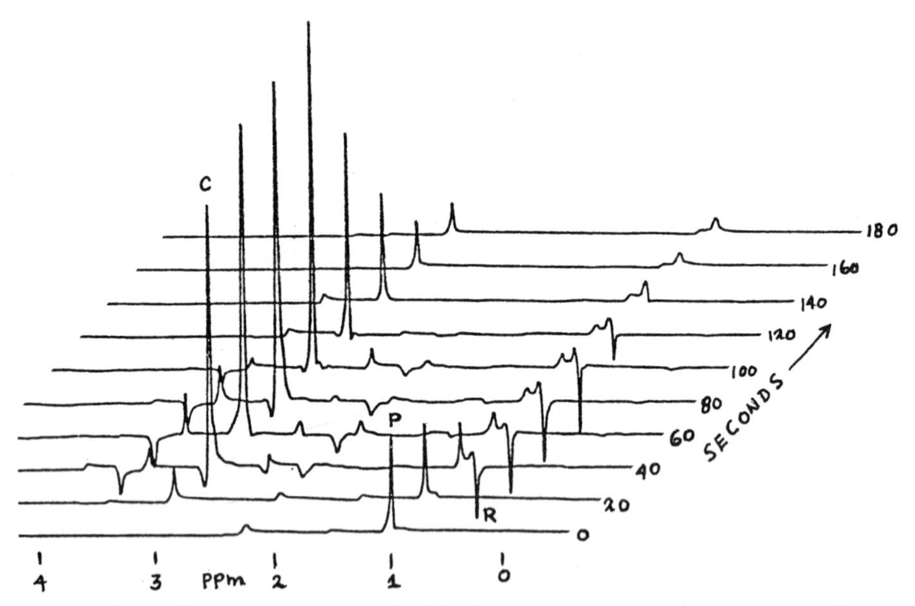

Figure 2. Time dependence of CIDNP from decomposing m-chlorobenzoyl-t-butylacetylperoxide in hexachloracetone at 120°. Peaks labelled C,P and R are discussed in the text.

From Figure 1 and Eqns.(1.3) and (1.4) we see that, except for very fast reactions where $k > T_1^{-1}$, the time constant for the initial rise in signal is T_1, typically a few seconds. This is the time required for the sample to reach a steady state in which any further change in signal intensity is a response to gradual depletion of the supply of reactant as the reaction proceeds. The steady state value of M_p^z then is simply

$$M^z_{p,ss}(t) = M_p^{z,o}(t) + V_p T_{1p} \dot{M}_p^{z,o}(t) \tag{1.5}$$

Thus the enhanced part of M_p^z becomes proportional to the reaction rate through $\dot{M}_p^{z,o}(t)$.

It is relatively easy to obtain pseudo-zero-order kinetics in photochemical reactions by using only brief periods of irradiation. Thermal reactions, however, are somehwat more difficult to control and it has become common practice to record the entire time evolution of thermally decomposing samples. This has become routine now that FT NMR spectrometers are widely available.

2. CIDNP FROM MANY-SPIN SYSTEMS

The time dependence of spectra exhibiting CIDNP from more than one nucleus may be described by a generalized master equation resembling the Bloch Equations [8].

2.1. Master Equation for Populations

The starting point for this approach is the proportionality between the intensity, Int_{nm}, of a transition between two spin states, n and m, and the population difference, $N_n - N_m$.

$$Int_{nm} = K\Omega_{nm}(N_n - N_m) \tag{2.1}$$

K is an instrumental constant which is the same for all lines and Ω_{nm} is the transition probability for the line in question. Denoting by P_n the contribution to the population of state n made by CIDNP and by r, as before, the rate of the reaction, we may write for the time-dependence of the population of state n

$$\dot{N}_n = rP_n - \sum_k W_{nk}[(N_n - N_k) - (N_n^o - N_k^o)] \tag{2.2}$$

where $W_{nk} = W_{kn}$ is the rate constant for relaxation-induced transitions between states n and k, the superscript zero indicates the population under conditions of equilibrium

$$N_n^o = \exp(-\varepsilon_n/kT)Z^{-1} \tag{2.3}$$

and Z is the total number of relevant nuclear spin states, e.g. for q spin I nuclei $Z = (2I+1)^q$.

2.2 AX Spin System

We shall illustrate the use of Eqns. (2.1) and (2.2) by describing some time-dependent effects expected in an AX CIDNP spectrum such as would be obtained from two weakly coupled protons or a proton and a carbon-13 in the same molecule. For simplicity we shall ignore the equilibrium populations, i.e. we shall assume that the polarized spectrum is much more intense than the unpolarized "background".

The energy levels and transition rates, W, for the AX spin system are shown schematically in Figure 3. The intensities of the four lines, $A_\pm$ and $X_\pm$, may be conveniently grouped and described by the following equations [9].

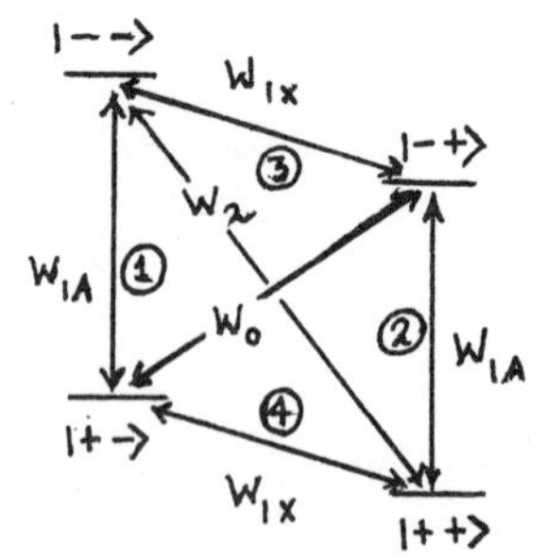

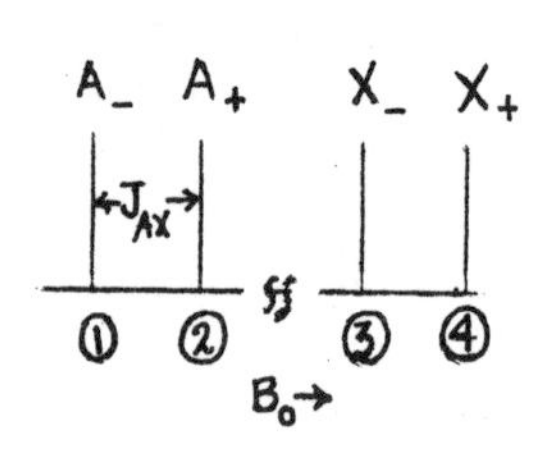

Figure 3. Energy levels, relaxation transition probabilities, W, and nmr transitions for an AX spin system. It is assumed that both nuclei have a positive magnetogyric ratio and that J_{AX} is positive.

$$(\dot{A}_+ + \dot{A}_-) = rP_A^+ - \rho_A (A_+ + A_-) - \sigma (X_+ + X_-) \quad (2.3)$$

$$(\dot{X}_+ + \dot{X}_-) = rP_X^+ - \sigma (A_+ + A_-) - \rho_X (X_+ + X_-) \quad (2.4)$$

$$(\dot{A}_+ - \dot{A}_-) = rP_A^- - \mu (A_+ - A_-) \quad (2.5)$$

$$(\dot{X}_+ - \dot{X}_-) = rP_X^- - \mu (X_+ - X_-) \quad (2.6)$$

where

$$P_A^{\pm} = [(P_{++} - P_{-+}) \pm (P_{+-} - P_{--})] \; ; \; \rho_A = (W_o + 2W_{1A} + W_2)$$

$$P_X^{\pm} = [(P_{++} - P_{+-}) \pm (P_{-+} - P_{--})] \; ; \; \rho_X = (W_o + 2W_{1X} + W_2)$$

$$\sigma = W_2 - W_o; \qquad \mu = 2(W_{1A} + W_{1X})$$

The sums and differences of transition intensities for each nucleus may be interpreted as a measure of the "net effect" and "multiplet effect" for that particular nucleus, respectively. For example, a pure AE multiplet effect for X would signify a negative value for $(X_+ - X_-)$ which would be twice the absolute magnitude of each of the oppositely enhanced X lines. We give below three limiting applications of Eqns. (2.3)-(2.6).

2.1.1 Net effects for both A and X, $\sigma = o$.

$$P_A^+ = P_A; \quad P_X^+ = P_X; \quad P_A^- = P_X^- = o$$

Under these conditions each line relaxes with a single rate constant, ρ_A or ρ_X, and exhibits a time dependence identical to that described by the modified Bloch Equation, Eqn.(1.1).

2.1.2 Pure multiplet effect.

$$P_A^+ = P_X^+ = o; \quad P_A^- = P_X^- = P$$

In this case all four lines relax with the same rate constant, μ, which is faster than the relaxation rate of either nucleus if W_o and W_2 are negligible. By comparison with the previous example we see that, in the absence of cross-relaxation, in a spectrum with mixed multiplet and net effects, the multiplet effect component of the enhancement will relax faster than the net effect.

2.1.3 Net effect for A only, $\sigma \neq o$.

$$P_A^+ = P_A; \quad P_X^+ = P_X^- = P_A^- = o$$

In the general case where cross-relaxation takes place between A and X, e.g. as in a dipole-dipole mechanism, the CIDNP intensities will be redistributed between A and X. A particularly striking effect of this type occurs when only one of the nuclei, e.g. A, is strongly polarized [10]. We will use as an extreme example the case where A and X relax solely by a dipole-dipole mechanism. Under those conditions [9]

$$W_o : W_{1A} : W_{1X} : W_2 = 2:3:3:12$$

If we assume zero-order kinetics, i.e. r = constant, we find

$$A_+(t)+A_-(t) = \frac{rP_A}{6\sigma}\{3[1-\exp(-\sigma t)] + [1-\exp(-3\sigma t)]\} \qquad (2.7)$$

$$X_+(t)+X_-(t) = \frac{rP_A}{6\sigma}\{3[1-\exp(-\sigma t)] - [1-\exp(-3\sigma t)]\} \qquad (2.8)$$

$$A_+(t) - A_-(t) = X_+(t) - X_-(t) = o$$

The time dependence of the A and X intensities is shown schematically in Figure 4. Two facts emerge from this analysis:

a) The steady state value of the X intensity is opposite in sign and half the magnitude of the steady state intensity of A, even though X was never directly pumped by the chemical reaction. This phenomenon has been called a Nuclear Overhauser Effect in a chemically pumped spin system [10]. Figure 2 shows such an effect in the emission line marked "R" which turns out to be the t-butyl protons of neopentylchloride enhanced by cross-relaxation with the line marked "C".

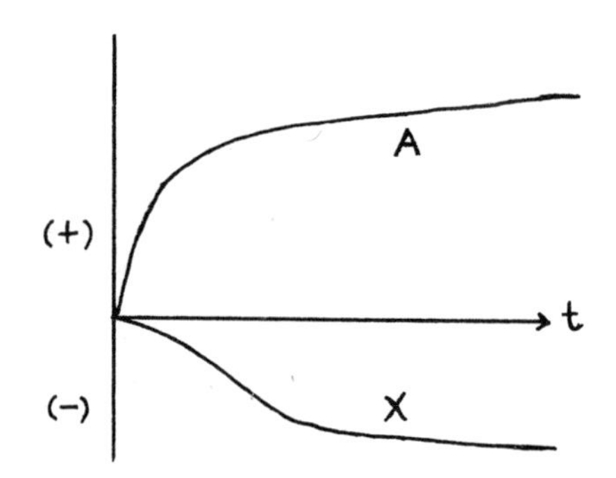

Figure 4. Time dependence of CIDNP intensities for two nuclei A and X which relax by the intramolecular dipole-dipole mechanism. It is assumed that A is chemically pumped to produce enhanced absorption and that the chemical reaction displays zero-order kinetics.

b) The enhancement of the X spins takes longer to develop than that in A since it is derived from the latter. It is thus possible to detect this effect by observing that the enhancement in X disappears if one looks at the spectrum a sufficiently short time after the reaction is started. Exactly this short of behavior is exhibited by the line marked "R" in Figure 1 if the spectrum is recorded at successively shorted times after saturation to remove residual CIDNP.

REFERENCES.

1. J. Bargon and H. Fischer, Z. Naturforsch., 22a, 1551 (1967).

2. S. Glarum, in "Chemically Induced Magnetic Polarization", Ed. G. L. Closs and A. R. Lepley, Wiley, New York, 1973; R. Kaptein and J. A. den Hollander, J. Am. Chem. Soc., 94, 6269 (1972); M. Lehnig and H. Fischer, Z. Naturforsch., 27a, 1300 (1972); H. R. Ward, R. G. Lawler, H. Y. Loken and R. A. Cooper, J. Am. Chem. Soc., 91, 4928 (1969); G. T. Evans and R. G. Lawler, Mol. Phys., 30, 1085 (1975).

3. S. Schaublin, A. Wokaun and R. R. Ernst, Chem. Phys., 14, 285 (1976).

4. S. Schaublin, A. Hohener and R. R. Ernst, J. Mag. Res., 13, 196 (1974).

5. J. A. Ferretti and R. R. Ernst, J. Chem. Phys., 65, 4283 (1976).

6. H. Fischer and G. P. Laroff, Chem. Phys., 3, 217 (1974).

7. A. L. Buchachenko and Sh.A. Markarian, Int. J. Chem. Kin., 4, 513, (1972).

8. G. L. Closs, Adv. Mag. Res., 7, 157 (1974); R. G. Lawler, Prog. NMR Spectros., 9, 145 (1975); R. Kaptein, J. Am. Chem. Soc., 94, 625 (1972).

9. C. L. Mayne, D. W. Alderman and D. M. Grant, J. Chem. Phys., 63, 2514 (1975).

10. G. L. Closs and M. S. Czeropski, Chem. Phys. Lett., 45, 115 (1977).

CHAPTER IV

CHEMICALLY INDUCED NUCLEAR SPIN POLARIZATION IN PHOTO-INITIATED RADICAL ION REACTIONS

Heinz D. Roth

Bell Laboratories, Murray Hill, New Jersey

Chemically induced nuclear spin polarization was discovered in thermal reactions[1,2] but its value as a tool in mechanistic photochemistry was soon recognized. The first observation of CIDNP in a photoreaction was reported by Cocivera in 1968;[3] since then, more than one hundred publications dealing with a wide range of photochemical systems have appeared. The first five years of photoinitiated CIDNP were reviewed in 1972.[4]

This paper will focus on CIDNP effects in radical ion reactions. We will begin, however, by discussing several examples of the observation of CIDNP effects in photoreactions involving neutral radicals, especially in systems which proved instrumental in developing the radical pair theory or which illustrate the application of the CIDNP technique in mechanistic photochemistry.

1. CIDNP IN PHOTOREACTIONS PROCEEDING VIA NEUTRAL RADICALS

The simplest mechanism of radical pair formation is bond homolysis. Unimolecular decomposition may result from thermal excitation, *i.e.* in a ground state reaction, or from a photoexcited state. For example, the photolysis of benzoyl peroxide leads to polarized scavenging products. The decomposition can also be achieved by photosensitization; in this case the signal direction is inverted suggesting a triplet precursor.[5,6]

$$^{1,3}C_6H_5\text{-}CO_2\text{-}O_2C\text{-}C_6H_5 \rightarrow {}^{1,3}\overline{C_6H_5\text{-}CO_2 \cdot\cdot C_6H_5} \rightarrow\rightarrow C_6H_5{}^{\dagger}X$$

L. T. Muus et al. (eds.), Chemically Induced Magnetic Polarization, 39-76.

(Here, as throughout this paper, a dagger, †, indicates nuclear spin polarization).

In the reaction of benzoyl peroxide, no polarization is observed for the coupling product and only one signal represents the protons in the ortho, meta, and para positions of the escape product. The photolysis of decafluorobenzoyl peroxide yields more information: because of more favorable spin lattice relaxation times polarization is observed for the coupling product as well as for the scavenging products; because of the wide range of ^{19}F chemical shifts, the ortho, meta, and para nuclei can be observed separately and all of them exhibit appreciable polarization.[7]

$$^{1}\overline{C_6F_5\text{-}CO_2\cdot \ \cdot C_6F_5} \rightarrow C_6F_5\text{-}CO_2\text{-}C_6F_5^{\dagger} + C_6F_5^{\dagger}X$$

The classic example of a homolysis reaction occurring exclusively upon photolysis is the Norrish Type I cleavage of ketones. The scope and mechanism of this reaction have been studied in great detail.[8] CIDNP studies confirmed that the cleavage in many cases occurs predominantly from the triplet state[9-12] as had been concluded from the results of standard photochemical experiments. The reactions of several aliphatic ketones show evidence for the involvement of excited-singlet and triplet precursors[11,13,14] which had been inferred previously from non-linear Stern-Volmer plots. The highly selective reaction of diisopropyl ketone with tetrachloromethane was explained by invoking an exciplex as an intermediate.[14]

A high degree of cage escape (and subsequent free radical encounters) was inferred from the selective use of free radical scavengers[11,12] and from the observation of cross coupling products.[13] For example, DoMinh photolyzed a mixture of benzoin methyl ether and its p,p′-dichloro derivative and observed four polarized benzoin ethers of comparable intensities (Figure 1).[13]

$$\begin{matrix} \text{Ar-CO-R} \\ \text{Ar}'\text{-CO-R}' \end{matrix} \rightarrow \left\{ \begin{matrix} \text{Ar-CO}\cdot \ \cdot\text{R} \\ \text{Ar}'\text{-CO}\cdot \ \cdot\text{R}' \end{matrix} \right\} \rightarrow \begin{matrix} \text{Ar-CO-R} & \text{Ar-CO-R}' \\ \text{Ar}'\text{-CO-R}' & \text{Ar}'\text{-CO-R} \end{matrix}$$

In contrast, a relatively high degree of cage return is indicated for the photolysis of dibenzyl ketone, since the polarization of this ketone was not affected by addition of the free radical quencher tri-n-butyl tin hydride.[12]

The photolysis of aromatic esters and ethers may result in intramolecular rearrangements generating substituted phenols. CIDNP studies suggest that the photo-Fries rearrangement of aromatic esters proceeds via radical pairs which are generated from the ester excited-singlet state.[15] A similar mechanism was assigned to the related rearrangement of phenyl ethers.[16,17] Of

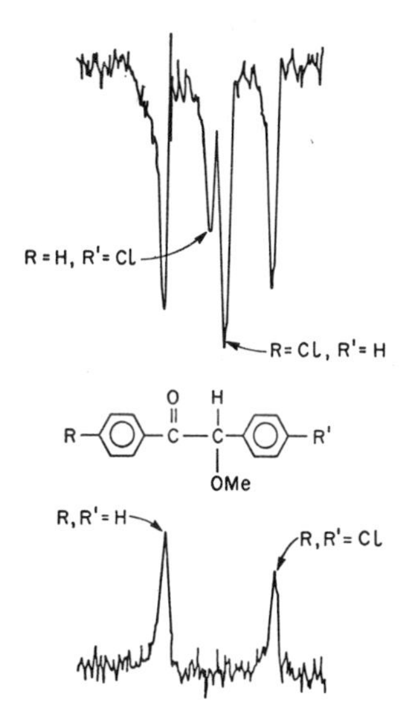

Figure 1. Pmr spectrum (benzylic protons) of benzoin methyl ether and its p,p'-dichloro derivative (bottom) and CIDNP spectrum observed during irradiation (top).[13]

special interest is the polarization of the meta coupling product in the photo-Claisen rearrangement of p-cresyl methallylether. Since this polarization was opposite to that of other coupling products, it was concluded that the meta product is formed by coupling of a triplet radical pair.[16] Examples of in-cage reactions of triplet pairs are rare since the conservation of spin angular momentum requires that the product is formed in the triplet state, which often lies considerably above the energy of the radical pair (cf. Section 7).

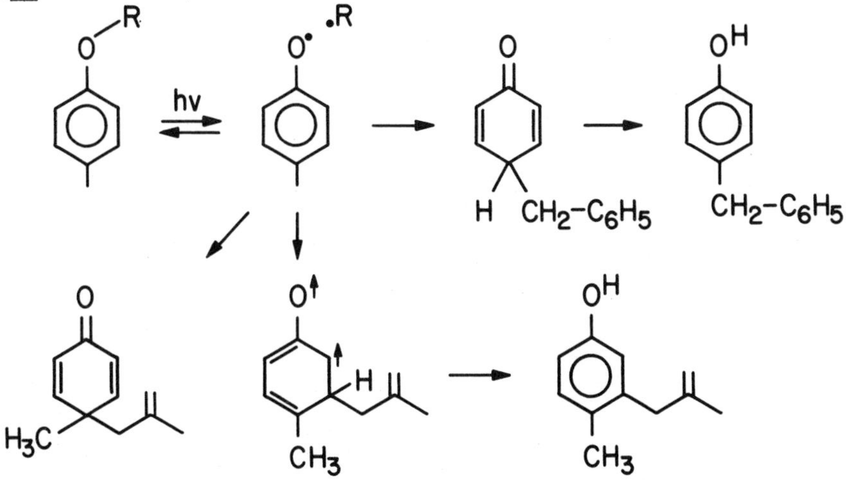

In these rearrangements, cyclohexadienones are primary coupling products, which undergo rapid hydrogen shifts to yield the phenolic products. However, under appropriate conditions the intermediates may be sufficiently long-lived to allow for the observation of their enhanced spectra. Thus, Benn and coworkers reported the CIDNP spectrum of 4-benzylcyclohexa-2,5-dienone during the photolysis of benzyl phenyl ether[17].

The identification of unstable diamagnetic intermediates is an attractive application of the CIDNP technique. The first intermediate identified this way was the enol of acetophenone,[18] which is formed during the photoreaction of acetophenone with phenol. The radical pair generated by hydrogen abstraction from the phenol reacts by hydrogen transfer from the ketyl methyl group to the phenoxyl radical. Additional enols have been generated by various chemical routes and their reactivity and magnetic parameters have been studied.[19-21] Another group of unstable intermediates, vinylamines, will be discussed in Section 5.

OH O• OH OH
CH3 → CH2

The reaction with phenols is just one of a large number of bimolecular reactions of carbonyl compounds involving radical pairs. Closs and coworkers systematically investigated the intermolecular photoreduction of carbonyl compounds by various substrates,[22] for example, of benzaldehydes by diphenylmethanes. This system played a key role in the development of the radical pair theory. The _g_ factor difference of the intermediate radical pair was varied by introducing para halogen substituents in carbonyl compound or substrate. In the extreme cases (Figure 2a,e) the hydroxybenzyl radical has either a considerably larger or a much smaller _g_ factor than the partner radical; therefore, the resulting spectra show net effects. For smaller _g_ factor differences (Figure 2b-d) the CIDNP spectra are dominated by multiplet effects.[22]

The reactions of divalent-carbon intermediates (carbenes) with many substrates proceed via radical pairs; therefore, it is not surprising that nuclear spin polarization has been observed in many carbene reactions. The first example, observed during the reaction of triplet diphenylmethylene with toluene illustrated an important feature of the radical pair theory, the dependence of the observed polarization upon precursor spin multiplicity.[23] The multiplet phase of the reaction product, triphenylethane (A/E),

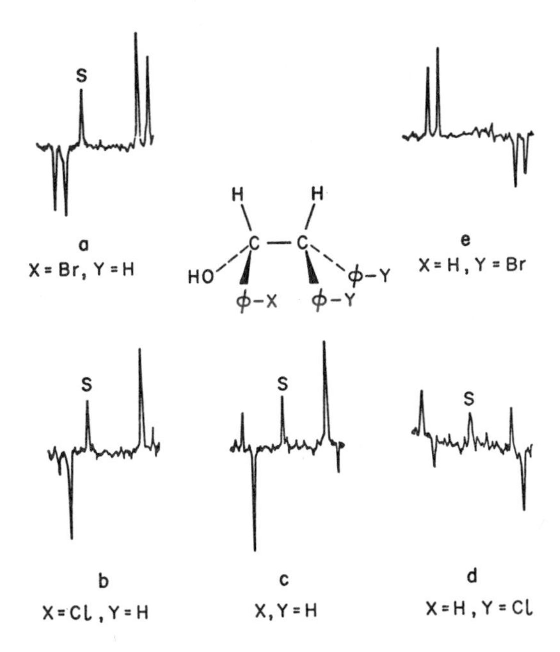

Figure 2. CIDNP spectra (benzylic protons) of coupling products obtained during the photoreaction of benzaldehydes with diphenylmethylenes. S indicates the low-field ^{13}C satellites of the substrate-solvent.

was opposite to that observed during the thermal cleavage of triphenylazomethane, which is generally considered to be a singlet reaction.[23]

$$^{3}\Phi_2C: + \Phi\text{-}CH_3 \xrightarrow{h\nu} \Phi_2CH\cdot \ \cdot H_2C\text{-}\Phi \xleftarrow{\Delta} {}^{1}\Phi_2CH\text{-}N{=}N\text{-}CH_2\text{-}\Phi$$

Subsequent CIDNP studies were aimed at the two standard mechanistic problems of carbene chemistry: to differentiate between "concerted" and radical pair reactions and to identify the spin multiplicity of the reacting carbene. In a systematic investigation of methylene and several monosubstituted derivatives, trichloromethane was found to be a unique probe for studying carbenes which may undergo intersystem crossing in competition with bimolecular reactions. The products formed by chlorine or hydrogen abstraction have characteristic nmr spectra so that the two reactions can be studied simultaneously and the precursor spin multiplicity can be assigned on the basis of the observed effects. These studies showed that singlet carbenes may undergo abstraction reactions and that they have a preference for chlorine abstraction whereas triplet carbenes preferentially abstract hydrogen.[24]

a) R = C_6H_5 b) R = $CO-CH_3$ c) R = $COOCH_3$ d) R = CN

The above selection of photochemical radical pair reactions is far from being complete but we hope that it conveys an impression of the wide range of reactions that may be studied by nuclear spin polarization techniques. In recent years, the CIDNP method has been applied to reactions of radical ion pairs which may be generated by photoinitiated electron transfer.[25] We will discuss the nuclear spin polarization observed in several reactions of this type in some detail.

2. PHOTO-INITIATED ELECTRON TRANSFER

The interactions between organic electron donors and acceptors cover a wide range of energies. Donors with low ionization potentials may react with acceptors of high electron affinities by electron transfer in the ground state; weaker interactions may be indicated by "charge transfer" bands in the electronic spectra; an even weaker association of poor donors with poor acceptors may be inferred from shifts in nmr line positions. For donor-acceptor pairs with weak associative forces in the ground state, electronic excitation of either reactant may result in strong interactions and lead to exciplex formation or radical ion pair generation. The classic criterion for an exciplex is the observation of a characteristic emission spectrum.[26] However, in recent years the intermediacy of exciplexes has also been inferred from specific chemical decay pathways.[27] The involvement of radical ions may be indicated by the observation of photoconductivity and confirmed by the identification of the transient species by electronic or esr spectroscopy.[29,30]

The solvent polarity is a major factor in determining the course of the reaction. The free energy of radical ion pair formation by the reaction of a photoexcited species with a ground state molecule is given by:[31]

$$\Delta G = E_{(D/D^{\cdot +})} - E_{(A^{\dot{-}}/A)} - E^{*}_{(0,0)} - e_0^2/\varepsilon a - T\Delta S$$

In this equation, $E_{(D/D^{\cdot +})}$ is the one-electron oxidation potential of the donor; $E_{(A^{\dot{-}}/A)}$ is the one-electron reduction potential of the acceptor; $E^{*}_{(0,0)}$ is the excitation energy (0,0 transition) of the photoexcited reactant; $e_0^2/\varepsilon a$ is a Coulomb term accounting for ion pairing and $T\Delta S$ is an entropy term. The first three of these parameters are known for many systems or can be measured relatively easily; the Coulomb term can be calculated for a given distance of separation from the dielectric constant of the solvent. The entropy term is rarely more than a minor correction since its upper limit, the entropy of formation of an exciplex, does not amount to more than 0.2 eV.

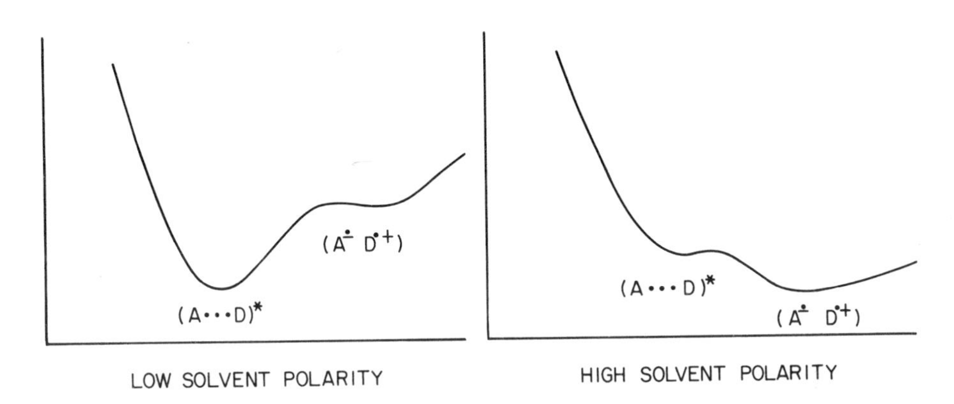

Figure 3. Schematic representation of the relative energies of an exciplex and of a radical ion pair in a low polarity solvent (left) and in a solvent of high polarity (right).

Figure 3 shows a schematic representation of the relative energies of an exciplex and of a radical ion pair in a polar and in a non-polar solvent. In non-polar solvents, the Coulomb term is important and exciplex formation predominates. In contrast, ion pair formation is favored in polar solvents because the energy of separation by diffusion becomes relatively small. These reaction conditions satisfy the chief prerequisite for the observation of CIDNP, since they allow the competition between the electron spin dependent reaction of a radical (ion) pair to form diamagnetic products and the electron spin independent separation of the individual radicals by diffusion. On the other hand, an exciplex has a relatively rigid geometry and the components do not have the mobility relative to each other that would be required for hyperfine-induced singlet-triplet mixing and, thus, for the generation of nuclear spin polarization. Consequently, one can expect to observe CIDNP effects in reactions involving radical ions but not from products formed directly from an exciplex.

3. REACTIONS OF RADICAL IONS

Radical ions, regardless of the mechanism of their formation, may undergo a variety of reactions. It is convenient to differentiate between unimolecular and bimolecular reactions. Among the unimolecular reactions of radical ions are rearrangement, *e.g.*, geometric isomerization or structural reorganization, and fragmentation, the cleavage of the radical ion into two or more fragments. The products of unimolecular rearrangements are always radical ions whereas a fragmentation reaction may generate a neutral, diamagnetic product plus a radical ion or it may form a neutral radical along with an ion.

In bimolecular reactions of radical ions several types of products may be formed. The encounter of two independently generated radical ions of opposite charge may result in the formation of an exciplex. In geminate radical ion pairs electron return regenerates the neutral reactants, most often in the ground state, but occasionally one of the reaction partners may be in the lowest triplet state (*vide infra*).

$$D^{\cdot +} + A^{\cdot -} \rightarrow D + A$$

In other donor-acceptor systems, the initial electron transfer may be followed by transfer of a proton or a different type of ion to the partner ion resulting in the formation of pairs of neutral radicals. This reaction, the bimolecular version of the fragmentation of a radical ion into a neutral radical and a proton, is important for the chemistry observed in reactions such as the photo-reduction of ketones or quinones by amines.

Frequently, degenerate electron exchange occurs between a radical ion and its diamagnetic parent. This exchange does not result in a net chemical change but it may give rise to line broadening or it may transfer polarization from a radical ion (with a short spin lattice relaxation time) to a neutral, diamagnetic product (with a considerably longer spin lattice relaxation time) so that it may be preserved. Degenerate exchange reactions play a major role in many systems that will be discussed in the following sections.

$$D^{\cdot +} + D \rightarrow D + D^{\cdot +}$$

The acidity of the solvent is of utmost importance in radical ion reactions. In basic solutions, the deprotonation of radical cations will be favored whereas in acidic solutions, radical anions may be protonated. In this case, the separation of the geminate pairs by diffusion will be facilitated. In a system undergoing rapid degenerate electron transfer, protonation may suppress the exchange, since the transfer of a hydrogen atom occurs at considerably slower rates than does electron exchange.

Another important reaction is the addition of radical ions to olefins resulting in an extended radical ion. This reaction is the reverse of the previously mentioned fragmentation of radical ions into a neutral diamagnetic fragment and a smaller radical ion. This reaction is an important step in the photo-sensitized dimerization of olefins (*vide infra*).

4. EFFECTS COMPETING WITH THE OBSERVATION OF RADICAL ION PAIR POLARIZATION

In many radical ion reactions giving rise to CIDNP effects, a slight change in one of the reaction parameters may result in line broadening effects in addition to or instead of the nuclear spin polarization. These broadening effects are often selective, *i.e.* the line widths of reaction partners and solvent are not affected. In addition, the lines representing different nuclei of the same reactant often show different degrees of broadening. Consequently, a general line broadening mechanism such as thermal paramagnetic relaxation is ruled out and a more selective mechanism is indicated.

Selective line broadening phenomena have been observed in a variety of reacting systems and have been explained by various chemical exchange mechanisms.[32] For example, the broadening of phenol spectra in the presence of phenoxyl radicals[33] and the broadening of the allyl iodide spectrum by transient allyl radicals[34] were explained by degenerate transfer of hydrogen and iodine atoms, respectively. In solutions containing N,N,N',N'-tetramethylphenylenediamine and its radical cation, Wurster's Blue, the observed broadening was explained by rapid electron exchange between those species.[35,36] Similarly, fast electron exchange can account for most cases of line broadening observed during photoreactions of donors with acceptors.[37-39] Formally, this exchange may involve a free radical ion

$$A^{\pm} + A \rightarrow A + A^{\pm}$$

or it may occur upon encounter of a radical ion pair with a neutral molecule, as a degenerate radical pair substitution

$$(A^{\dot{-}}D^{\cdot+}) + A \rightarrow A + (D^{\cdot+}A^{\dot{-}})$$

For the exchange contribution to the nmr line width, ΔT_2^{-1}, two limiting cases have been formulated,[32] the strong-pulse (slow-exchange) case

$$\Delta T_2^{-1} = k_e[P]$$

and the weak-pulse (fast-exchange) case

$$\Delta T_2^{-1} = \underline{a}_n^2[P]/4k_e[D]^2$$

where [D] and [P] are the concentrations of the diamagnetic species and its radical ion, respectively, k_e is the rate constant of electron exchange and $\underline{a}_n$ is the hyperfine coupling constant of the nucleus whose spectrum is being observed. These equations offer three criteria to differentiate between fast and slow exchange. Since k_e should increase with temperature, the broadening should show a negative temperature coefficient in the fast-exchange limit ($\Delta T_2^{-1} \propto 1/k_e$) whereas a positive temperature coefficient is expected for slow exchange ($\Delta T_2^{-1} \propto k_e$). In addition, the magnitude of the hyperfine coupling constants and the concentration of the diamagnetic precursor should have a marked influence on the degree of broadening in the fast-exchange limit but not in the case of slow exchange. We illustrate these relations with spectra observed during the electron transfer quenching of photo-excited decafluorobenzophenone (DFB) by 1,4-dimethoxybenzene (DMB) and N,N-diethyl-p-toluidine (DET), respectively.[38,40]

The role of the magnitude of the hyperfine coupling constants is illustrated by the broadened pmr spectra observed during the electron-transfer quenching of ^{3}DFB* by DET. In this system, the degree of broadening reflects the spin density distribution in the radical ion derived from DET. The methylene quartet ($\underline{a} \sim 12G$) was substantially broadened whereas the corresponding methyl triplet ($\underline{a} < 1G$) remained essentially unchanged; the aromatic region showed pronounced broadening for the ortho protons ($\underline{a}_o = 5.3G$) and somewhat weaker broadening of the meta protons ($\underline{a}_m = 1.35G$). These effects show the dependence expected for electron exchange between DET and the N,N-diethyl-p-toluidinium radical ion in the rapid-exchange limit.[38]

Figure 5 shows the temperature dependence and the effect of quencher concentration on the pmr spectrum of the methoxy signal of DMB during the electron transfer quenching of ^{3}DFB*. Clearly, the degree of broadening shows a negative temperature coefficient and increases with decreasing quencher concentration. These results are compatible with rapid electron exchange between DMB and its radical cation.[40]

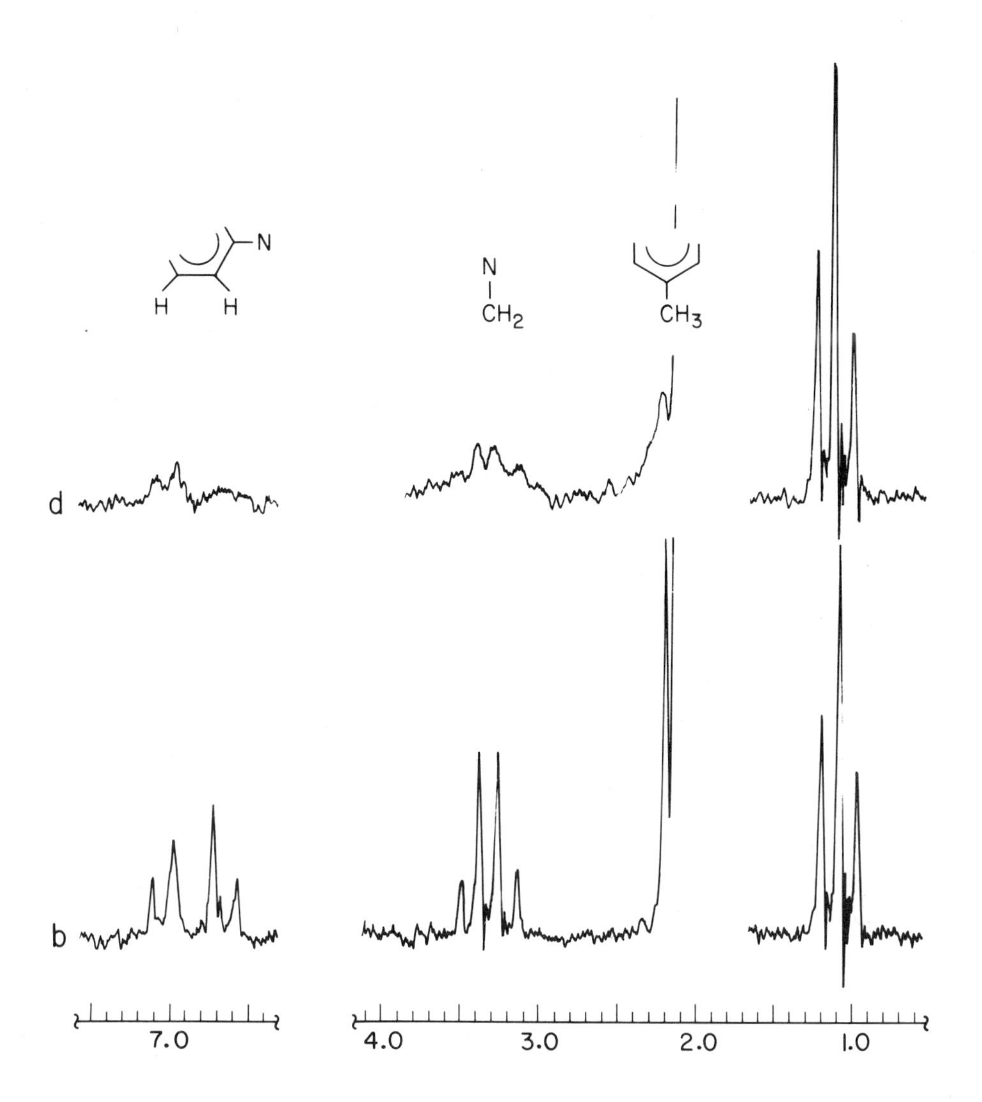

Figure 4. Pmr spectra (60 MHz) of a solution of 0.02-M N,N-diethyl-p-toluidine and 0.02-M decafluorobenzophenone in acetonitrile-d_3 in the dark (bottom) and during uv irradiation (top). The broadened aromatic, aminomethylene, and benzylic methyl resonances are labelled by the appropriate structural fragments.

The variations in the degree of line broadening as a function of temperature and hyperfine coupling constants appear to be valid criteria for the differentiation between rapid and slow electron

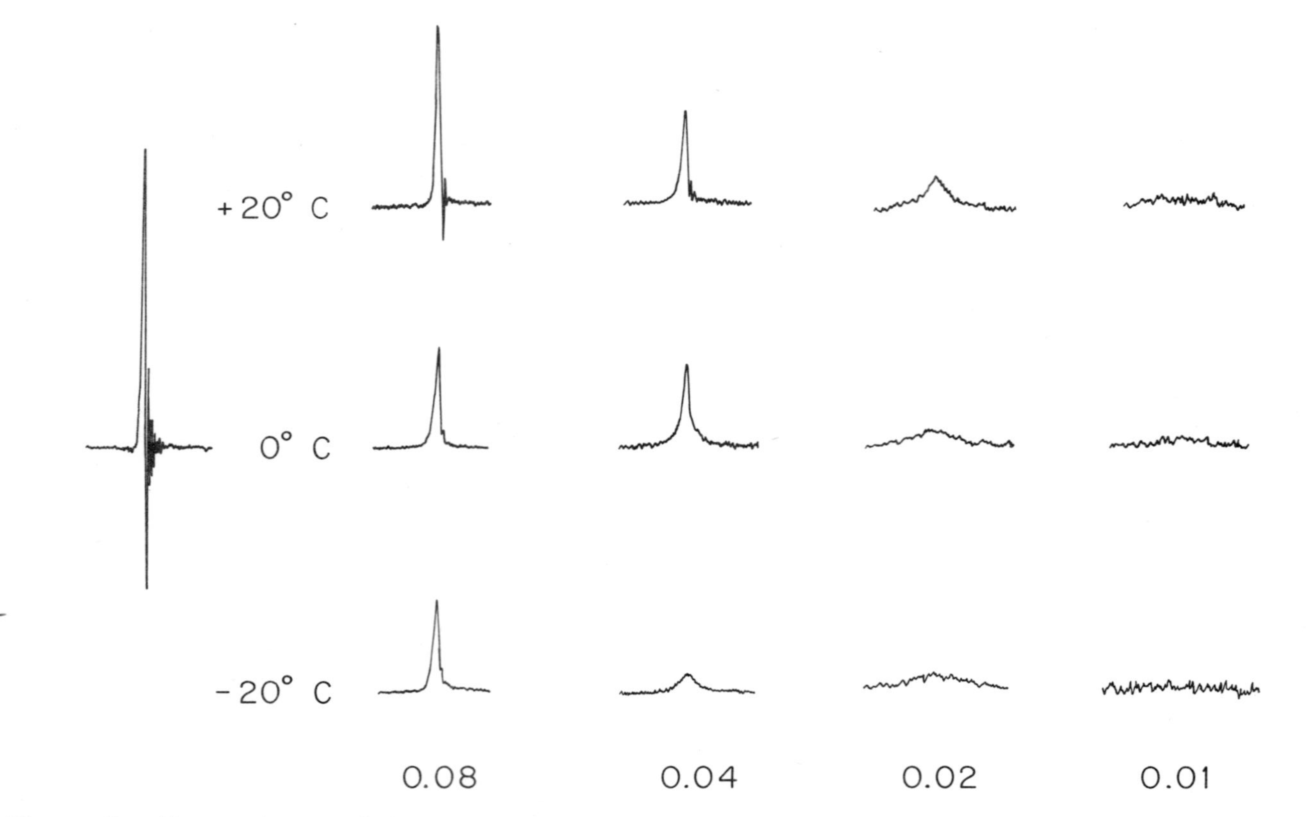

Figure 5. Temperature and concentration dependence of the line broadening observed for the methoxy signal of 1,4-dimethoxybenzene (DMB) during the electron transfer quenching of triplet decafluorobenzophenone by DMB in acetonitrile-$\underline{d}_3$.

exchange. The dependence upon reactant concentration is less reliable as a criterion since the concentration of either reaction partner may have a marked influence on the concentration of radical ions produced in a photoreaction and since this concentration affects the degree of broadening in both exchange limits.

Degenerate electron exchange accounts for the observation of line broadening but we have yet to explain the failure to observe nuclear spin polarization, which one may expect for reactions initiated by the generation of radical ion pairs. In order to observe pure line broadening, as discussed so far, there has to be a provision either to prevent the generation of nuclear spin polarization or to cancel any polarization that may have been induced in these pairs. The generation of spin polarization will be inhibited in radical ion pairs with marginal free energies of formation. In these cases, the separation of the radical ions by diffusion is inefficient and the line broadening would have to be ascribed to degenerate pair substitution. In the case of radical ion pairs, which diffuse apart efficiently, nuclear spin polarization is induced, but may be cancelled _via_ either of two mechanisms. First, cancellation may occur if the radical ions are relatively stable so that the pair lifetime is longer than the spin lattice relaxation time of the radical ions. Secondly, cancellation may occur because the reaction partners may be regenerated _via_ two independent pathways. In-cage electron return will regenerate the reactants and transfer to them a characteristic polarization pattern; degenerate electron exchange between "free" radical ions and their diamagnetic precursors will transfer the opposite polarization to the reactants. In most cases, the in-cage polarization, which is transferred to the reactant-product without loss, will outweigh the escape polarization. However, if the lifetime of an escaping radical is considerably shorter than its spin lattice relaxation time, the escape polarization will be transferred to the starting material-product with an insignificant loss and, therefore, effectively cancel the in-cage polarization of opposite sign. Only line broadening will be observed under these conditions.

However, the complete cancellation of the in-cage polarization can be avoided. Since the lifetime of a polarized radical ion depends on the reactant concentration but the spin lattice relaxation is independent of it, it is possible to vary the lifetime of the intermediate and, thereby, the degree of relaxation before exchange. Sufficiently high reactant concentrations will allow for complete cancellation of nuclear spin polarization and, thus, result in pure line broadening; at lower reactant concentrations, relaxation of the escape polarization will occur and the in-cage polarization will be observed in addition to line broadening. The result is a broadened CIDNP spectrum. Effects of this type have been observed, for example, during the photoreactions of porphyrins and chlorins with benzoquinones and benzohydroquinones.[41] We consider both line broadening and broadened CIDNP spectra characteristic phenomena for many photo-initiated radical ion reactions. To date, electron exchange induced line broadening has been observed for ketones and quinones, amines and aromatic electron donors. We see no reason why additional substrates should not show this effect under appropriate reaction conditions.

However, degenerate electron exchange is not the only possible cause for line broadening during the irradiation of organic materials. As early as 1968, Cocivera reported broadening of the pyrene spectrum during uv irradiation in deuteriobenzene and ascribed this phenomenon to degenerate excitation energy transfer.[42] The observed broadening was uniform for all signals, thus indicating that the strong pulse limit had been reached. Under these conditions, the rate constant of exchange can be determined from the degree of broadening ($k_e = 4 \times 10^7$ $lmol^{-1}sec^{-1}$) but no information is available concerning the spin density distribution in the triplet state.

$$^3Py^* + {}^1Py_0 \rightarrow {}^1Py_0 + {}^3Py^*.$$

More recently, Boxer and Closs were successful in observing degenerate singlet-triplet exchange in the weak-pulse limit.[43] Under these conditions, the nuclei are not completely dephased and the degree of broadening reflects the spin density distribution in the triplet state ($\Delta T_2^{-1} \propto \underline{a}^2$). The uv irradiation of anthracene in bromobenzene resulted in differential line broadening which increased with decreasing anthracene concentration. From these results, the rate constant of exchange ($k_e = 3 \times 10^9$ $lmol^{-1}sec^{-1}$) was determined. In the case of anthracene, the spin density distribution was known from ENDOR experiments, but for other organic materials line broadening due to singlet-triplet exchange should be of value in characterizing the spin density distribution of the triplet state. For example, this method has been applied to photosynthetic pigments. As a process interfering with the generation of radical ion pair polarization, excitation energy transfer can be avoided. Since the rate constant for singlet-triplet exchange are either substantially (pyrene) or marginally slower (anthracene) than the diffusion controlled rate, the exchange can be eliminated by choosing an efficient quencher at sufficient concentrations to ensure that the rate of quenching is larger than the rate of degenerate exchange.

$$k_q[Q] \gg k_e[K]$$

Occasionally, an alternative polarization mechanism competes with radical ion pair polarization, specifically in photoreactions of quinones and fluorine-substituted aryl alkyl ketones. A detailed description of this mechanism and its kinetic requirements is given in Chapter XXI.

5. PHOTOREDUCTION OF KETONES AND QUINONES BY AMINES

Photoexcited aryl ketones undergo fast reactions with amines which may result in ketone reduction or in quenching of the triplet ketone without net chemical change. A large number of experimental studies have been carried out in an effort to elucidate the detailed mechanism of this reaction. The results indicate that charge transfer from the amine to the excited ketone must be considered in any mechanistic formulation.[44] The following generalized scheme includes the pathways which have been proposed: the formation of an exciplex, electron transfer producing a radical ion pair, and hydrogen abstraction leading to a pair of neutral radicals. The neutral intermediates are necessary to explain any reaction products; their formation may occur via direct hydrogen abstraction or via a two-step process: electron transfer followed by proton transfer.

ELECTRON TRANSFER
ELECTRON RETURN
DEEXCITATION
RADICAL ION PAIR
EXCIPLEX
PROTON TRANSFER
RADICAL PAIR
PRODUCTS

As noted above, nuclear spin polarization cannot be generated in an exciplex. Accordingly, the CIDNP method does not provide any information concerning the intermediacy of excited-state complexes. However, this technique can be used to identify the nature of the radical intermediates because the g factors and the hyperfine coupling patterns of aminium radical ions and of aminoalkyl radicals may be substantially different. In addition, indirect evidence has been used to identify the intermediates.

For example, the photoreduction of aromatic ketones with primary anilines may proceed via anilinyl radicals (hydrogen abstraction) or via anilinium radical ions (electron transfer). If both these intermediates are π radicals, i.e. if the spin density is fully delocalized into the benzene ring, they should give rise to identical or very similar spectra. Obviously, the polarization pattern observed for two species, which differ only by protonation or deprotonation at the site of unpaired spin density, should not be very different.

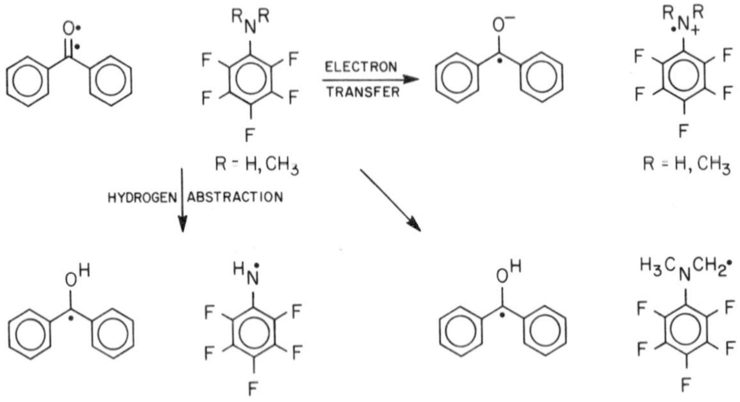

ELECTRON TRANSFER

R = H, CH_3

R = H, CH_3

HYDROGEN ABSTRACTION

H_3C N CH_2•

The mechanism was assigned on the basis of results obtained during the irradiation of the ketones in the presence of N,N-dialkylanilines. The tertiary anilines may form anilinium radical ions in a one-step process but cannot produce anilinyl radicals directly. Since the N,N-dialkylanilines did not show CIDNP effects and since the energetics of anilinium radical ion formation should be somewhat more favorable for tertiary than for primary anilines, it appears that the polarization of the primary anilines cannot be due to a radical ion pair but is induced in the pair of neutral radicals generated by hydrogen abstraction.

Several other ketone - amine systems provide direct evidence for the nature of the intermediates. For example, the quenching of benzophenone and several p,p'-substituted derivatives by the bicyclic diamine dabco results in strogly enhanced ketone spectra.[37] In this system, the g factor difference is the key to the

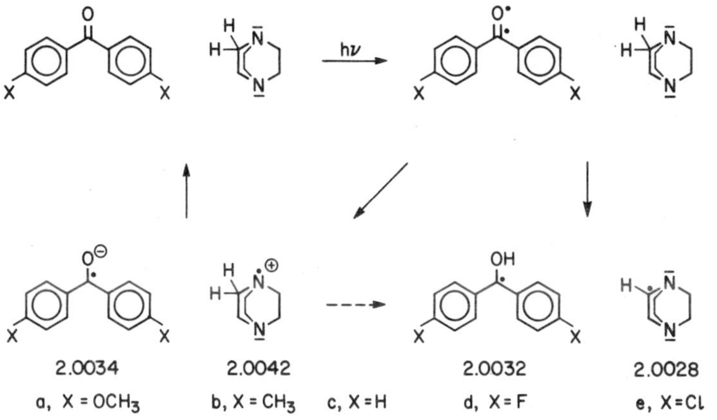

2.0034 2.0042 2.0032 2.0028

a, X = OCH_3 b, X = CH_3 c, X = H d, X = F e, X = Cl

identification of the intermediates. The radical cation derived from dabco has a substantially larger $\underline{g}$ factor (2.0040) than the ketyl radical anion (~2.0032) whereas the aminoalkyl radical has a slightly smaller $\underline{g}$ factor (~2.0028) than the diphenylhydroxymethyl radical (~2.0030). Three of the polarization determining parameters can be assigned readily: only the triplet state of the photoexcited ketones can be quenched at low amine concentrations (5×10^{-2}-$\underline{M}$) since the singlet state is too short-lived ($\mu > 0$); the ketones are most likely regenerated by an in-cage reaction ($\varepsilon > 0$); the proton hyperfine coupling constants of the intermediates are those of π radicals with alternating carbon spin densities ($\underline{a}_{ortho} < 0 < \underline{a}_{meta}$). Given these parameters, the observed signal directions, A for the ortho protons, E for the meta protons (and for the methyl group of $\underline{p},\underline{p}$'-dimethylbenzophenone), are compatible only with $\underline{g} < 0$ and, thus, indicate that the quenching reaction proceeds via electron transfer.[37]

Reaction parameters affecting the stability of the intermediate radical ion pairs, e.g. the reduction potential of the ketone, the oxidation potential of the amine, or the dielectric constant of the solvent, may have a pronounced effect on the observed phenomena. For example, the formation of radical ions should be more favorable in polar than in non-polar solvents. As illustrated in Figure 6, a and b, the CIDNP enhancements of the aromatic signals of $\underline{p},\underline{p}$'-dimethylbenzophenone are considerably stronger in acetonitrile ($\varepsilon = 36$) than in hexafluorobenzene ($\varepsilon \sim 2$). Similarly, electron transfer to the photoexcited ketone should be more favorable for ketones with electron-withdrawing substituents (Cl or F) than for ketones with electron-donating substituents (CH_3 or OCH_3). At the same time, the reverse electron transfer regenerating the reactants should be less exothermic for $\underline{p},\underline{p}$'-dichloro than for $\underline{p},\underline{p}$'-dimethylbenzophenone. The resulting longer lifetimes of the radical ion pairs and their higher steady-state concentrations should give rise to stronger

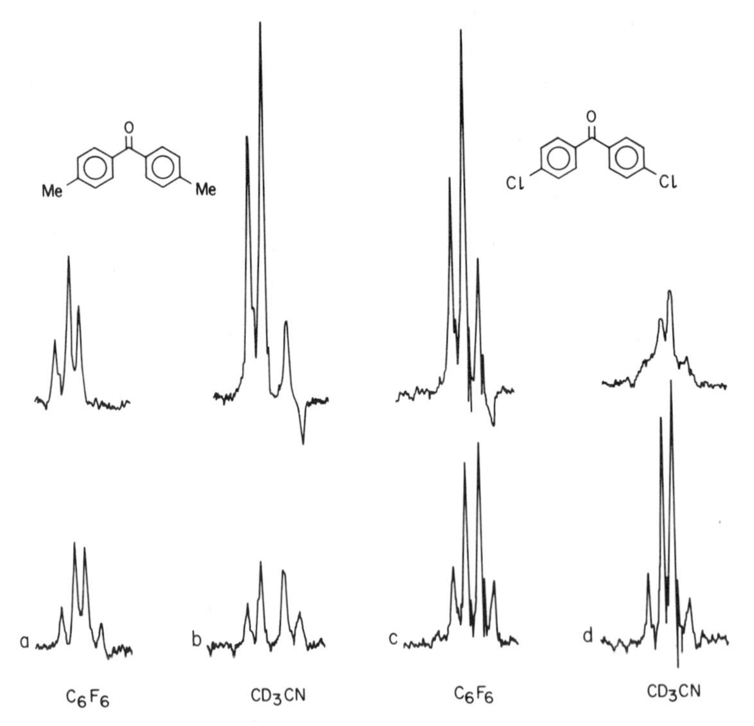

Figure 6. Pmr spectra (60 MHz) observed during the quenching of p,p'-dimethylbenzophenone and p,p'-dichlorobenzophenone by dabco in C_6F_6 and CD_3CN, respectively. The dark spectra are added for comparison.

enhancements. This consideration is born out by a comparison of Figure 6, a and c.

In contrast, a comparison of Figure 6, b and d, does not show the expected stronger enhancement for the chlorine-substituted ketone. Instead, line broadening is observed. This phenomenon indicates that in acetonitrile the radical ion concentrations are sufficiently high and their lifetimes are sufficiently long to permit spin lattice relaxation. As a result, only the line broadening due to degenerate electron exchange between ketyl anion and parent ketone can be observed.[37]

In the interaction with dabco, benzophenone is quenched efficiently but no products are formed in appreciable quantum yields. This result is understandable in view of the rigid geometry of both dabco and its radical ion. For both species, the axes of the carbon hydrogen bonds are nearly orthogonal to the orbitals containing the non-bonding electron pair and the unpaired electron, respectively.

Therefore, neither the direct hydrogen abstraction from the amine nor the proton transfer from the aminium radical ion derive any assistance from the interaction with these orbitals despite their proximity.

For conformationally mobile amines, this restriction does not apply. Therefore, the formation of the neutral radicals should be facilitated. The photoreactions of benzophenone and several of its derivatives with triethylamine result in strongly enhanced spectra, which have been assigned to the vinyl protons of diethylvinylamine. This unsaturated amine is unstable under the reaction conditions and has not been isolated. In this system, the intermediates can be identified on the basis of the observed polarization patterns. The neutral aminoalkyl radicals resulting from hydrogen abstraction have sizeable hyperfine coupling constants for both α and β protons whereas the aminium radical cation has an appreciable hyperfine coupling constant only in the α position.

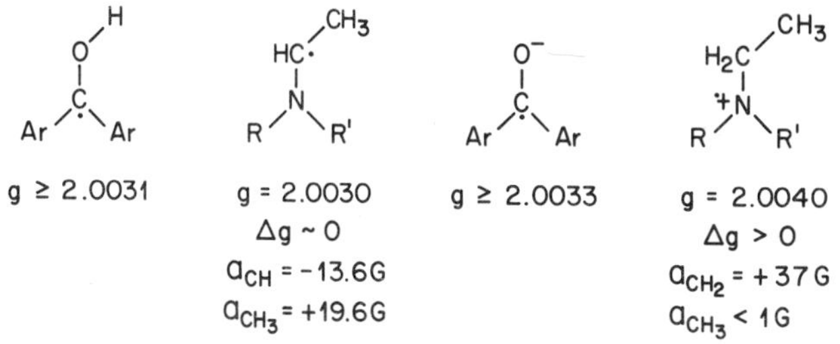

$Ar_2\dot{C}$–O–H	$R(R')N$–$\dot{C}H$–CH_3	$Ar_2\dot{C}$–O^-	$R(R')\overset{\cdot+}{N}$–CH_2–CH_3
$g \geq 2.0031$	$g = 2.0030$	$g \geq 2.0033$	$g = 2.0040$
	$\Delta g \sim 0$		$\Delta g > 0$
	$a_{CH} = -13.6G$		$a_{CH_2} = +37G$
	$a_{CH_3} = +19.6G$		$a_{CH_3} < 1G$

The CIDNP spectra observed in these systems show polarization for both α and β protons. For example, the reaction of p,p'-dichlorobenzophenone with triethylamine produced emission for the doublet of doublets representing the internal proton, and enhanced absorption for the overlapping doublets representing the terminal protons. For acetone or p,p'-dimethylbenzophenone, the photoreaction with triethylamine resulted in A/E multiplet effects for all three vinyl protons. These results are compatible only with the neutral aminoalkyl radical as precursor of the vinylamine.[39]

With N,N-diethyl-p-toluidine as quencher, nuclear spin polarization was observed for the starting amine as well as for a reaction product. Three A/E multiplets with comparable enhancements can be assigned to the three protons of a vinyl group. Obviously, this polarization is induced in a neutral aminoalkyl radical. At the same time, net polarization was found for the methylene quartet (strong emission) and for the ortho protons (enhanced absorption) of the starting amine but not for its methyl triplet. Since these effects are different in type and pattern from those observed for the vinyl compound, they must originate in a different precursor pair. Signal direction and polarization pattern are compatible with the radical ion pair that can be generated by electron transfer. Thus, the experimental observations suggest the occurrence of both, net hydrogen abstraction and electron transfer. Neutral aminoalkyl radicals are intermediates in the formation of products whereas the aminium radical ions regenerate the starting reactants by reverse electron transfer.[39]

These results demonstrate that the CIDNP method can differentiate between aminoalkyl radicals and aminium radical ions, in favorable cases even in the presence of each other. However, the question has yet to be answered whether these intermediates are formed independently from a common precursor or whether the aminoalkyl radicals are formed in a two-step process initiated by electron transfer. The results obtained in the ketone - amine systems discussed above do not lend themselves to such an assignment. However, the irradiation of benzoquinone in the presence of triethylamine gives rise to nuclear spin polarization effects, which shed some light on this question. The observed polarization patterns are incompatible with either the aminium radical ion or the aminoalkyl radical, but they may be interpreted as a superposition of effects originating in both species. These results can be considered evidence for a two-step mechanism of aminoalkyl radical formation.[46]

Figure 7 shows three different polarization patterns for the vinyl group generated by the photoreaction of quinones with triethylamine. One of these (Figure 7c), observed with anthraquinone as reaction partner, is similar to the spectrum observed during the reaction of p,p'-dichlorobenzophenone; it reflects the spin density distribution of an aminoalkyl radical. The reaction of benzoquinone with triethylamine gives rise to substantially different spectra. In acetonitrile, the all-absorption spectrum of Figure 7a is observed whereas in acetone only the terminal protons show polarization (Figure 7b). These observations are incompatible with any chemically reasonable intermediate but they could be the result of two contributions, one from the aminium radical ion, the other one from the aminoalkyl radical. Since the product vinylamines cannot be formed from the radical ion pair in a simple direct reaction, the polarization generated in a radical ion pair must be transferred to the product amines via the neutral aminoalkyl radical.

2.0045 2.0040 → 2.0038 2.0030

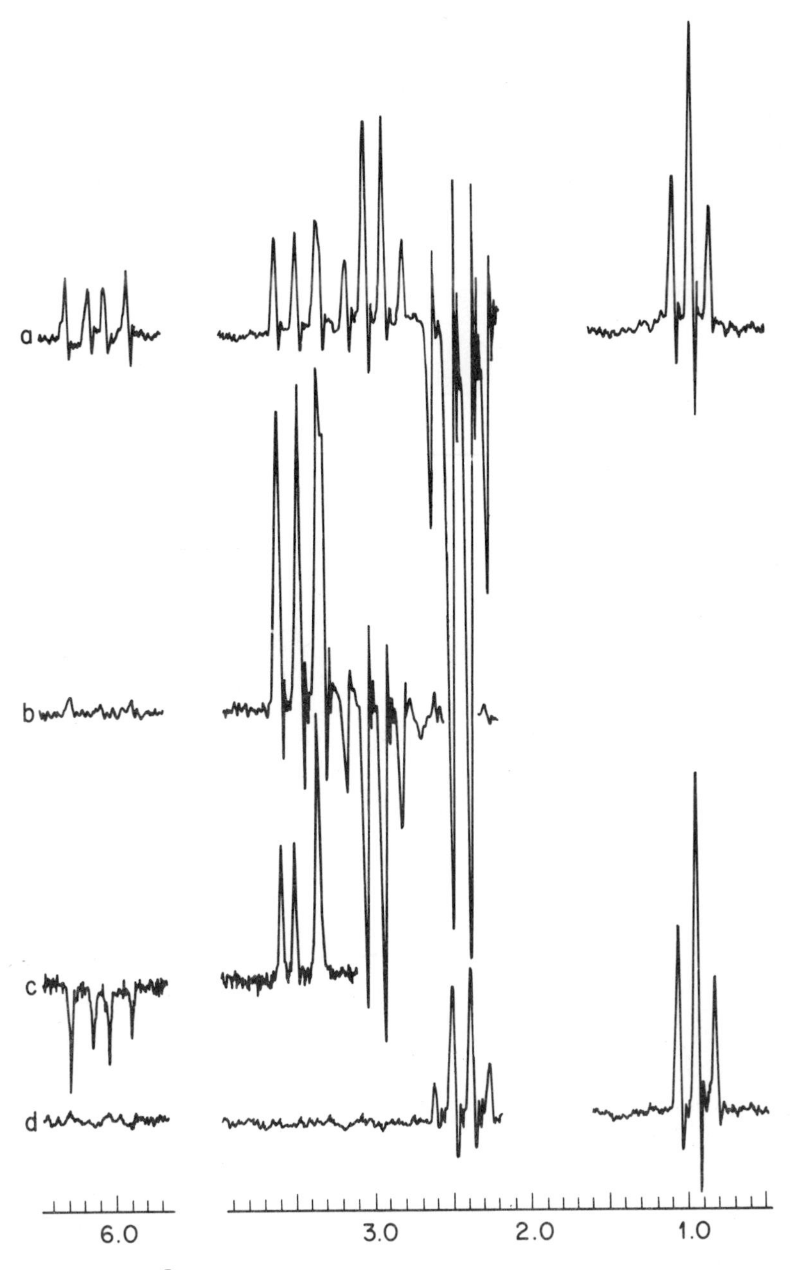

Figure 7. ^{1}H CIDNP spectra (60 MHz) observed during the photoreaction of benzoquinone (a,b) and anthraquinone (c), respectively, with triethylamine in acetonitrile-$\underline{d}_3$ (a,c) or acetone-$\underline{d}_6$. The dark spectrum is shown in trace d.

The assignment of an observed polarization to an intermediate other than the immediate precursor is not without precendent. Several authors have reported that the plarization generated in a geminate radical pair may be observed in products derived from a subsequent pair with different values of g or a. These effects are known as "cooperative effects."[47,48] This model implies that the precursor spin multiplicity and the mechanism of product formation (the parameters μ and ε in Kaptein's formalism) are the same for the contributions from both pairs. In the case discussed here, an examination of the g factors shows that Δg has the same sign for both pairs. Consequently, the contributions of the two pairs to the polarization of the internal proton should be of opposite sign; the signal direction of the terminal protons is, of course, solely determined by the neutral intermediate.

The relative contribution of the radical ion pair should depend on its lifetime prior to proton transfer and should be stronger in a more polar solvent than in a less polar one. The observed results are fully consistent with this expectation. In acetonitrile, the contribution of the ion pair dominates a weak contribution due to the neutral radical, whereas in acetone the two contributions are of equal magnitude and, thus, cancel each other. The results observed in the photoreduction of anthraquinone are consistent with this explanation. Anthraquinone has a less favorable reduction potential (0.94V) than benzoquinone (0.52V); therefore, the corresponding radical ion pair is less stable and does not contribute appreciably to the vinylamine polarization (Figure 7 c). Nevertheless, some involvement of electron transfer in the anthraquinone photoreduction is suggested by the broadening of the anthraquinone spectrum and by the somewhat decreased absorption of the methylene quartet. However, this effect is not nearly as strong as the pronounced emission observed in the more favorable case of benzoquinone (Figure 7 a).

The assumption that the aminium radical cation contributes to the vinylamine polarization implies that the neutral aminoalkyl radical, the only chemically reasonable link between the two species, is formed *via* proton transfer from the radical anion. Consequently, the CIDNP results observed in the system benzoquinone-triethylamine elucidate an important mechanistic facet: they reveal that, in this system, net hydrogen abstraction is a two-step process.[46]

In conclusion, the CIDNP effects observed during the photoreduction of ketones or quinones by amines reveal a wide range of reactivities. In several cases, the esr parameters of the potential intermediates are sufficiently different to allow the differentiation between electron transfer, which is dominant, for example, in the quenching of ketones by dabco, and net hydrogen abstraction, which accounts for the photoreduction of several ketones by acyclic tertiary amines. Of special interest are systems such as benzophenone - N,N-diethyltoluidine, where neutral radical and radical ion reactions are detected simultaneously, and the case benzoquinone-triethylamine, where the aminium radical ion was indicated as a precursor of the aminoalkyl radical.

6. PHOTOSENSITIZED ISOMERIZATION AND DIMERIZATION OF OLEFINS

Olefins are efficient quenchers for the excited states of various photoreactive materials. The quenching interaction may occur without net chemical change or it may initiate the dimerization or geometric isomerization of the olefins. Two mechanistic pathways are frequently discussed for these reactions. One mechanism involves triplet energy transfer and subsequent reaction of the triplet olefins; alternatively an electron transfer process followed by a radical ion reaction may be discussed. A variety of reactions of this type have been studied by nmr techniques and strong CIDNP effects have been observed in a number of systems. The CIDNP data are consistent with radical ion pair reactions and suggest an electron transfer process as primary quenching reaction.[49]

The irradiation of electron acceptors, mainly quinones, in the presence of donor olefins, such as N-vinylcarbazol, phenylvinylether or dimethylindene, led to strong polarization of the olefinic protons and weaker polarization of the aromatic protons of the starting materials. At low conversion to dimers, no polarization was observed for any dimer signal. In these systems, three of the parameters determining the polarization can be assigned readily. Regardless of the nature of the olefin it is reasonable to assume that the triplet state of the quinone is involved in the reaction ($\mu > 0$) since the singlet state is too short-lived. The most likely mechanism for the regeneration of the olefins is in-cage electron return ($\varepsilon > 0$). For all three cases, the $\underline{g}$ factors of the olefin-derived radical cations should be smaller than that of the semiquinone radical anion ($\Delta \underline{g} < 0$). Consequently, the observed CIDNP effects allow some conclusions concerning the spin density distribution in the intermediate radical cations.

The polarization patterns of N-vinylcarbazole and phenylvinylether are quite similar. The terminal olefinic protons appear in enhanced absorption indicating negative hyperfine coupling constants whereas the internal protons, represented by a doublet of doublets, show emission and, thus, indicate positive hyperfine coupling constants. These results suggest structures where the positive charge is localized on the heteroatom whereas the unpaired spin is delocalized between the terminal carbon, the heteroatom, and the aromatic moiety.

The effects observed in the reaction of dimethylindene are different. Two doublets representing the α and the β proton appear in enhanced absorption, but the α doublet is considerably weaker. This finding indicates that the unpaired spin density is higher in the homobenzylic than in the benzylic position and that the radical ion derives optimal stabilization from the more favorable delocalization of the positive charge into the benzene ring.

The radical ion pairs undergoing in-cage electron return represent one fraction of the initially produced pairs which, aside from the dissipation of the excitation energy, do not result in a net chemical change. Another fraction of ion pairs, carrying polarization of equal intensity but of opposite sign, will separate by diffusion and undergo free radical ion reactions. Among these, the interactions of the radical cations with unreacted olefin are important for the formation of two types of products. These encounters may result in degenerate electron exchange or in addition as shown in the Scheme.[49]

1,3(S⁻· A B) → S A B

S⁻· A B → A B → (e⁻) A B

A B A B → B A⁺· B A → (e⁻) B A B A

The electron exchange does not, per se, generate new products. However, in the case of donor olefins capable of geometric isomerization, the radical cation may isomerize before the encounter with the unreacted olefin, so that the electron exchange may produce a polarized isomeric olefin. The CIDNP spectra observed during the irradiation of quinones in the presence of propenylbenzene or anethol are typical examples for the polarization of isomerized olefins. During the reaction of trans-anethol, the β protons (emission) and the allylic methyl group (enhanced absorption) of the cis-isomer were polarized in addition to the corresponding nuclei of the starting olefin (cf. Figure 8). Generally, the polarization of the rearranged olefin is of lower intensity than that of the starting material and of opposite sign, as can be expected for products formed by an escape mechanism. The degree of polarization of the rearranged olefin depends on the lifetime of the free ion before exchange, τe, relative to the spin lattice relaxation time (T_1) of the nuclei in the free radical ion. Since relaxation is a first-order process, whereas exchange is a second-order reaction, the olefin concentration will influence the degree of polarization.

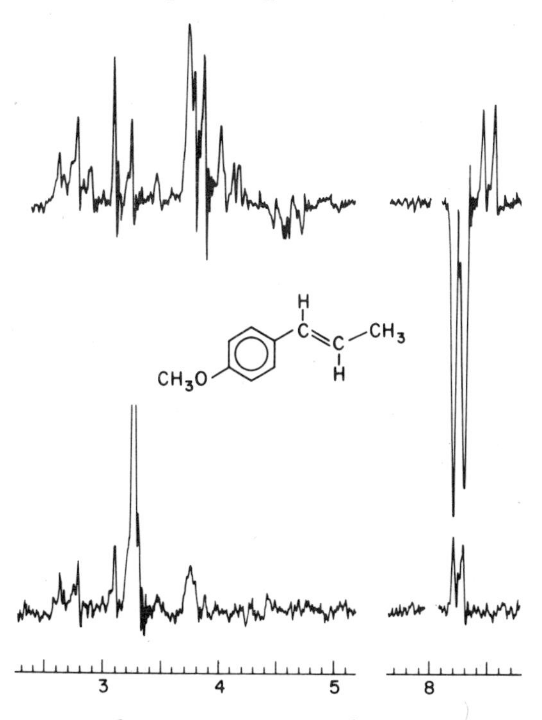

Figure 8. Pmr spectra (60MHz) during the photoreaction of tetrachlorobenzoquinone with anethol (top) and of the same solution in the dark (bottom).

Instead of undergoing electron exchange, radical ions and olefins may react by addition. This reaction is important since it initiates the formation of dimers, which is completed via two additional steps, ring closure to a seondary adduct and electron transfer from a suitable reaction partner such as the olefin. The stability of the primary adduct will depend to a large degree on the delocalization of the unpaired spin and of the positive charge; therefore a head-to-head adduct is usually the most stable one.[50-52] The relative thermodynamic stability of the primary adducts is the key to the high selectivity of electron-transfer induced dimerization of olefins. In contrast, the triplet sensitized reaction is much less selective.[53]

During the acceptor-sensitized dimer formation from donor olefins (N-vinylcarbazol, phenylvinylether, dimethylindene) the monomers showed CIDNP effects but no polarization was observed for any of the dimers. This finding is quite reasonable in view of the mechanism suggested above. As shown in the case of propenyl benzene or anethol, an escaped radical ion has suffered substantial relaxation by the time it encounters an olefin. It appears reasonable that the remaining polarization will suffer a similar degree of relaxation before the adduct encounters an electron donor which converts it to a neutral diamagnetic species.[49]

S
cyclobutane radical ion
$S^{\bullet+}$
closed radical ion
S
S
open radical ion
S
S

That nuclear spin polarization from such a reaction can, in principle, be observed, was shown for the reaction of acceptor triplets with the trans-head-to-head dimer of dimethylindene.[54] During this reaction, both the dimer and the monomer showed CIDNP effects. The dimer polarization is quite unusual, since the benzylic (H_α) and the homobenzylic cyclobutane protons (H_β) show emission spectra of comparable intensities. At the same time, the methyl groups, which are also homobenzylic, showed no polarization at all. Since the spin lattice relaxation times of the two types of cyclobutane protons are similar, these results appear to suggest an intermediate in which H_α and H_β have comparable hyperfine coupling constants. The two radical cations discussed as intermediates in the dimer formation do not meet this requirement, and a cyclobutane radical cation, which is consistent with the observed results, appears energetically unfavorable, since it fails to utilize any element of benzylic stabilization. Therefore, the polarization is explained as the sum of two contributions, one from the "closed" radical cation, ($a_\alpha > a_\beta > 0$), the other from the "open" form ($a_\alpha < 0 < a_\beta$).[40,49]

This assignment implies an equilibrium between the two isomeric radical cations and requires that both species have reasonable lifetimes before the electron return from the semiquinone anion, or before cleavage to a monomeric olefin and a monomer radical cation. In other cases of electron transfer induced dimer cleavage, the balance between these processes may be shifted substantially. For example, the dimer of dimethylthymine (T-T) is rapidly and efficiently cleaved upon quenching triplet anthraquinone ($^3A^*$). During this reaction, three signals of the monomer showed abnormal signal intensities but no spin polarization was observed for any of the dimer signals. Apparently the cleavage of the initially formed dimer cation is faster than the intersystem crossing in the radical ion pair so that essentially no electron return to the dimer cation is possible. These results are important since they provide guidelines for understanding the light-requiring step in photoreactivation, the photoenzymatic reversal of pyrimidine dimer formation in DNA. Historically, it is of interest that the nuclear spin polarization observed in this system was the first example of CIDNP effects in a photoinitiated electron transfer reaction.[25]

So far in this Section, we have dealt exclusively with nuclear spin polarization of donor olefins. In the following, we discuss the interactions of photo-excited electron donors, mainly aromatic hydrocarbons, with acceptor olefins such as 1,2-dicyanoethylene, cinnamonitrile, or its derivatives. These systems are of interest because the relative energies of the reactant excited states and of the intermediate radical ion pair provide for a unique spin selection principle. In the cases discussed so far, it was tacitly assumed that only pairs of singlet multiplicity can undergo reverse electron transfer. This assumption is valid for systems where the energy of the radical ion pair

$$\Delta G = E_{(D/D^{\cdot +})} - E_{(A^{\dot{-}}/A)} - e^2/\varepsilon a - T\Delta S$$

is lower than the combined energies of ground state donor plus triplet acceptor and of ground state acceptor plus triplet donor ($\Delta G < E_T$). However, the triplet energies of cyanoolefins may lie below the energies of radical ion pairs generated by their reaction with aromatic hydrocarbons ($\Delta G > E_T$). Therefore, reverse electron transfer can occur in both singlet and triplet pairs. Singlet pairs, in regenerating the reactant olefin, transfer to it the "in-cage" polarization; triplet pairs populate the triplet state of the olefin and transfer to it what in other systems would be the "escape" polarization. The polarized triplet state may decay directly to the ground state or suffer rearrangement and decay to a rearranged olefin. The "escape" polarization will be observed in rearranged olefin, if the triplet lifetime is short compared to the spin lattice relaxation time of the nuclei in the triplet species.

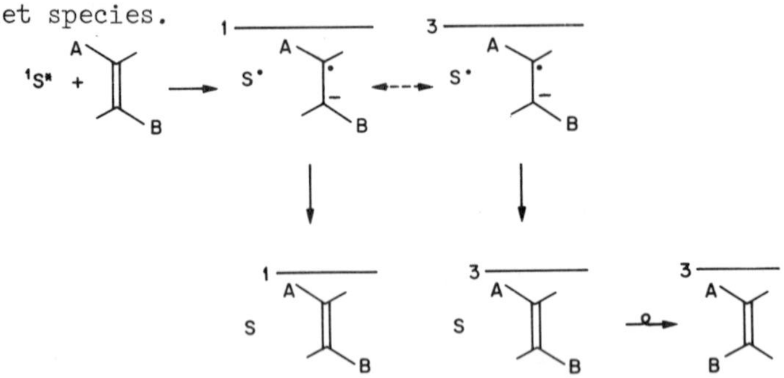

The CIDNP spectra observed in these systems have the characteristic feature that the polarizations of the reactant olefins and of the rearranged ones are of equal intensity (though of opposite sign). Taylor observed the first example of such an effect in the irradiation of naphthalene in the presence of _trans_-1,3-dicyanoethylene. During this experiment, the _trans_-olefin showed strongly enhanced absorption, whereas the _cis_-isomer showed equally strong emission. Further examples are found in the reactions of cinnamonitrile and several of its derivatives. In contrast to the dicyanoethylenes, the cinnamonitriles have two non-identical olefinic protons appearing as doublets so that the polarization for the two positions can be monitored separately. The results do not indicate a dramatic difference for the spin densities in the two positions. The doublets show multiplet effects, which are more pronounced for cinnamonitrile and the _p_-methoxy derivative than for the _p_-chloro derivative (_cf_. Figure 11).[40]

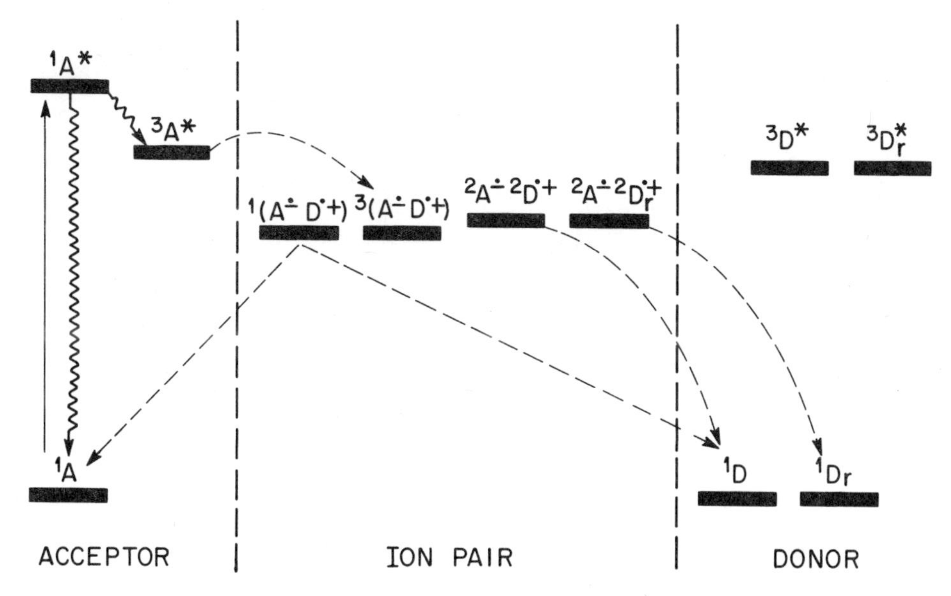

Figure 9. Schematic energy diagram for the reaction of a triplet acceptor with a donor olefin. The free energy of formation of the ion pair lies below the triplet energy of the reactants ($\Delta G < E_T$).

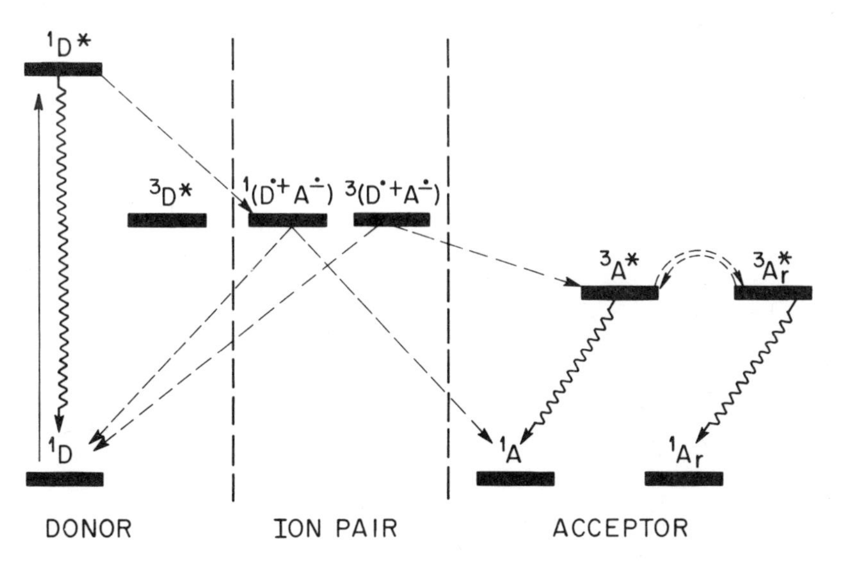

Figure 10. Schematic energy diagram for the reaction of an excited-singlet donor with an acceptor olefin. The free energy of formation of the ion pair lies above the triplet energy of one reaction partner ($\Delta G > E_T$).

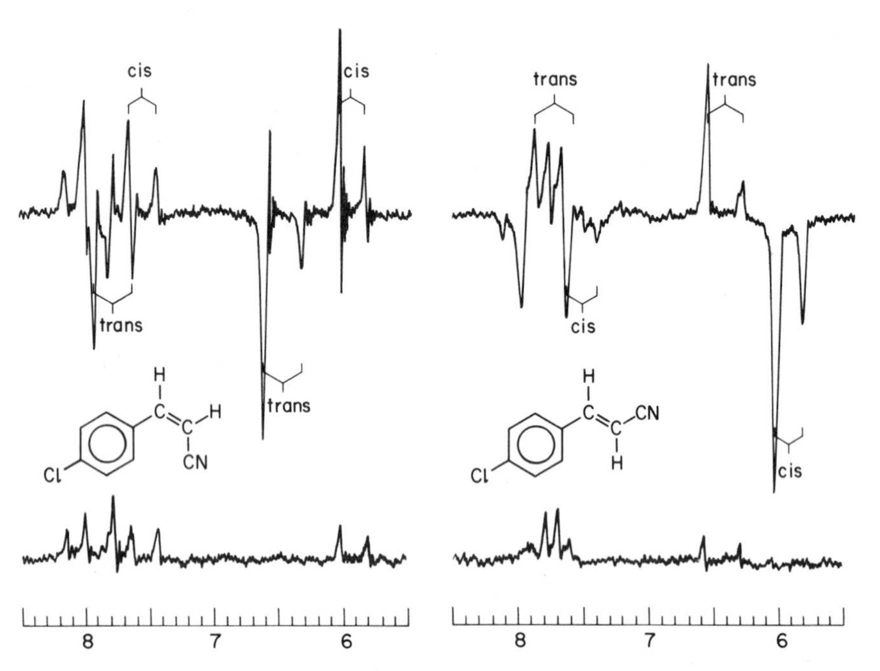

Figure 11. Pmr spectra (60 MHz) observed during the photoreaction of phenanthrene with cis (left) and trans-p-chlorocinnamonitrile.

In conclusion, the interaction of photo-excited materials with olefins may give rise to interesting CIDNP phenomena. The results contribute to the understanding of two mechanisms of geometric isomerization. In the case of donor olefins, the rearrangement occurs in the radical cations and is facilitated by the reduced double bond character in these species. In the case of acceptor olefins, the rearrangement occurs in the triplet state. However, the detailed pathway is different from triplet-sensitized isomerization since the triplet state is not populated by triplet energy transfer but by reverse electron transfer from a pair of radical ions initially generated from a singlet precursor. Electron transfer is likely to be involved in other types of photoreactions of numerous other olefins. We are confident that the CIDNP technique will be a valuable mechanistic tool in many of these reactions.

7. PHOTOREACTIONS OF PORPHYRINS AND CHLORINS

The reactions of photoexcited organic pigments such as porphyrins and chlorins with electron donors and acceptors have been studied widely because of their relation to the primary process in photosynthesis.[55] The initially excited singlet states of these pigments decay by fluorescence and by intersystem crossing to the lower lying triplet states; however their lifetimes are sufficiently long that they may be intercepted by efficient quenchers of sufficiently high concentration. The lifetimes of the corresponding triplet states are substantially longer so that they are quenched even at lower reactant concentrations.

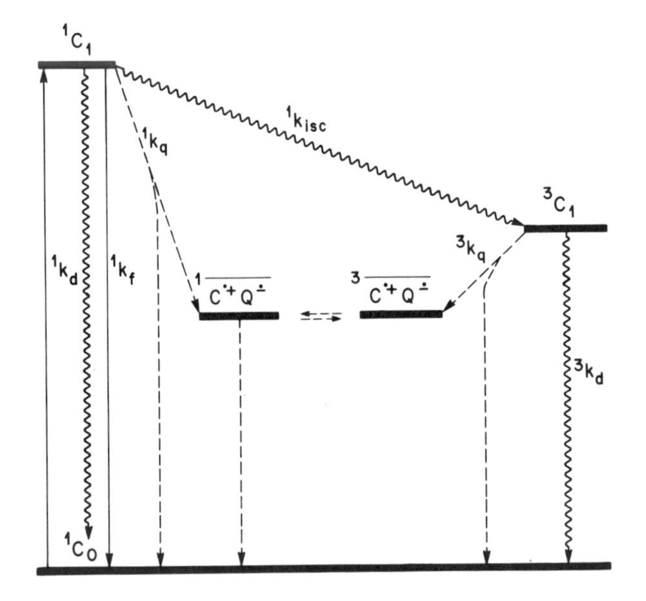

Figure 12. Schematic energy diagram of chlorophyll excited states and of the chlorophyll cation - semiquinone anion pair.

The first CIDNP study of chlorophyll photooxidation, with 1,4-benzoquinone or 2,6-dimethyl-1,4-benzoquinone as reaction partners was carried out by Tomkiewicz and Klein, who reported enhanced absorption for the ring protons and emission for the methyl group of the quencher, but did not find any polarization for the chlorophyll. The reported polarizations can be interpreted in terms of a singlet precursor.[56] However, subsequent investigations did not confirm this report. Instead, two different effects were observed depending on the acidity of the solvent. In commercially available methanol-$\underline{d}_4$, which was contaminated with up to 10^{-4}-$\underline{M}$ base, the irradiation resulted in pronounced line broadening whereas in carefully neutralized solutions CIDNP effects were observed: emission for the ring protons, enhanced absorption for the methyl groups. These effects suggest that the intermediate pair is generated by a triplet precursor.[38]

These effects can be explained as follows: photoexcited chlorophyll is quenched by electron transfer to the quinones (7.1); in the resulting pair, the electron spin dependent electron return (7.2) competes with the electron spin independent separation of the radical pairs by diffusion (7.3); degenerate electron exchange (7.4) is the likely cause of the observed line broadening. The failure to observe any chlorophyll polarization might be ascribed to the relatively small hyperfine coupling constants[57] or to relatively short spin lattice relaxation times. However, Closs and Boxer were able to show that line broadening due to a degenerate exchange reaction (7.5) prevents the observation of chlorophyll polarization.[58]

$$^{1,3}\mathrm{Chl}^{*} + \mathrm{Q} \rightarrow {}^{1,3}\overline{\mathrm{Chl}^{\cdot +} + \mathrm{Q}^{\cdot -}} \qquad (7.1)$$

$$^{1}\overline{\mathrm{Chl}^{\cdot +} \quad \mathrm{Q}^{\cdot -}} \rightarrow \mathrm{Chl} + \mathrm{Q}^{\dagger} \qquad (7.2)$$

$$^{3}\overline{\mathrm{Chl}^{\cdot +} \quad \mathrm{Q}^{\cdot -}} \rightarrow {}^{2}\mathrm{Chl}^{\cdot +} + {}^{2}\mathrm{Q}^{\cdot -} \qquad (7.3)$$

$$^{2}\mathrm{Q}^{\cdot -} + \mathrm{Q} \rightarrow \mathrm{Q} + {}^{2}\mathrm{Q}^{\cdot -} \qquad (7.4)$$

$$^{2}\mathrm{Chl}^{\cdot +} + \mathrm{Chl} \rightarrow \mathrm{Chl} + {}^{2}\mathrm{Chl}^{\cdot +} \qquad (7.5)$$

The line broadening of the quinone can be avoided in neutral solutions were partial protonation of the semiquinone radical ions slows down the degenerate exchange to rates too low for broadening. This observation and the fact that CIDNP can be observed only in protic solvents suggest that the protonation of the semiquinone radicalions is essential.

The two contradictory assignments of precursor spin multiplicity are not by necessity irreconcilable. Conceivably, the observed polarizations could be the net result of two large contributions of opposite sign due to quenching excited-singlet or triplet chlorophyll, respectively. The potential importance of these contributions was evaluated by a kinetic analysis of singlet and triplet quenching as a function of quencher concentration.[38]

The probability of quenching a primary photoexcited state (excited-singlet chlorophyll) depends on the rate of quenching relative to the rate of decay in the absence of quenching ($^{1}\tau$-1) and should increase monotonously with increasing quencher concentration (cf. Figure 13).

$$S = \frac{^{1}k_q[\mathrm{Q}]}{^{1}k_q[\mathrm{Q}] + {}^{1}\tau^{-1}}$$

In contrast, the probability of quenching a secondary excited state (triplet chlorophyll) formed by the decay of a primary excited state shows a different concentration dependence. In addition to the competition between quenching and triplet decay ($^3\tau^{-1}$), this probability depends on the unquenched fraction of the precursor state and on the quantum yield of formation in the absence of quenching (Φisc).

$$T = \frac{^3k_q[Q]}{^3k_q[Q] + {^3\tau^{-1}}} \times (1 - S) \times \Phi isc$$

$$= \frac{^3k_q[Q]}{^3k_q[Q] + {^3\tau^{-1}}} \times \frac{^1\tau^{-1}}{^1k_q[Q] + {^1\tau^{-1}}} \times \Phi isc$$

At sufficiently low quencher concentrations only the triplet should be quenched and the degree of quenching should increase with increasing quencher concentration. However, as soon as singlet quenching becomes significant, the triplet function should begin to decrease (cf. Figure 13). The rate parameters used to calculate the numerical values for the probabilities S and T as follows:

$$^1\tau^{-1} = 1.6 \times 10^8 \text{ sec}^{-1}$$

$$^1k_q = 7 \times 10^9 \; \ell \text{ mol}^{-1}\text{sec}^{-1}$$

$$^3\tau^{-1} = 2 \times 10^5 \text{ sec}^{-1}$$

$$^3k_q = 1 \times 10^9 \; \ell \text{ mol}^{-1}\text{sec}^{-1}$$

$$\Phi = 0.87$$

The CIDNP effects generated by the interaction of triplet born pairs will be opposite to those of singlet born pairs and the intensities will be proportional to the quantum yield of pair formation in the quenching reaction((Φs, Φt) and to the relative enhancements generated by these pairs (V_s, V_t). Therefore, the CIDNP intensities due to singlet and triplet quenching can be expressed as

$$I_S = \Phi_s V_s S \quad \text{and} \quad I = \Phi_t V_t T$$

and the overall CIDNP intensity will be

$$I = \Phi_s V_s S + \Phi_t V_t T$$

The concentration dependent CIDNP intensities observed during the quenching of chlorophyll by 2,5-dimethyl-1,4-benzoquinone are shown in Figure 14. These results show clearly that the

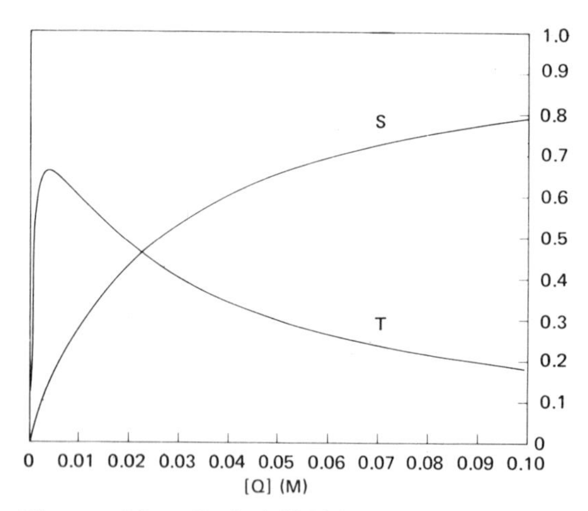

Figure 13. Probabilities of quenching the excited-singlet state (S) or the triplet state (T) of chlorophyll by a quinone, as a function of quencher concentration.[38]

quenching probability of the triplet state describes the relative CIDNP intensity quite satisfactorily over the entire concentration range and eliminate any significant contribution of singlet quenching to the CIDNP effects.[38] The results of Closs and Boxer obtained during the photoreduction of several chlorins with a series of hydroquinones are fully consistent with this assignment.[58] All CIDNP effects observed in these systems are derived from a triplet precursor.

The photoinduced interactions of porphyrins with quinones have also been probed by the CIDNP method. For example, Glazkov *et al*. studied the reactions of tetraphenylporphins with 1,4-benzoquinone in alcoholic solvents and found a dependence of CIDNP signal direction upon the *g* factor of the porphyrin radical cation.[59] In systems where the *g* factor of the radical cation is smaller than that of the semiquinone (2.0047), as is the case for tetraphenylporphin (g_{D^+} = 2.0030) and its tetrakis-*p*-chloro derivative (g_{D^+} = 2.0036), the spectrum of the quinone appeared in emission, whereas enhanced absorption was observed in the reaction of the tetrakis-*p*-bromo derivative (g_{D^+} = 2.0060).[59]

The reactions of tetraphenylporphin and octaethylporphyrin with several quinones (2,5-dimethyl-1,4-benzoquinone, tetrafluoroquinone) and with hydroquinones (hydroquinone, tetrafluorohydroquinone) resulted in CIDNP effects, which were in all cases compatible with a triplet precursor. The ring protons showed emission whereas the methyl groups and the ^{19}F signals appeared in enhanced oabsorption. The signal intensities increased with

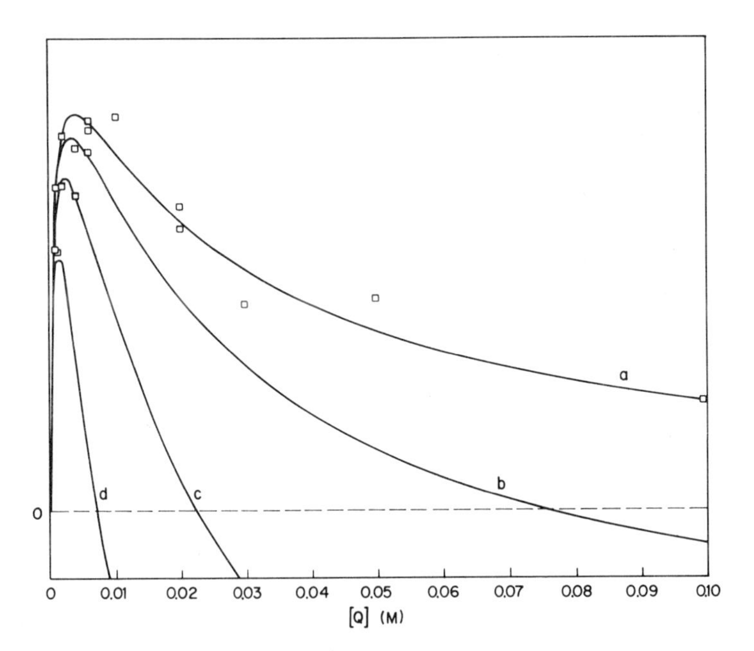

Figure 14. Experimental CIDNP intensities (methyl signal of 2,5-dimethyl-1,4-benzoquinone) and predicted CIDNP intensities for different values of $\Phi_s V_s$ relative to $\Phi_t V_t$. a) $\Phi_s V_s = 0$; b) $\Phi_s V_s = \Phi_t V_t/3$; c) $\Phi_s V_s = \Phi_t V_t$; d) $\Phi_s V_s = 3\Phi_t V_t$.

quencher concentration and reached maxima between 10^{-3} and 10^{-2} M quencher. Significantly, there was no inversion of CIDNP direction at high quencher concentrations. These results suggest, as in the case of chlorophyll photooxidation, that only the quenching of triplet porphyrins leads to the formation of radical ion pairs with appreciable lifetimes.[41]

The results obtained in the reactions of porphyrins and chlorins illustrate the importance of reactant concentrations for CIDNP studies. When analyzing a concentration dependence, it is absolutely essential that the sample is irradiated at a wavelength where only one reaction partner has an absorption band. Otherwise, the quencher may act as a filter and, at increasing concentrations may cause a decrease in signal intensity. In addition, these experiments cannot be carried to concentrations where the quencher changes the properties of the solvent appreciably. Nevertheless, when evaluated judiciously, the concentration dependence of CIDNP intensities will provide valuable information. They should be considered in all photoreactions.

8. CONCLUSION

Photoinitiated electron transfer reactions and radical ion induced nuclear spin polarization are not limited to the three reaction types discussed in this Chapter. Electron transfer reactions have been established in many additional systems and nuclear spin polarization effects have been observed in several of them and can be expected to be observed in many more. The examples discussed here were chosen to illustrate the effects that can be observed, the difficulties that may be encountered in interpreting them, the detrimental effects of even small amounts of impurities, and the type of information that may be derived for carefully studied systems. Looking ahead, one can expect interesting results in systems where radical pair polarization competes with polarization due to electron-nuclear cross relaxation. Only a few examples of these effects have been observed to date but more can be expected to surface once appropriate systems are investigated systematically.[60,61]

REFERENCES

1. J. Bargon, H. Fischer and U. Johnsen, Z. Naturforsch., A, 22, 1551 (1967).
2. H. R. Ward and R. G. Lawler, J. Amer. Chem. Soc., 89, 5518 (1967).
3. M. Cocivera, J. Amer. Chem. Soc., 90, 3261 (1968).
4. H. D. Roth, Mol. Photochem., 5, 91 (1973).
5. R. Kaptein, J. A. den Hollander, D. Antheunis and L. J. Ooosterhoff, Chem. Commun., 1687 (1970).
6. S. R. Fahrenholtz and A. M. Trozzolo, J. Amer. Chem. Soc., 93, 251 (1971).
7. H. D. Roth and M. L. Kaplan, J. Amer. Chem. Soc., 95, 262 (1973).
8. N. J. Turro, "Energy Transfer and Organic Photochemistry", A. A. Lamola and N. J. Turro, Eds., Wiley Interscience, New York, 1969, Chapter 3.
9. M. Cocivera and A. M. Trozzolo, J. Amer. Chem. Soc., 92, 1772 (1970).
10. G. L. Closs and D. R. Paulson, J. Amer. Chem. Soc., 92, 7229 (1970).
11. M. Tomkiewicz, A. Groen and M. Cocivera, J. Chem. Phys., 56, 5850 (1972).
12. B. Blank, P. G. Mennitt and H. Fischer, Special Lectures, 23rd Congr. Pure Appl. Chem., Boston, Mass., July 1971, 4, 1, Butterworths, London.
13. T. DoMinh, Chem. Ind. Belge, 36, 1080 (1971).
14. J. A. den Hollander, R. Kaptein, and P. A. T. M. Brand, Chem. Phys. Letters, 10, 430 (1971).

15. W. Adam, J. A. de Sanabia and H. Fischer, J. Org. Chem., 38, 2571 (1973).
16. W. Adam, H. Fischer, J.-J. Hansen, H. Heimgartner, H. Schmid and H.-R. Waespe, Angew, Chem., 85, 669 (1973).
17. R. Bausch, H.-P. Schuchmann, C.von Sonntag, R. Benn and H. Dreeskamp, Chem. Commun., 418 (1976).
18. S. M. Rosenfeld, R. G. Lawler, and H. R. Ward, J. Amer. Chem. Soc., 95, 946 (1973).
19. B. Blank and H. Fischer, Helv. Chim. Acta., 56, 506 (1973); G. P. Laroff and H. Fischer, ibid., 56, 2011 (1973).
20. J. Bargon and K. G. Seifert, Ber. Bunsenges. Phys. Chem., 78, 187 (1974).
21. S. A. Sojka, C. F. Poranski, Jr., and W. B. Moniz, J. Magn. Resonance, 23, 417 (1976).
22. G. L. Closs and L. E. Closs, J. Amer. Chem. Soc., 91, 4550 (1969); G. L. Closs, C. E. Doubleday, and D. R. Paulson, ibid., 92, 2185 (1970); G. L. Closs and A. D. Trifunac, ibid., 92, 2186 (1970).
23. G. L. Closs and L. E. Closs, J. Amer. Chem. Soc., 91, 4549 (1969); G. L. Closs and A. D. Trifunac, ibid., 91, 4554 (1969).
24. H. D. Roth, Accounts Chem. Res., 10, 85 (1977).
25. A. A. Lamola and H. D. Roth, J. Amer. Chem. Soc., 94, 1013 (1972).
26. H. Leonhardt and A. Weller, Ber. Bunsenges. Phys. Chem., 67, 791 (1963); A. Weller, Pure Appl. Chem., 16, 115 (1968).
27. Cf. e.g., N. C. Yang, D. M. Shold, and J. K. McVey, J. Amer. Chem. Soc., 97, 5004 (1975); N. C. Yang, K. Srinivasachar, B. Kim, and J. Libman, ibid., 97, 5006 (1975).
28. Cf. e.g., Y. Taniguchi, Y. Nishina, and N. Mataga, Bull. Chem. Soc. Jap., 45, 764 (1972).
29. K. H. Grellman, A. R. Watkins, and A. Weller, J. Lumin., 1, 2, 678 (1970).
30. R. S. Davidson and S. Santhanam, J. Chem. Soc., Perkin Trans. 2, 2351, 2355 (1972).
31. H. Knibbe, D. Rehm, and A. Weller, Ber. Bunsenges. Phys. Chem., 72, 257 (1968).
32. E. deBoer and H. van Willigen, Progr. Nucl. Magn. Resonance Spectrosc., 2, 111 (1967).
33. R. W. Kreilick and S. I. Weissman, J. Amer. Chem. Soc., 88, 2645 (1966).
34. R. G. Lawler, H. R. Ward, R. B. Allen, and P. E. Ellenbogen, J. Amer. Chem. Soc., 93, 791 (1971).
35. C. R. Bruce, R. E. Norberg, and S. I. Weissman, J. Chem. Phys., 24, 473 (1956).
36. C. S. Johnson, J. Chem. Phys., 39, 2111 (1963).
37. H. D. Roth and A. A. Lamola, J. Amer. Chem. Soc., 96, 6270 (1974).
38. A. A. Lamola, M. L. Manion, H. D. Roth, and G. Tollin, Proc. Nat. Acad. Sci., USA, 72, 3265 (1975).

39. H. D. Roth and M. L. Manion, J. Amer. Chem. Soc., 97, 6886 (1975).
40. M. L. Schilling and H. D. Roth, unpublished results.
41. H. D. Roth and M. L. Schilling, Abstracts, VI IUPAC Symp. Photochem., Aix-en-Provence, June 1976, p 279.
42. M. Cocivera, Chem. Phys. Lett., 2, 529 (1968).
43. S. G. Boxer and G. L. Closs, J. Amer. Chem. Soc., 97, 3268 (1975).
44. S. G. Cohen, A. Parola, and G. H. Parsons, Chem. Rev, 73, 141 (1973).
45. M. L. Kaplan, M. L. Manion, and H. D. Roth, J. Phys. Chem., 78, 1837 (1974).
46. M. L. M. Schilling, R. S. Hutton, and H. D. Roth, unpublished results.
47. J. den Hollander, Chem. Commun., 352 (1975); Chem. Phys., 10, 167 (1975).
48. R. Kaptein, Chapter XIV.
49. H. D. Roth and M. L. Manion, Abstracts, VIII Int. Conf. Photochem., Admonton, Alberta, 1975, U2.
50. A. Ledwith, Accounts Chem. Res., 5, 133 (1972).
51. S. Farid and S. E. Shealer, Chem. Commun., 677 (1973).
52. S. Kuwata, Y. Shigemitsu, and Y. Odaira, J. Chem. Soc., D, 2 (1972); J. Org. Chem., 38, 3803 (1973).
53. Cf. e.g., J. J. McCullough, Can. J. Chem., 46, 43 (1968).
54. G. N. Taylor, private communication; quoted in ref. 4.
55. Cf. e.g., G. Tollin, J. Bioenerg. 6, 69 (1974).
56. M. Tomkiewicz and M. P. Klein, Proc. Nat. Acad. Sci. USA, 70, 143 (1973).
57. J. K. M. Sanders and J. C. Waterton, Chem. Commun., 247 (1976).
58. G. L. Closs, Chapter XX.
59. Yu. V. Glazkov, A. G. Zhuravlev, A. M. Shulga, and L. A. Khilmonovich, Zh. Priklad. Spectr., 25, 133 (1976).
60. F. J. Adrian, H. M. Vyas, and J. K. S. Wan, J. Chem. Phys., 65, 1454 (1976).
61. M. J. Thomas, P. J. Wagner, M. L. Manion Schilling, J. Amer. Chem. Soc., 99, 3842 (1977).

CHAPTER V

RADICAL PAIR MECHANISM OF CHEMICALLY INDUCED MAGNETIC POLARIZATION*

F. J. Adrian

The Johns Hopkins University Applied Physics
Laboratory, Laurel, Maryland, USA

1. INTRODUCTION

Chemically induced magnetic polarization (CIMP) refers to various processes whereby free radical reactions in liquids lead to nonequilibrium population of the nuclear spin states of the products and reactants, and similar anomalies in the electron spin state population of the free radical intermediates. The nuclear and electron polarizations are referred to as chemically induced dynamic nuclear polarization (CIDNP) and chemically induced electron polarization (CIDEP), respectively. The first case of CIMP is probably the 1963 observation by Fessenden and Schuler of CIDEP in the electron spin resonance (ESR) spectra of H and D atoms produced by electron bombardment of liquid CH_4 and CD_4 [1]. This observation undoubtedly stimulated some theoretical work; however, no satisfactory explanation resulted. In hindsight, given the limited experimental data and the then unknown complexity of the CIDEP mechanism, this failure is quite understandable. It was the numerous observations of CIDNP in nuclear magnetic resonance (NMR) spectra, beginning in 1967 with the work of Bargon, Fischer and Johnson [2] and Ward and Lawler [3], that provided both stimulus and a sizable body of experimental data to work which led to satisfactory understanding of most CIMP phenomena [4]. The first theories postulated that a small deviation from equilibrium in the electron spin states of the chemically generated radicals (such as all states equally populated) could be converted by

*This work has been supported by the U. S. Naval Sea Systems Command under Contract N00017-72-C-4401.

L. T. Muus et al. (eds.), Chemically Induced Magnetic Polarization, 77-105.

electron-nuclear-spin cross relaxation processes into a large deviation of the nuclear spin states from equilibrium [2b, 3b]. This model was analogous to the dynamic polarization or Overhauser effect in which microwave pumping of the electron spin states of a paramagnetic species leads to large nuclear polarizations, and suggested the name CIDNP [5]. It failed, however, to explain many observed features of CIDNP, most notably the occurrence of both emission and enhanced absorption lines in many polarized spectra, and in 1969 Closs [6] and, independently, Kaptein and Oosterhoff [7] proposed an alternative explanation, the radical pair mechanism (RPM), often called the CKO model after its originators. (Somewhat paradoxically, the Overhauser mechanism may be making a limited comeback in certain systems where intersystem crossing to a triplet precursor leads to strong electron polarization in the resulting radicals, cf. the chapter of this book entitled "The Triplet Overhauser Mechanism of CIDNP".) The RPM proposed that the cage reactions of a radical pair could be affected by magnetic interactions within the radicals, which produce nuclear-spin-dependent mixing of the reactive singlet and unreactive triplet electron spin states of the pair. This model, discussed in this chapter, immediately provided a qualitative understanding of the then available CIDNP effects, as well as a sizable portion of CIDEP phenomena, and its modification by Adrian [8] and Kaptein [9] to include the role of the diffusive behavior of radical pairs in liquids, made the model more realistic and quantitative.

2. QUALITATIVE RADICAL PAIR MODEL

2.1 Overview of a free radical reaction

The photochemical radical reaction, shown in Fig. 1, illustrates various factors involved in CIMP. Similar considerations apply to thermally induced radical reactions except involvement of a triplet state precursor is less likely. A reactant ^1R is photoexcited to the state $^1R^*$ which may initiate a reaction in several ways. $^1R^*$ may dissociate or, as shown in Fig. 1, react with a substrate forming, in either case, a radical pair in a singlet electronic state. Alternatively, $^1R^*$ may make an intersystem crossing to a triplet state $^3R^*$, which then reacts to form a radical pair in a triplet electronic state. These pairs due to simultaneous generation of two radicals are known as geminate singlet (S) or triplet (T) pairs, respectively. As shown in Fig. 1, the radical pair may undergo various in-cage reactions to restore the original reactants or to form new products. Alternatively, the pair will separate, yielding two independent or free radicals. These free radicals will eventually recombine in pair-

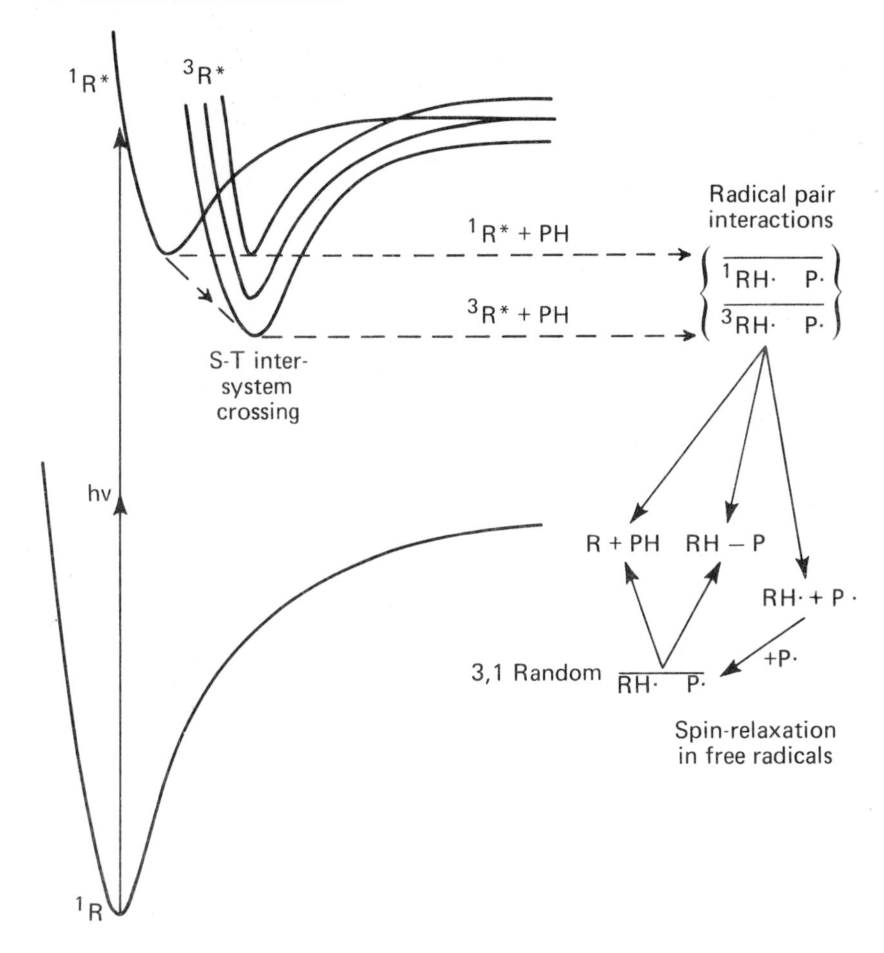

Fig. 1. Typical photochemical free radical reaction

wise encounters, but meanwhile they may be involved in other reactions such as fragmentation, abstraction, etc. A radical pair formed by the chance encounter of two free radicals is known as a random pair or an F pair as distinguished from the geminate pairs.

Since the electron and all nuclei, such as H, F, ^{13}C, ^{31}P, etc., for which CIMP has been observed are spin ½ particles, and thus have only a magnetic moment, all polarization processes must involve a magnetic interaction involving that spin. There are a number of stages in this reaction involving spin-altering magnetic interactions. The singlet-triplet intersystem crossing is due to the spin-orbit interaction, i.e., the interaction between the magnetic moment of an electron and the magnetic field generated by its orbital motion in the electrostatic field of the molecule. This interaction is usually electron-spin selective and leads to a strong electron polarization of the triplet which can be transferred to the radicals formed by reaction of the triplet. This process, known as the triplet mechanism, is discussed in detail elsewhere in this book. In the radical pair the magnetic interactions of the unpaired electrons of the radicals with the external magnetic field (Zeeman interaction) and the magnetic nuclei (hyperfine structure or hfs interactions), combined with the electron-spin-dependent valence bonding or exchange interaction between the radicals, are the basis of the RPM discussed in this chapter. Finally, magnetic interactions in the isolated radicals induce spin transitions which drive the various spin systems toward thermal equilibrium. During the approach to equilibrium those relaxation transitions which involve simultaneous flips of two spins can transfer polarization from one spin system to another, e.g., from electrons to nuclei. Interestingly, it will be seen that relaxation processes which destroy polarization in free radicals sometimes play an important indirect role in the RPM.

2.2 Singlet-triplet splitting and mixing in a radical pair

For a qualitative description of the RPM consider Fig. 2 which shows the bonding singlet (S) and antibonding triplet (T) energies of a radical pair in an external magnetic field, H_0, as a function of separation, R. As the pair separates after formation either as a geminate or an F pair the short-range exchange interaction, J(R), decreases rapidly to zero, and degeneracies occur in the S and T levels, enabling mixing of these levels by the hfs interactions and difference terms in the electron Zeeman interactions. Indeed, if the radicals were to separate very slowly, which they do not, then the S-T mixing would split these degeneracies and the separation would take place along the adiabatic energy levels shown by the dotted lines in Fig. 2. Such adiabatic dissociation would lead to very large polarizations; for example, the T_- Zeeman level of a triplet would go over completely to the singlet leaving the electron spin system strongly T_+ polarized. In fact, however, the radical pair separation is normally so rapid that S-T mixing interactions have no effect during the initial separation, which thus follows the diabatic energy levels shown by the solid curves in Fig. 2.

An approximate criterion governing whether pair separation will take place along the adiabatic or diabatic energy curves can be obtained from the Heisenberg uncertainty relation

$$\Delta\omega\Delta t \geq 1 \quad (1)$$

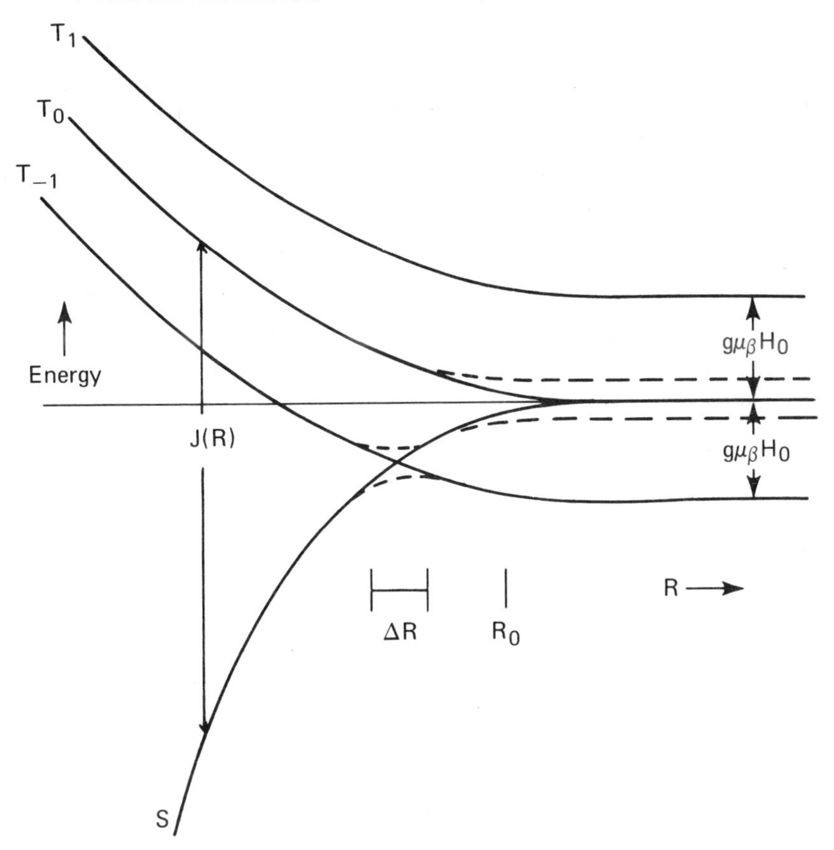

Fig. 2. Singlet and triplet energy levels of a separating radical pair in an external magnetic field. —— diabatic separtion. --- adiabatic separation.

As applied here, the polarizations resulting from adiabatic separation are related to the splitting of the adiabatic energy levels, $\Delta\omega$. Therefore, it is possible, at least in principle, to measure $\Delta\omega$ from the observed polarizations, in which case the time, Δt, of the adiabatic separation, which is the relevant process in this case, must satisfy the Heisenberg relation. The magnetic interactions responsible for $\Delta\omega$, are typically of the magnitude 10^8 rad/sec, as, for example, the proton hfs in the CH_3 radical. Therefore, $\Delta t \cong 10^{-8}$ sec, which is much longer than the time involved in pair separation either by dissociative vibration or simple diffusion. Thus, we may expect the initial separation will follow the diabatic energy curves and not lead immediately to polarization.

Consequently, $S-T_{-1}$ mixing is strongly inhibited at high fields because the system passes rapidly through the narrow region, ΔR, where these levels are degenerate into a region where the Zeeman splitting precludes $S-T_{\pm 1}$ mixing by typical hfs interactions, although polarizations due to mixing of these states can occur in low magnetic fields. $S-T_0$ mixing is possible, however, because the degeneracy of these levels begins after the radicals have separated a few molecular diameters and continues on out to infinite separation. Even here, however, one encounters the problem of how the radical pair retains its identity during the S-T mixing process once the radicals have separated to a point where the exchange interaction, which makes the radicals "aware" of each other, has become very small. The answer to this problem will be shown to lie in the Noyes concept of diffusional separation and reencounter of the components of the pair [10]; however, we postpone discussion of this point for the sake of presenting a simplified qualitative picture of the RPM and the rules it predicts.

Assuming that adequate $S-T_0$ mixing can occur, it is straightforward to envision a process whereby a geminate T pair can form diamagnetic cage reaction products only insofar as it acquires S character by $S-T_0$ mixing. Because this mixing involves the hfs interactions it is faster for some nuclear spin states than others, and these states are favored in the pair reaction products. For the opposite case of an initial S pair those nuclear spin states which give the fastest $S-T_0$ mixing will be converted most rapidly to the unreactive T state and thus will be disfavored in the cage reaction products. An F pair formed by random encounter of two free radicals will behave qualitatively as a triplet, because the higher the triplet character of a given pair the more likely it is to survive the formative encounter without reacting. These considerations lead to the first rule of radical pair CIDNP, which has been verified experimentally [11]: The sign of the polarization of T and F pairs is opposite to that of S pairs.

Another important principle comes from the fact that only $S-T_0$ mixing is allowed in high fields. Neither of these states have a net magnetization with respect to $\underline{H}_0$; thus, this process cannot produce a net magnetization of the pair. The polarization, therefore, must consist of an excess of "up" spins in certain products (net effect) or certain spin states of a given product (multiplet effect) with a corresponding excess of "down" spins in other products or other spins states of the given product. This type of polarization is often referred to as entropy polarization as distinguished from energy polarization where there is a net excess of "up" or "down" spins in the entire system. This explains why many CIDNP and CIDEP spectra exhibit multiplet effects with some lines appearing in emission and other lines in enhanced absorption. Furthermore, if the product (P) of a pair reaction has a net polarization then the escaping radicals must have a polarization equal and opposite to that of P, which polarization may be transferred to reaction products of these radicals. This polarization of the escape radicals and their products is often reduced by relaxation processes within the radicals, and offset by F-pair polarization in the case of F-pair products involving this radical. One especially noteworthy consequence of this process is that if all the radicals eventually yield the same product P either during geminate reaction or in subsequent F pair encounters, then P will be unpolarized, unless the lifetime of the free radicals is long enough for loss of some of the polarization via relaxation processes in the radicals [12].

2.3 Vector model of S-T_0 mixing and CIDNP rules

To proceed further we consider the Hamiltonian for the S-T-mixing magnetic interactions within the individual radicals [8]:

$$\mathcal{H}^{(M)} = \mu_\beta(g_1\underset{\sim}{S}_1+g_2\underset{\sim}{S}_2)\cdot\underset{\sim}{H}_0+\sum_n A_{1n}\underset{\sim}{I}_{1n}\cdot\underset{\sim}{S}_1+\sum_m A_{2m}\underset{\sim}{I}_{2m}\cdot\underset{\sim}{S}_2, \tag{2}$$

where μ_β is the Bohr magneton, $\underset{\sim}{S}_1$ and $\underset{\sim}{S}_2$ are the electron spins of the two radicals, g_1 and g_2 are the electronic g factors of radicals 1 and 2, A_{1n} and $\underset{\sim}{I}_{1n}$ are, respectively, the isotropic hfs constant and nuclear spin of the nth nucleus of radical 1, and A_{2m} and $\underset{\sim}{I}_{2m}$ are the analogous quantities for radical 2. We have neglected the anisotropic terms in $\mathcal{H}^{(M)}$ because we anticipate that the time interval between separation and return of the radicals will average these terms to zero. The g factors of the two radicals differ slightly from the free spin value, g_e=2.0023, because spin-orbit interactions in the individual radicals provide a small orbital contribution to the net electron magnetic moment. In high fields, where the nuclear spins are quantized with respect to $\underset{\sim}{H}_0$, it can be shown that $\mathcal{H}^{(M)}$ is equivalent to each radical experiencing a different nuclear-spin-dependent magnetic field, where both fields are parallel to $\underset{\sim}{H}_0$. After the radicals separate to a point where the electron-spin-dependent exchange interaction is negligible, each spin will precess independently about its own effective magnetic field; the difference in precession frequencies for a particular nuclear spin state (ab) being

$$w_{ab}=w_{1a}-w_{2b}=\tfrac{1}{2}\mu_\beta H_0\Delta g+\tfrac{1}{2}\sum_n A_{1n}M_{1n}^{(a)}-\tfrac{1}{2}\sum_m A_{2m}M_{2m}^{(b)}, \tag{3}$$

where $\Delta g=g_1-g_2$, $M_{1n}^{(a)}$ is the magnetic quantum number of the nth nucleus of radical 1 in the overall nuclear-spin state a, and $M_{2m}^{(b)}$ is the magnetic quantum number of the mth nucleus of radical 2 in the overall nuclear-spin state b.

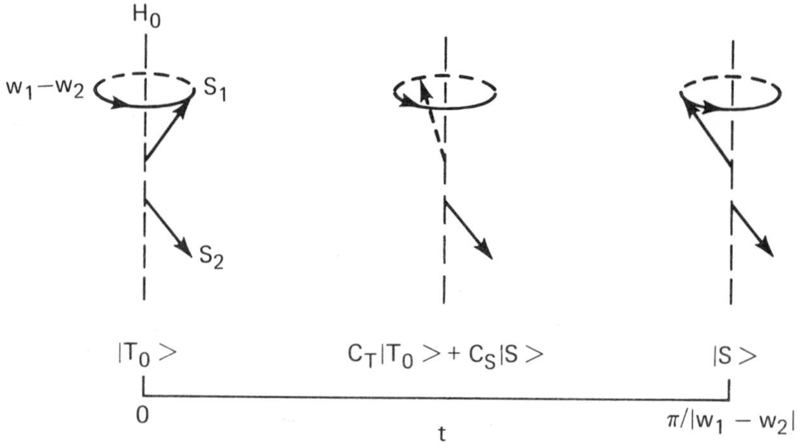

Fig. 3. Vector model of S-T_0 mixing.

This leads to the vector model of S-T_0 mixing, illustrated in Fig. 3. A difference in electron spin precession rates for a given nuclear spin state will cause the electron spin system to oscillate between the S state, where the two electron spins are antiparallel, and the T_0 state where the projection of the two spins along $\underset{\sim}{H}_0$ is zero but is maximized along an axis perpendicular to $\underset{\sim}{H}_0$. The period of this oscillation is $2\pi/|w_{ab}|$. Thus, an initial T_0 state will evolve in time $\pi/|w_{ab}|$ into a pure S state, with a mixture of T_0 and S states at intermediate times, and vice versa for an initial S state. This qualitative picture of the rate of S-T_0 mixing, together with the fact that the time available for S-T_0 mixing (i.e., the lifetime of the pair as an entity before it either undergoes cage raction or separates completely) is invariably less than π/w_{ab}, enables derivation of the rules for the signs of nuclear spin polarization by the radical pair mechanism.

For the case of net CIDNP effects, which arises when $\Delta g \neq 0$, consider the pair $\overline{R_1H\cdot \; R_2\cdot}$ where R_1H has one nucleus of spin $\frac{1}{2}$ and R_2 has no magnetic nuclei. Let the pair be formed as a triplet and assume that $\Delta g = g_1 - g_2 > 0$ and $A_1 > 0$. Finally, assume, as is usually the case, that the nuclear magnetic moment is positive so that the ground nuclear Zeeman state has the nuclear spin parallel to the external field. This case is illustrated in Fig. 4. Clearly, the S-T_0 mixing rate $|w_{ab}|$ for this case is greater in the $M_1 = \frac{1}{2}$ or α_1 nuclear spin state than in the $M_1 = -\frac{1}{2}$ or β_1 state. Thus, the former state will acquire S character faster than the latter and will be favored in the recombination product giving

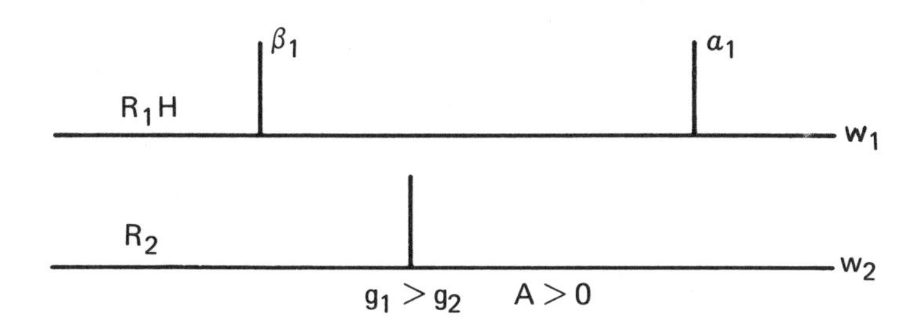

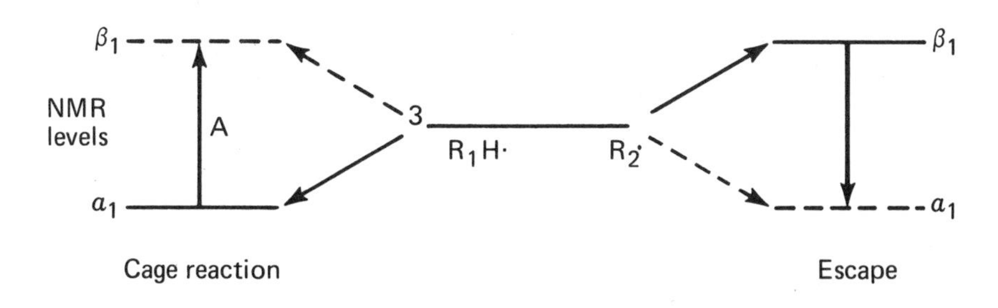

Fig. 4. Radical pair mechanism giving net effect polarization. Solid arrow indicates favored state.

enhanced absorption (A) CIDNP for this product. Conversely, the β_1 state, which retains antibonding triplet character longer will be favored in the products formed from the escaped R_1H radical, which thus have emissive (E) CIDNP. Starting with an S instead of a T pair, or changing the sign of either of the quantities Δg or A_1 will reverse this pattern giving E CIDNP for the recombination product and A CIDNP for the escape products.

If $\Delta g=0$ there can be no net polarization in any product but there can be a mixture of emissive and enhanced absorptive polarization known as a multiplet effect. For this case consider the pair $\overline{R_1H\cdot \quad R_2H\cdot}$ where each radical has one nucleus of spin ½. Again let the pair be formed as a triplet and assume that $A_1, A_2 > 0$. This case is illustrated in Fig. 5. Obviously, $|w_{ab}|$ is the same for the nuclear spin states $\alpha_1\beta_2$ and $\beta_1\alpha_2$ and greater than the $|w_{ab}|$ value common to the states $\alpha_1\alpha_2$ and $\beta_1\beta_2$. Thus, the former states will be favored if the recombination product R_1H-R_2H is formed by cage reaction of the radical pair and disfavored if it is formed from escaped radicals. For the common case where the nuclear spin-spin coupling constant (J_{12} in $J_{12}\underset{\sim}{I}_1\cdot \underset{\sim}{I}_2$) of the recombination product is small compared to the difference in chemical shift between the nuclei, $\hbar\gamma H_0(\sigma_1-\sigma_2)$, we have an AX type

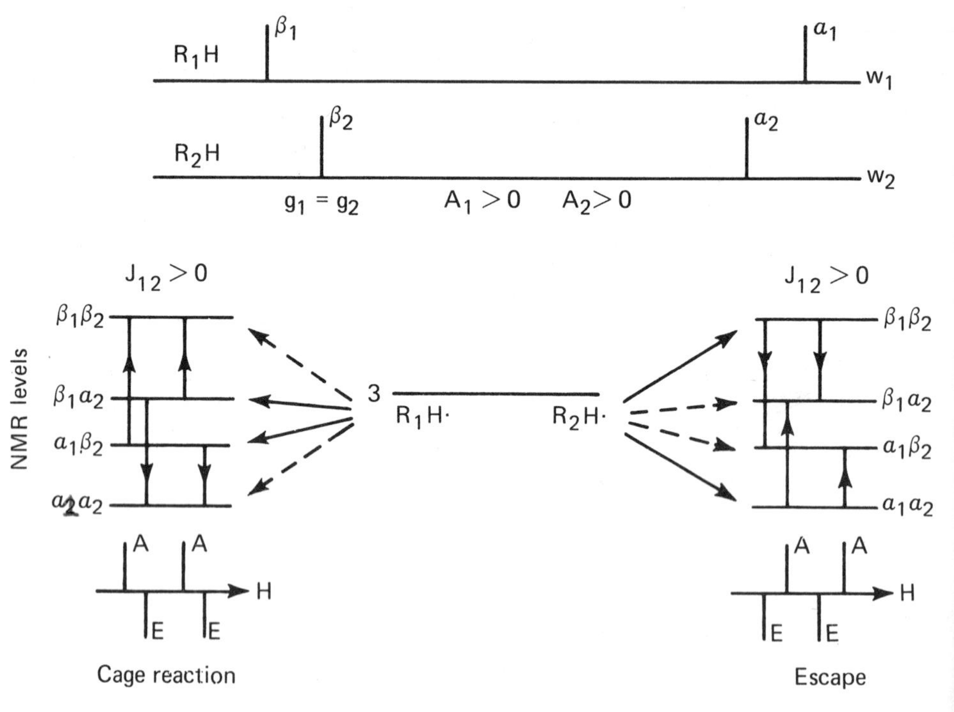

Fig. 5. Radical pair mechanism giving multiplet effect polarization. One magnetic nucleus on each radical. Solid arrows indicate favored states.

NMR spectrum consisting of two doublets resulting from the separate NMR transitions of $\underset{\sim}{I}_1$ and $\underset{\sim}{I}_2$, with each transition split by the spin-spin interaction with the other nucleus. If $J_{12}>0$ the $\alpha_1\alpha_2$ and $\beta_1\beta_2$ levels are raised relative to the $\alpha_1\beta_2$ and $\beta_1\alpha_2$ levels by the $J_{12}\underset{\sim}{I}_1\cdot\underset{\sim}{I}_2$ interaction, with the result that the NMR transitions with $\alpha_1\alpha_2$ as the lower level involve a smaller energy difference, and thus appear at higher field in an NMR spectrum run at constant frequency, than do the corresponding transitions with $\beta_1\beta_2$ as the upper level [13]. This combined with preferential population of the $\alpha_1\beta_2$ and $\beta_1\alpha_2$ levels of the cage-reaction product gives each multiplet an absorption-emission (A/E) CIDNP. Similarly, the escape product has an emission-absorption (E/A) spectrum. It is readily shown that if $J_{12}<0$ then the $\alpha_1\alpha_2$ and $\beta_1\beta_2$ levels are then lowered relative to the $\alpha_1\beta_2$ and $\beta_1\alpha_2$ levels without changing the spin selective CIDNP process so that the previous A/E multiplets become E/A and vice versa. Changing the sign of $(\sigma_1-\sigma_2)$ has no effect because this merely interchanges the energies of the equally populated $\alpha_1\beta_2$ and $\beta_1\alpha_2$ levels. Changing the sign of either A_1 or A_2 makes $|w_{ab}|$ greatest for the $\alpha_1\alpha_2$ and $\beta_1\beta_2$ levels, and thus changes an A/E pattern to an E/A and vice versa.

The foregoing analysis can be extended to show that these multiplet effect rules often apply to the AB type NMR spectrum [13], where J_{12} is comparable to or greater than the chemical shift difference. There can, however, be exceptions where the simple rules break down, as, for example, an A_nB NMR spectrum in which second-order effects are important. These cases are discussed elsewhere [4,9].

Finally, if we consider multiplet effects for the recombination product of the pair $\overline{R_1HH'\cdot \quad R_2\cdot}$ where both magnetic nuclei are on the same radical then the NMR levels are the same as before, assuming the parameters J_{12} and $(\sigma_1-\sigma_2)$ remain the same, but w_1 and w_2 are different, as shown in Fig. 6. Thus, for the case analogous to Fig. 5, i.e., A_1, $A_2>0$, $|w_{ab}|$ is greatest for the $\alpha_1\alpha_2'$ and $\beta_1\beta_2'$ states, and this will lead to E/A multiplets in the cage recombination product $R_1HH'-R_2$ of a T pair.

These qualitative predictions of the radical pair theory of CIDNP have been summarized by Kaptein in the following rules involving the signs of two quantities, Γ_{ne} for the net effects and Γ_{me} for multiplet effects [4,9].

$$\Gamma_{ne} = \mu\varepsilon\Delta g A_i \qquad \begin{cases} + & : \ A \\ - & : \ E \end{cases} \tag{4}$$

$$\Gamma_{me} = \mu\varepsilon A_i A_j J_{ij}\sigma_{ij} \qquad \begin{cases} + & : \ E/A \\ - & : \ A/E \end{cases} \tag{5}$$

where the NMR of nucleus i of radical 1 is being considered and the symbols are defined as follows:

$$\mu \quad \begin{cases} + \text{ for T and F precursor} \\ - \text{ for S precursor} \end{cases}$$

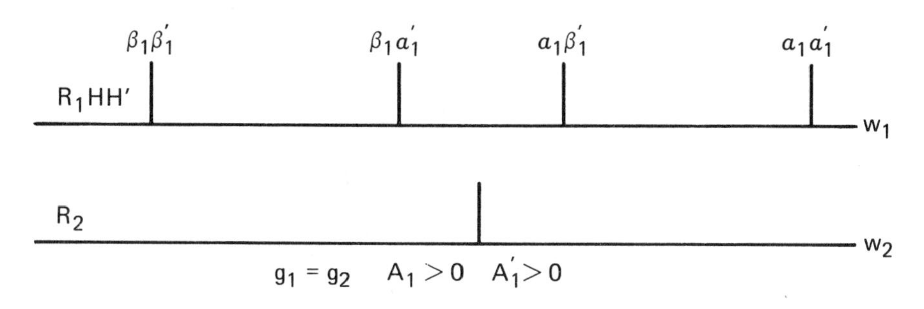

Fig. 6. Radical pair mechanism giving multiplet effect polarization. Both magnetic nuclei on same radical.

$$\varepsilon \begin{cases} + & \text{for cage recombination (P) products} \\ - & \text{for escape (D) products} \end{cases}$$

$$\sigma_{ij} \begin{cases} + & \text{when i and j reside in the same radical} \\ - & \text{when i and j reside in different radicals} \end{cases}$$

These rules have been found to be obeyed in almost all cases where the reaction kinetics are unambiguous. Thus, they are very useful in determining precursor multiplicities or the signs of hyperfine constants or other parameters from the observed CIDNP spectra. There are many examples of the application of these rules in the literature [4] and others will undoubtedly be given elsewhere in this book.

In radical pairs with $g_1 \neq g_2$ and several magnetic nuclei there can be various combinations of the net and multiplet effects. These CIDNP spectra have a number of interesting features including strong magnetic field dependences because of the term $\frac{1}{2}\mu_B H_0 \Delta g$ in Eq. (3) for w_{ab}. A useful qualitative picture of these spectra often can be obtained by treating them as a mixture of appropriate pure net and pure multiplet spectra with adjustable mixing coefficients. This procedure is illustrated in Fig. 7 for the product

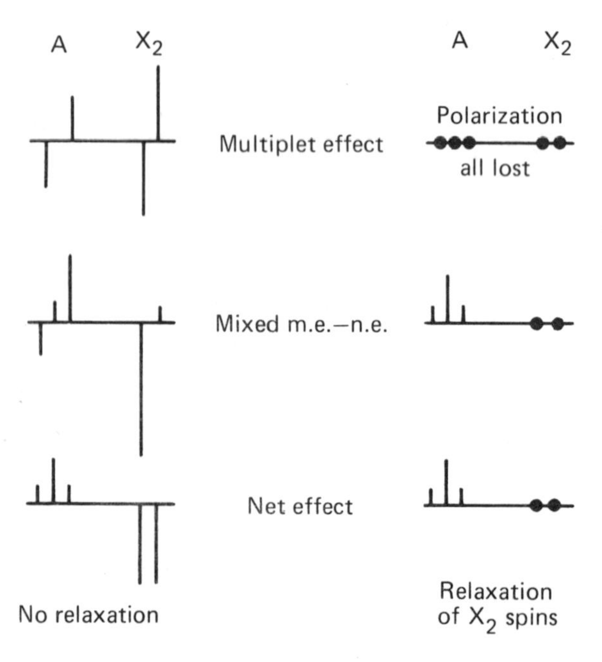

Fig. 7. Combination of net and multiplet effects, and effect of Selective nuclear spin relaxation.

RCH_2-CHR'R" formed from the pair $\overline{RCH_2\cdot \quad CHR'R''\cdot}$. A good example of this effect is given by Closs and coworkers [11,14] who studied CIDNP effects in a number of phenylethanes where the g factor difference was varied by attaching chemically inert halogen atoms in para positions of various phenyl groups. As discussed by Lawler [4c] selective relaxation effects can also play a role in determining the CIDNP spectra observed in these systems. A number of relaxation effects are possible; however, the common effect, illustrated in Fig. 7, is that a single magnetic nucleus bound to a given atom (e.g., the CH proton in $RCH_2CHR'R''$) has a much longer relaxation time than two magnetic nuclei bound to the same atom, because the latter can relax by intramolecular dipole-dipole interaction. The rapid relaxation of the CH_2 protons removes all multiplet effect polarization, as can be seen by noting that rapid $\alpha_1\alpha_2 \leftrightarrow \beta_1\alpha_2$ and $\alpha_1\beta_2 \leftrightarrow \beta_1\beta_2$ relaxation transitions in Fig. 5 will completely destroy the population differences responsible for the multiplet effect. This relaxation only removes any net polarization from the CH_2 NMR levels, however, with the obvious results as shown in Fig. 7. If the faster relaxing nuclei achieve only partial equilibration then the multiplet effect is altered but not completely removed. Effects such as this have been detected by Müller and Closs by studying the growth of CIDNP at the start of a photolytic reaction, the steady state CIDNP, and the decay of the CIDNP after the light is turned off [15].

Quantitative methods for predicting CIDNP spectra usually involve computer calculation of the effect of the RPM on the reactivity of various nuclear spin levels of the radical pair and correlation of these levels with the nuclear spin states of the reaction products [4,8,16]. Closs has given some graphical procedures which are useful, although less accurate, and have the advantage of giving a physical picture of how the CIDNP spectra change with various parameters [4b]. Further discussion of these procedures is beyond the scope of this presentation except to note that they require the quantitative picture of S-T mixing in radical pairs, as governed by the time dependent Schrodinger equation, combined with the diffusive behavior of the pairs which determines the time available for S-T mixing. This is discussed in the next section.

3. QUANTITATIVE RADICAL PAIR MODEL

3.1 Time dependent radical pair wave function

A quantitative description of the RPM is based on the time-dependent Schrodinger equation [17] for the development of the electronic wave function of the radical pair. If all energies are in units of $\hbar$ this equation is [8]

$$[-J(R)(2\underset{\sim}{S}_1\cdot\underset{\sim}{S}_2+\tfrac{1}{2})+\mathcal{H}^{(M)}]\psi(t) = \partial\psi(t)/\partial t \tag{6}$$

where $2J=(E_S-F_T)$ is the exchange interaction which splits the singlet and triplet states, and $\mathcal{H}^{(M)}$ has been described in Eq.(2). This equation neglects the dipolar interaction between $\underset{\sim}{S}_1$ and $\underset{\sim}{S}_2$ and spin-orbit interactions. The dipolar interaction can introduce an additional singlet-triplet splitting; however, this anisotropic interaction is small and tends to be averaged out by the diffusive motions of the radicals. The spin-orbit interaction can mix singlet and triplet states; however it is very small when the unpaired electrons reside on separate molecules [9b].

Exact solution of Eq.(6) is not feasible because the variation of R in the diffusing radical pair makes J(R) a random function of time. Fortunately, as we shall see, a good approximation for the case of CIDNP can be obtained by assuming that J=0. A solution can also be obtained for the case of constant J. Since this case is useful and instructive, and includes the J=0 case, it is given below

$$\psi(t) = [C_{S,ab}(t)|S\rangle+C_{T,ab}(t)|T_0\rangle]\phi_{ab}{}^{(N)}, \tag{7}$$

where $|S\rangle$ is the singlet function $2^{-\frac{1}{2}}(\alpha\beta-\beta\alpha)$, $|T_0\rangle$ is the triplet function $2^{-\frac{1}{2}}(\alpha\beta+\beta\alpha)$ (mixing of these functions with the other triplet spin functions $\alpha\alpha$ and $\beta\beta$ is unimportant in a strong magnetic field), and $\phi_{ab}{}^{(N)}$ is a nuclear-spin function which denotes the nuclear-spin states of radicals 1 and 2 as a and b, respectively. The singlet and triplet coefficients are given by the equations

$$C_{S,ab}(t) = C_S(0)[\cos wt-(iJ/w)\sin wt]-(iw_{ab}/w)C_T(0)\sin wt, \tag{8a}$$

and

$$C_{T,ab}(t) = C_T(0)[\cos wt+(iJ/w)\sin wt]-(iw_{ab}/w)C_S(0)\sin wt, \tag{8b}$$

where w_{ab} is given by Eq.(3) and

$$w = (w_{ab}^2+J^2)^{\frac{1}{2}}. \tag{9}$$

These equations show that S-T mixing cannot begin until the radicals have separated to a point where $|J(R)|\cong|w_{ab}|$. Because J(R) is a short-range interaction the radicals do not have to separate very far to reach this point, and once it is reached the diffusing radicals will tend to separate further so that $|J| \ll |w_{ab}|$. Thus, we can set J=0 in the foregoing equations. This approximation has been supported both by comparison of observed and computed CIDNP spectra [8,18] and by computer calculations of S-T mixing in a diffusing radical pair using the stochastic Liouville method [19]. The need for a finite J, indicated in some comparisons of theory with experiment, was removed when contributions to w_{ab} from all magnetic nuclei were included rather than including just the nuclei actually involved in the NMR spectrum [18]. Qualitatively, including all nuclei has the effect of giving a spread to the w_{ab} values for various states of the observed spins, thereby minimizing somewhat the polarizing effect of differences in w_{ab} between the observed spin states. This reduction of the polarization is effectively mimicked in at least some cases by using a finite J in Eqs. (7) - (9). If J=0, the singlet character at time t of a radical pair "born" as a triplet at t=0 is

$$|C_{S,ab}(t)|^2 = \frac{1}{3}\sin^2 w_{ab}t, \tag{10}$$

where account has been taken of the fact that 1/3 of the triplets will be in the T_0 state, with the remainder in unmixed $T_{\pm 1}$ states. Similarly, if the radical pair was initially a pure singlet we have

$$|C_{S,ab}(t)|^2 = 1 - \sin^2 w_{ab}t. \tag{11}$$

3.2 Role of diffusion controlled reaction

Clearly, for typical w_{ab} values of 10^8 rad/sec the rate of S-T mixing is far too slow for even slightly significant changes in the wave function during the 10^{-10} to 10^{-11} sec lifetime of the newly created radical pair before it either undergoes cage reaction or separates. As suggested by Adrian [8] and independently by Kaptein [9], the Noyes concept of secondary cage recombination [10] enables the radical pair to retain its identity during the relatively lengthy S-T mixing process. This process, illustrated in Fig. 8, starts with those pairs which fail to undergo primary cage reaction and diffuse apart to a point R_0 such that $|J(R_0)| \cong |w_{ab}|$. At this point S-T mixing begins and proceeds as the radicals diffuse freely. A number of these diffusing pairs will separate forever, but others will reencounter each other, where reencounter is defined as the pair separation diminishing to a point, R_σ, where the pair again has a chance to react. In the Appendix the probability of a reencounter in the time interval t to (t+dt) is shown to be

$$dP(R\to R_\sigma, t|R_0,0) = \sqrt{\frac{3}{4\pi}}\,\frac{R_0-R_\sigma}{R_0}(t/\tau)^{-3/2} \times \exp\left[-3(R_0-R_\sigma)^2\tau/4R_\sigma^2 t\right]d(t/\tau) \tag{12}$$

where τ is the time between diffusive displacements of the radicals.

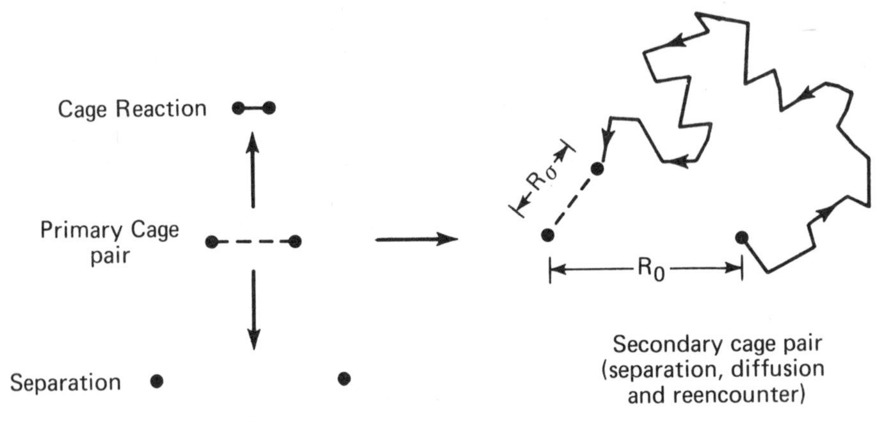

Fig. 8. Radical Pair Separation and Reencounter

To estimate the distances in Eq.(12) we need estimates of J(R). It is reasonable to approximate this short range function as

$$J(R) = \frac{1}{2}(E_S - E_T) = J_0 e^{-\lambda R}. \tag{13}$$

Two models for estimating the parameters J_0 and λ, described elsewhere [20], use the fact that J(R) depends on the overlap of the nonbonded electron charge clouds around the interacting molecules. Thus, the dependence of J on R scales approximately as the extent of these charge clouds, which also determines the molecular collision diameter R_σ. This suggests it is reasonable to assume that the dimensionless quantity λR_σ is independent of the nature of the radicals, which permits an estimate of J(R) by scaling the theoretical calculations for a pair of hydrogen atoms [21]. In the first model J(R) is simply taken to be the direct interaction between the radicals, which gives:

$$J_0 = -2.1(10)^{18} \text{ rad/sec}; \quad \lambda R_\sigma = 8.3. \tag{14a}$$

The second model considers the possibility that in dense media such as a liquid there may also be a transferred exchange interaction due to successive transfer of small amounts of the unpaired electron densities of the two radicals into the orbitals of adjacent molecules until the migrating unpaired electron charge densities reside on the same molecule or nearest neighbor pairs of molecules. This model gives:

$$J_0 = -1.0(10)^{16} \text{ rad/sec}; \quad \lambda R_\sigma = 4.1. \tag{14b}$$

With these estimates of J(R) the condition $|J(R_0)| \cong |w_{ab}| \cong$

10^8 rad/sec yields $R_0=2.9R_\sigma$ and $4.5R_\sigma$, respectively. The corresponding values of the coefficient $(3/4\pi)^{\frac{1}{2}}(R_0-R_\sigma)/R_0$ in Eq.(12) are 0.32 and 0.38, which we approximate henceforth as 1/3. These values agree satisfactorily with the value 0.24 deduced by Noyes [10], given the approximate nature of the calculations. The important thing is that this coefficient does not depend strongly on the rather approximate quantity R_0. Also, the typical values of R_0-R_σ are such that the exponential term in Eq.(12) is approximately unity at the reencounter times needed for adequate S-T mixing in most CIMP processes. Thus, a useful approximation to Eq.(12) is

$$dP(R\rightarrow R_\sigma,t\,|\,R_0,0) = \frac{1}{3}\frac{d(t/\tau)}{(t/\tau)^{3/2}} \; . \tag{15}$$

3.3 Diffusion-controlled radical pair model of CIDNP

The probability of radical pair reaction at the second encounter is proportional to $|C_{S,ab}(t)|^2$. Thus, Eqs.(10) and (15) give the following equation for the probability that a radical pair which separated in a triplet state will reencounter and react:

$$p_R^{(T)}(ab) = \frac{1}{9}k_R\int_0^\infty \frac{\sin^2 w_{ab}t}{(t/\tau)^{3/2}}\,d(t/\tau) \tag{16}$$

where k_R is the rate constant expressed as the probability that a radical pair in a pure singlet state will react when $R<R_\sigma$. If we neglect the possibility that the radicals separate and return more than once before recombination, which assumption cannot change the results significantly because the probability that a pair will return after separation is only 0.32 [this is determined by integrating Eq.(12) over all time], then the rate of formation of product in nuclear-spin state ab is proportional to $p_R^{(T)}(ab)$. To evaluate the integral in Eq.(16) we introduce the variable $x = w_{ab}t$, and integrate the resulting integral by parts to obtain a known definite integral of $\sin(x)/\sqrt{x}$. The result is

$$p_R^{(T)}(ab) = 0.20k_R(|w_{ab}|\tau)^{\frac{1}{2}} \tag{17}$$

If the radical pair is initially in a pure singlet state, then a similar calculation using Eq.(11) instead of Eq.(10) gives

$$p_R^{(S)}(ab) = 0.32k_R - 0.58k_R(|w_{ab}|\tau)^{\frac{1}{2}}. \tag{18}$$

As noted previously, an F-pair formed by the random encounter of two free radicals will behave qualitatively as a triplet, because the higher the triplet character of the pair, the more likely it is to survive the initial encounter and begin the process of S-T mixing and reencounter. For a quantitative treatment we note that setting J=0 in Eqs. (8a) and (9) gives the following equation for the singlet character of a radical pair with uncorrelated spins at time t after a nonreactive collision

$$|C_{S,ab}(t)|^2=|C_S(0)|^2+\left[|C_T(0)|^2-|C_S(0)|^2\right]\sin^2|w_{ab}|t$$
$$+(iw_{ab}/2|w_{ab}|)\ \left[C_T{}^*(0)C_S(0)-C_T(0)C_S{}^*(0)\right]$$
$$\times \sin 2|w_{ab}|t. \qquad (19)$$

The last term in Eq. (19) can be neglected because both the magnitude and phase of $C_S(0)$ and $C_T(0)$ are random and this term vanishes when averaged over the random phase factor. Because of reaction involving the singlet component of the wave function, the components after the initial encounter [$C_S(0)$ and $C_T(0)$] are related to those before the encounter by the equations [9,20]

$$|C_S(0)|^2 = (1-k_R)|C_S(0)|^2_{before}\ ;\ |C_T(0)|^2=|C_T(0)|^2_{before} \qquad (20)$$

Because all components, including the $|T_{\pm1}\rangle$, are equally likely in the random pair wave function before the first collision, the average values of these quantities are

$$\langle|C_S(0)|\rangle^2 = \frac{1}{4}(1 - k_R);\ \langle|C_T(0)|\rangle^2 = \frac{1}{4} \qquad (21)$$

Inserting these values in Eq. (19) and averaging over the distribution of return times given by Eq. (15) gives the result

$$p_R^{(F)}(ab) = 0.08k_R(1-k_R)+ 0.14k_R^2(|w_{ab}|\tau)^{\frac{1}{2}} \qquad (22)$$

As in previous qualitative models, the diffusion model contains the dependences on radical pair multiplicity and the S-T mixing rate w_{ab} which led to the CIDNP rules given in Eqs. (4) and (5). The significant result of the diffusion model is the dependence of the polarizations on the factor $(|w_{ab}|\tau)^{\frac{1}{2}}$ which is important in quantitative work and because it shows that large polarizations can be produced even if $|w_{ab}\tau|$ is small. For example, if w_{ab} and its differences among the various nuclear spin states are of the order of 10^8 rad/sec, $\tau = 10^{-11}$ sec, and $k_R=0.5$, then the nuclear spin selectivity in the reaction probabilities for T, S and F pairs is $3(10)^{-3}$, $9(10)^{-3}$ and $1(10)^{-3}$, by Eqs. (17), (18) and (22), respectively. These correspond to nuclear spin enhancement factors of 300, 900 and 100 relative to the equilibrium population difference between spin levels separated by 60 MHz, which is given by the Boltzman factor $w_n/kT=9(10)^{-6}$.

3.4 Diffusion controlled radical pair model of CIDEP

The same S-T_0 evolution of the radical pair wave function described in Eqs.(7)-(9) leads to radical pair CIDEP; however, the process is more complicated and involves the combined effects of S-T_0 mixing by the magnetic interactions, and S-T_0 splitting by the exchange interaction [7a,22,23]. Therefore, it is not possible in this case to assume that J=0 once the radical pair separates. As in the nuclear spin case, the S-T_0 mixing cannot produce an absolute electron spin polarization in the radical pair; thus, in all cases $\langle S_{1z}+S_{2z}\rangle=0$ where S_{1z} and S_{2z} are the components of $\underset{\sim}{S}_1$ and $\underset{\sim}{S}_2$ along $\underset{\sim}{H}_0$. It can lead to net polarization of one radical which is balanced by equal and opposite polarization of the other radical, and to multiplet effects wherein different hyperfine levels of a given radical are polarized equally and oppositely.

The electron spin polarizations of the two radicals, denoted as ρ_1 and ρ_2, are given by the equation

$$\rho_{1,ab}(t) = -\rho_{2,ab}(t) = 2\langle\psi^*(t)|S_{1z}|\psi(t)\rangle. \tag{23}$$

Since the electron spin is parallel to $\underset{\sim}{H}_0$ in the upper Zeeman level, $\rho_{1,ab}>0$ and <0 correspond to emission and enhanced absorption, respectively, in the ESR spectrum. Use of Eq. (7) for $\psi(t)$ gives

$$\rho_{1,ab}(t) = C_T(t)C_S^*(t) + C_T^*(t)C_S(t) \tag{24}$$

The physical significance of this result is that the polarization is proportional to the product of the in-phase parts of the $|S\rangle$ and $|T_0\rangle$ components of the wave function, where the phases are given by the complex exponential factors in the wave function coefficients, i.e., $C_S=|C_S|\exp(i\lambda_S)$, $C_T=|C_T|\exp(i\lambda_T)$. The S-$T_0$ mixing process tends to produce out-of-phase components; for example, starting with a $|T_0\rangle$ state with $\lambda_T=0$ (pure real), Eq.(8a) predicts the admixed $|S\rangle$ state will have $\lambda_S=\pi$ (pure imaginary). No amount of S-T_0 mixing can produce radical pair CIDEP unless there is an accompanying singlet-triplet splitting interaction which changes the phases of the $|T_0\rangle$ and $|S\rangle$ components by the equal and opposite energy-phase factors exp(iJt) and exp(-iJt), thereby partially aligning the phases [24].

For a more quantitative treatment [22] we use Eqs.(8) and (9) for C_S and C_T in Eq.(24), obtaining the result

$$\rho_{1,ab}(t) = [C_T(0)C_S^*(0)+C_T^*(0)C_S(0)][\cos 2wt+2(w_{ab}/w)^2\sin^2 wt]$$
$$+(iJ/w)[C_T(0)C_S^*(0)-C_T^*(0)C_S(0)]\ \sin 2wt$$
$$+(2w_{ab}J/w^2)[|C_S(0)|^2-|C_T(0)|^2]\ \sin^2 wt. \qquad (25)$$

The last term on the right side of Eq.(25) can lead to spin polarization among the radicals, and this was the term considered in the first radical-pair treatment of this problem [7a]. Because of its dependence on the quantity $2w_{ab}J/w^2$, this term is significant only in the very narrow range of R for which $|w_{ab}|\cong|J|\cong w/\sqrt{2}$. Since the time spent by the radicals at this or any other particular separation will be of the order of the correlation time for Brownian motions in liquids (i.e., 10^{-10}-10^{-12} sec) and since the w_{ab} are typically 10^8 rad/sec, the contribution of this term to the spin polarization will be vanishingly small because of its dependence on $\sin^2 wt$.

It is also readily shown that none of the other terms in Eq. (25) can lead to a spin polarization among the radicals following their first encounter. At an initial encounter the radical pair is either pure S, pure T, or its spins are uncorrelated. In the first two cases the terms in $C_T(0)C_S(0)$ are obviously zero. In the case of uncorrelated spins both the magnitude and phase of the S and T_0 components of the radical pair wavefunction are random, and the quantities $[C_T^*(0)C_S(0)\pm C_T(0)C_S^*(0)]$ vanish when averaged over this random phase factor.

This situation is changed, however, if we consider the development of polarization following a second encounter of the radicals at time t', an expression for which is obtained by making the substitutions $C_S(0)\to C_S(t')$, $C_T(0)\to C_T(t')$, and $t\to t-t'$ in the right-hand side of Eq.(25). It is readily shown for J=0 in the interval 0 to t' that $C_T(t')C_S^*(t')+C_T^*(t')C_S(t')=0$, so there is still no contribution to the polarization from the first term in Eq.(25). Equations (8)-(9) show, however, that

$$C_T(t')C_S^*(t')-C_T^*(t')C_S(t')$$
$$= -i(w_{ab}/|w_{ab}|)[|C_S(0)|^2-|C_T(0)|^2]\sin 2|w_{ab}|t'. \qquad (26)$$

This quantity is not zero so inserting it in Eq.(25) gives the following equation for spin polarization in the radicals following their second encounter:

$$\rho_{1,ab}(t,t') = (w_{ab}J/|w_{ab}J|)[|C_S(0)|^2-|C_T(0)|^2]$$

$$\times \sin 2|w_{ab}|t' \sin 2|J|(t-t'), \tag{27}$$

where we have assumed that $|J|>>|w_{ab}|$ during the second encounter when the radicals are close together. This equation, which is very similar to the last term on the right-hand side of Eq.(25), also predicts electron spin polarization in the radical pair. It is free, however, from the aforementioned difficulties with the last term in Eq.(25), because in this case the effects of singlet-triplet mixing by w_{ab} and singlet-triplet splitting by J are not constrained to occur simultaneously. Rather, the polarization in this case is the combined result of the weak separation-independent singlet-triplet mixing during the relatively long interval between the first and second encounters of the radicals (0 to t'), and the strong short-range singlet-triplet splitting during the relatively short time when the radicals are close together for the second time (t' to t).

Averaging Eq.(27) over the distribution of reencounter times given by Eq.(15) gives the result

$$\bar{\rho}_{1,ab}(|J|\tau_C) = 1.2\ \mathrm{sgn}(w_{ab}J)(|w_{ab}|\tau)^{\frac{1}{2}}$$

$$\times [|C_S(0)|^2-|C_T(0)|^2]\sin 2|J|\tau_C, \tag{28}$$

where $\tau_C=(t-t')$ is the duration of a typical collision between the radicals, and

$$\mathrm{sgn}(x) = \begin{cases} 1 & x>0 \\ 0 & x=0 \\ -1 & x<0 \end{cases} \tag{29}$$

A working definition of the duration of the collision is that interval during which $J>w_{ab}$. Now J(R) is a short range interaction which is effectively zero when the radicals are separated by 3-4 molecular diameters, and τ is the average time between diffusive displacements of the order of a molecular diameter; therefore, τ_C and τ should be roughly equal. There is no problem about the function $\sin 2|J|\tau_C$ in Eq.(28) always being very small, because J can be very large when the radicals are close together. On the contrary, many collisions will be "strong" in the sense that $J \geq kT/\hbar = 4(10)^{13}$ rad/sec at T=300 K. Even if the radicals do not recombine in a "strong" collision no spin polarization is produced because $2|J|\tau_C >> 1$ for $\tau_C \simeq \tau > 10^{-12}$ sec, and, thus, the average of the function $\sin 2|J|\tau_C$ over the duration of a "strong" collision is zero. (This, of course, is just the spin exchange process wherein a collision between radicals that produces a large splitting of their singlet and triplet states randomizes their electron spins [25].) On the other hand, there will be some "weak" collisions ($2|J|\tau_C \approx 1$) in which the radicals do not approach as closely. This suggests the following rough approximation [22], which is supported by a more sophisticated calculation that includes the variation of J(R) in a diffusing pair [20]:

$$\langle \sin 2|J|\tau_C \rangle (\text{collision Av}) \cong 0.1. \tag{30}$$

Use of this estimate in Eq.(28) gives the following expression for the expectation value of the electron spin polarization in a pair of radicals that are known to have been together at time t=0:

$$\langle \bar{\rho}_{1,ab} \rangle = 0.12 \ \text{sgn}(w_{ab}J)(|w_{ab}|\tau)^{\frac{1}{2}}[|C_S(0)|^2 - |C_T(0)|^2]. \tag{31}$$

For an estimate of the spin polarization produced by this mechanism, we take $\tau = 10^{-11}$ sec and $|w_{ab}| = 10^8$ rad/sec. Then Eq.(31) gives $|\langle \bar{\rho}_{1,ab} \rangle| \approx 4.0(10)^{-3}$ or $1.3(10)^{-3}$ depending on whether the radical pair was initially pure singlet or pure triplet, cf. Eqs. (10,11). The thermal equilibrium population difference between electron Zeeman levels separated by 9000 MHz is $1.4(10)^{-3}$. Thus, the estimated polarization is in qualitative agreement with observations that the differences between polarized spin levels are comparable to the thermal equilibrium values [1,23,26,27].

For the case of an F pair the considerations are the same as in the CIDNP case. The arguments applied there, cf. Eqs.(19)-(22) and accompanying discussion, give

$$\langle \bar{\rho}^{(F)}_{1,ab} \rangle = -0.03 \ k_R \ \text{sgn}(w_{ab}J)(|w_{ab}|\tau)^{\frac{1}{2}}. \tag{32}$$

As expected, F pairs behave similarly to geminate triplet pairs.

These results, as does the original treatment of Kaptein and Oosterhoff [7a], provide a set of qualitative rules of radical pair CIDEP, analogous to the CIDNP rules in Eqs.(4) and (5). A net effect in which one radical has an excess of "up" spins and the other radical has an equal excess of "down" spins requires $\Delta g \neq 0$. If $\Delta g > 0$, Eq.(3) shows that $\langle w_{ab} \rangle > 0$. Then for radical 1 of a T or F pair and taking $J<0$, Eqs.(31) and (32) show that $\rho_1 > 0$, which corresponds to emissive (E) CIDEP. The sign of ρ_1 is reversed for an S pair or by changing the sign of either J or Δg.

If $g_1 = g_2$ there can only be multiplet effects. Again, consider a T or F pair with a single magnetic nucleus of spin ½ on radical 1, and take $J<0$. Taking $A_1 > 0$, then $w_{ab} > 0$ for the $M_1 = \frac{1}{2}$ nuclear spin state; thus, this state will be emissive and the $M_1 = -\frac{1}{2}$ state will be absorptive. In this $A_1>0$ case the ESR transition energy is greatest for the $M_1=\frac{1}{2}$ hyperfine level; consequently, in an ESR spectrum run at constant frequency and varying field this transition will occur at low field and the $M_1=-\frac{1}{2}$ transition occurs at high field, giving an emission-absorption (E/A) CIDEP spectrum. Changing the sign of A_1 does not affect the CIDEP spectrum, because, although it makes $M_1=-\frac{1}{2}$ the emissive state, it also makes $M_1=-\frac{1}{2}$ the low field transition. Changing to an S pair or changing the sign of J changes the sign of ρ, and converts an E/A spectrum to A/E, and vice versa. Furthermore, if radical 1 has a number of magnetic nuclei and a corresponding multiline ESR spectrum, it can be shown, by expanding the foregoing arguments, that in an E/A spectrum all hyperfine lines below the center of the spectrum will be emissive and all those above the center will be absorptive, and vice versa for an A/E spectrum. These rules may be summarized in terms of the two quantities, Λ_{ne} for net effects, and Λ_{me} for multiplet effects:

$$\Lambda_{ne} = \mu J \Delta g \qquad \begin{cases} + : \mathrm{A} \\ - : \mathrm{E} \end{cases} \tag{33}$$

$$\Lambda_{me} = \mu J \qquad \begin{cases} + : \mathrm{A/E} \\ - : \mathrm{E/A} \end{cases} \tag{34}$$

where the multiplicity factor μ is defined following Eqs.(4)-(5).

To date there is no known case of a pure net CIDEP effect due to the radical pair mechanism; all net effects are probably due to the triplet mechanism [28]. Trifunac and Avery have observed a case of combined net and multiplet effects dependent on Δg which can be varied by placing a halogen atom in a chemically inert region of one of the radicals [29]. Most other radical pair CIDEP spectra appear to be E/A multiplets if allowance is made for distortion of the pure E/A pattern by relaxation effects [1,23,26,27]. Because it is unlikely that the radical pair precursors invariably will be triplets, it appears that most radical pair CIDEP involves F pairs [30]. This requires that $J<0$, which is reasonable because the bonding state of a radical pair is a singlet.

Quantitatively, the theory described here enables calculation of the relative polarization intensities of various hyperfine components of a CIDEP spectrum, and several comparisons of these predictions with experiment have yielded good agreement [22,31]. There have been several attempts to improve this theory, particularly the treatment of the exchange interaction. Adrian has treated J(R) as a first-order perturbation whose time variation is determined by the diffusion of the radicals [20]. Although some diffusion trajectories are certain to make J too large to be treated as a first order perturbation, it is argued that because spin-exchange destroys all polarization in these pairs, this difficulty can be overcome by selecting out and recycling these pairs. Evans, Fleming and Lawler have used a δ-function approximation to J(R) which permits an exact solution of the S-T mixing equation [32]. Pedersen and Freed have applied the stochastic Liouville method to this problem and have obtained, insofar as this model is applicable, exact computer solutions for various forms of J(R) and other parameters [33]. Generally, these theories predict greater polarizations than does the highly approximate theory discussed here. Otherwise, at the comparison level allowed by presently available experimental data all theories agree substantially; however, there are differences between them which hopefully will be probed by future experiments [34].

4. LOW FIELD RADICAL PAIR CIDNP

Unfortunately, space limitations and the objective of as simple a presentation as possible preclude any substantial discussion of the radical pair mechanism at low fields where the splitting of the electron Zeeman levels is comparable to the hyperfine interaction. In this region, it is no longer possible to consider only S-T_0 mixing, but it is necessary to consider the $T_{\pm 1}$ states both regarding their effect on the S and T_0 energy levels and wave functions and how this affects S-T_0 mixing, and the effects of actual spin transitions between the S and T_0 states and the $T_{\pm 1}$ states. The result is a number of interesting departures from high field CIDNP, most notably, the production of net effects in a number of systems which do not exhibit net effects at high field. Fortunately, for the interested reader there are a number of interesting experimental and theoretical studies of this aspect of the problem [35].

5. APPENDIX: ASPECTS OF DIFFUSION IMPORTANT IN CIMP

We present here a derivation of the diffusion equation and a solution of it which is especially pertinent to the RPM. Emphasis will be on presenting a physical picture of the process rather than mathematical rigor [36]. A molecule moving randomly in a liquid under the influence of its own thermal energy and collisional interactions with other molecules will in a short time τ be displaced a small amount $\underset{\sim}{\rho}$. The displacement $\underset{\sim}{\rho}$ will be randomly distributed in size and direction, with the probability that the magnitude of a given displacement, ρ', lies between ρ and $\rho+d\rho$ given by the equation

$$dP[\rho<\rho'<\rho+d\rho] = f(\rho)d\rho. \qquad \text{(A1)}$$

In an isotropic fluid the directions of the diffusional displacements will be uniformly distributed; thus the probability that the direction of a given displacement lies in the element of solid angle $d\Omega = \sin\theta d\theta d\phi$ is given by the equation

$$dP(d\Omega) = \frac{1}{4\pi}\sin\theta d\theta d\phi \tag{A2}$$

We are neglecting here effects associated with the instantaneous orientation of the diffusing molecule and the surrounding molecules which might lead to time-varying preferences for certain directions of diffusional displacement. Molecular reorientations, which take about 10^{-10} to 10^{-12} sec for the size molecules involved in most CIMP reactions, will average out these orientational effects over the time scale of the diffusion processes involved in CIMP.

The location of a given molecule at a time t after diffusion begins can be specified only in terms of a probability distribution function $P(\underset{\sim}{R},t)$, where

$$dP\begin{Bmatrix}\text{Particle in volume element } dV\\ \text{centered on } \underset{\sim}{R} \text{ at time } t\end{Bmatrix} = P(\underset{\sim}{R},t)dV \tag{A3a}$$

Of course, $P(\underset{\sim}{R},t)$ will depend on the initial probability distribution of the diffusing particle, $P(\underset{\sim}{R},0)$, which quantity is a boundary condition of the partial differential diffusion equation. It should be noted that $P(\underset{\sim}{R},t)$ can also describe the density distribution of an ensemble of identical diffusing particles, i.e.,

$$dN\begin{Bmatrix}\text{Particles in volume element } dV\\ \text{centered on } \underset{\sim}{R} \text{ at time } t\end{Bmatrix} = NP(\underset{\sim}{R},t)dV \tag{A3b}$$

where N is the total number of particles. It will be convenient in future discussion to use these definitions of $P(\underset{\sim}{R},t)$ interchangeably.

A difference equation for $P(\underset{\sim}{R},t)$ in terms of $\underset{\sim}{\rho}$ and τ may be established by noting that if the linear dimensions of dV are small compared to $\underset{\sim}{\rho}$ then a particle reaching $\underset{\sim}{R}$ at time $t+\tau$ by a diffusional displacement $\underset{\sim}{\rho}$ must have been at $\underset{\sim}{R}-\underset{\sim}{\rho}$ at time t. Considering the possible magnitudes and orientations of the displacement $\underset{\sim}{\rho}$ and averaging over these, using their probabilities as given by Eqs.(A1) and (A2), gives the equation

$$P(x,y,z,t+\tau) = \frac{1}{4\pi}\int_0^{\infty} d\rho f(\rho)\int_0^{2\pi} d\phi\int_0^{\pi} d\theta\sin\theta \tag{A4}$$

$$\times\ P(x-\rho\sin\theta\cos\phi, y-\rho\sin\theta\sin\phi, z-\rho\cos\theta)$$

where $\underset{\sim}{R}$ has been expanded in Cartesian components. Expanding this expression in the small quantities $\underset{\sim}{\rho}$ and τ and carrying out the integrations, many of which vanish, gives

$$\frac{\partial P}{\partial t} = \frac{1}{6}\frac{\langle\rho\rangle^2_{rms}}{\tau}\left[\frac{\partial^2 P}{\partial x^2}+\frac{\partial^2 P}{\partial y^2}+\frac{\partial^2 P}{\partial z^2}\right] \tag{A5}$$

where

$$\langle\rho\rangle_{rms} = \left\{\int_0^\infty f(\rho)\rho^2 d\rho\right\}^{\frac{1}{2}} \tag{A6}$$

Of course, Eq.(A5) is the partial differential equation for diffusion, where the diffusion constant is

$$D = \langle\rho\rangle^2_{rms}/6\tau \tag{A7}$$

Treatment of the diffusion processes involved in CIMP by this diffusion equation is somewhat more convenient than the original treatment of Noyes, which used a difference equation approach [10]. Use of the continuous partial differential equation is permissible so long as the time interval τ required for a significant displacement, $\langle\rho\rangle_{rms}$, is small compared to the time required for the S-T mixing processes involved in CIMP. This requirement is clearly fulfilled: if $\langle\rho\rangle_{rms}$ is taken to be of the order of a molecular diameter ($\sim$4 Å) then τ is approximately 10^{-10} to 10^{-11} sec yielding typical liquid phase diffusion constants of $2.6(10)^{-5}$ to $2.6(10)^{-6}$ cm^2/sec, whereas the S-T mixing processes take around 10^{-8} sec.

In all our work the exchange interaction and the related singlet-state reaction probability are taken to be isotropic functions of radical separation. Thus, we are interested only in the radial separation of a diffusing radical pair, the equation for which is obtained by transforming Eq.(A5) to spherical coordinates and dropping the angular terms. This gives

$$D_{RP}\left(\frac{\partial^2 P}{\partial R^2}+\frac{2}{R}\frac{\partial P}{\partial R}\right) = \frac{\partial P}{\partial t} \tag{A8}$$

Also, we are concerned here with the relative diffusion of two radicals, which is treated by regarding one radical stationary and the other diffusing with the diffusion constant

$$D_{RP} = D_{R1}+D_{R2}\,. \tag{A9}$$

The basic diffusion process involved in CIMP, as illustrated in Fig. 8, is separation of the radical pair to a point R_0 where

S-T mixing may begin $[|J(R_0)| \simeq |w_{ab}|]$, a period of diffusion during which S-T mixing proceeds, and then, return <u>for the first time</u> to a separation of the order of a molecular diameter R_σ where the singlet component of the radical pair may enter into cage reactions. We are thus interested in the distribution of times at which a diffusing particle, originally at R_0, first reaches R_σ where $R_\sigma < R_0$. This problem is solved by a method, first used in CIDNP by Deutsch [37], where one considers that particle distribution function $P(R,t|R_0,0;R>R_\sigma)$ which is a solution of the diffusion equation (A8), with the initial state boundary condition

$$P(R,0) = (1/4\pi R_0^2)\,\delta(R-R_0) \tag{A10}$$

and an absorbing barrier introduced at $R=R_\sigma$ by the boundary condition

$$P(R_\sigma,t) = 0 \tag{A11}$$

The solution of this equation is well known; it has been given by Carslaw and Jaeger [38] for the analogous problem of heat conduction:

$$P(R,t|R_0,0;\ R>R_\sigma) = (4\pi RR_0)^{-1}(4\pi Dt)^{-\frac{1}{2}}\Big\{\exp[-(R-R_0)^2/4Dt] - \exp[-(R+R_0-2R_\sigma)^2/4Dt]\Big\} \tag{A12}$$

The physical significance of this result is that the particles, be they material particles, heat "particles", or whatever, originating at R_0 are absorbed at the barrier R_σ. Thus, we may calculate the number of particles arriving at R_σ, for what must be the first time, because they do not survive to arrive again, by calculating the flux of particles at R_σ. This is given by the product of the area of the sphere of radius R_σ, the diffusion constant, and the gradient of the particle distribution function. Thus, the probability that a particle originating at R_0 reaches R_σ for the first time during the time interval t to t+dt is:

$$\begin{aligned} dP(R\to R_\sigma,t|R_0,0) &= 4\pi R_\sigma^2 D[\partial P(R,t|R_0,0;R>R_\sigma)/\partial R]_{R=R_\sigma}dt \\ &= [R_\sigma(R_0-R_\sigma)/R_0](4\pi Dt^3)^{-\frac{1}{2}}\exp[-(R_0-R_\sigma)^2/4Dt]dt \end{aligned} \tag{A13}$$

Taking the average diffusive displacement equal to the collision diameter in accordance with previous discussion ($\langle\rho\rangle_{rms}=R_\sigma$), and using Eqs.(A7) and (A9) assuming $D_{R1}=D_{R2}$ gives $D=(1/3)R_\sigma^2/\tau$. Substituting this result into Eq.(A13) gives

$$dP(R\to R_\sigma,t|R_0,0)=\sqrt{\frac{3}{4\pi}}\,\frac{R_0-R_\sigma}{R_0}(\tau/t)^{\frac{3}{2}}\exp\left[-3(R_0-R_\sigma)^2\tau/4R_\sigma^2 t\right]d(t/\tau) \tag{A14}$$

References

1. R.W. Fessenden and R.H.Schuler, J. Chem. Phys. 39, 2147 (1963).
2.(a) J. Bargon, H. Fischer and U. Johnsen, Z. Naturforsch. 22a, 1551 (1967);
(b) J. Bargon and H. Fischer, ibid., p. 1556.
3.(a) H.R. Ward and R.G. Lawler, J. Am. Chem. Soc. 89, 5518 (1967);
(b) R.G. Lawler, Ibid., p. 5519.
4. Some reviews are:
(a) Chemically Induced Magnetic Polarization, Eds. A.R. Lepley and G.L. Closs (Wiley, New York 1973).
(b) G.L. Closs, Advances in Magnetic Resonance, Ed. J.S. Waugh, Vol. 7, p. 157 (Academic Press, New York 1974);
(c) R.G. Lawler, Progress in Nuclear Magnetic Resonance Spectroscopy, Eds. J.W. Emsley, J. Feeney and L.H. Sutcliffe, Vol. 9, p. 147 (Pergamon, New York 1975);
(d) R. Kaptein in Advances in Free-Radical Chemistry, Ed. G.H. Williams, Vol.5, p. 319 (Elek Science, London 1975).
5. K.H. Hauser and D. Stehlik, Adv. Magn. Reson. 3, 79 (1968).
6.(a) G.L. Closs, J. Am. Chem. Soc. 91, 4552 (1969);
(b) G.L. Closs and A.D. Trifunac, J. Am. Chem. Soc. 92, 2183 (1970).
7.(a) R. Kaptein and L.J. Oosterhoff, Chem. Phys. Letters 4, 195 (1969);
(b) R. Kaptein and L.J. Oosterhoff, ibid., p. 214.
8. F.J. Adrian, J. Chem. Phys. 53, 3374 (1970); J. Chem. Phys. 54, 3912 (1971).
9.(a) R. Kaptein, Doctoral Dissertation, University of Leiden (1971), Chap. VIII;
(b) R. Kaptein, J. Am. Chem. Soc., 94, 6251 (1972).
10. R.M. Noyes, J. Chem. Phys. 22, 1349 (1954).
11. G.L. Closs and A.D. Trifunac, J. Am. Chem. Soc. 92, 2186 (1970).
12. G.L. Closs, Chem. Phys. Lett. 32, 277 (1975).
13. E.D. Becker, High Resolution NMR (Academic Press, New York, 1969), Chap. 7.
14. G.L. Closs, C.E. Doubleday and D.R. Paulson, J. Am. Chem. Soc. 92, 2185 (1970).
15. K. Möller and G.L. Closs, J. Am. Chem. Soc. 94, 1002 (1972).
16. R. Kaptein, J. Am. Chem. Soc. 94, 6262 (1972).
17. H. Eyring, J. Walter and G.E. Kimball, Quantum Chemistry (Wiley, New York 1949), Chap. III.
18. R. Kaptein and J.A. den Hollander, J. Am. Chem. Soc. 94, 6269 (1972).
19. J.B. Pedersen and J.H. Freed, J. Chem. Phys. 61, 1517 (1974).
20. F.J. Adrian, J. Chem. Phys. 57, 5107 (1972).
21. C. Herring and M. Flicker, Phys. Rev. A134, 362 (1964).

22. F.J. Adrian, J. Chem. Phys. 54, 3918 (1971).
23. Some reviews are
(a) J.K.S. Wan, S.K. Wong and D.A. Hutchinson, Accts Chem. Research 7, 58 (1974);
(b) P.W. Atkins and G.T. Evans, Advances in Chemical Physics, Vol. XXXV, edited by I. Prigogine and S.A. Rice (Wiley, New York, 1976), p. 1.
24. Reference 17, p. 29.
25. E.M. Purcell and G.B. Field, Astrophys. J. 124, 542 (1956).
26. B. Smaller, J.R. Remko and E.C. Avery, J. Chem. Phys. 48 5174 (1968).
27. H. Paul and H. Fischer, Z. Naturforsch 25a, 443 (1970).
28.(a) S.K.Wong, D.A. Hutchinson and J.K.S. Wan, J. Chem. Phys. 58, 985 (1973);
(b) P.W. Atkins and G.T. Evans, Mol. Phys. 27, 1633 (1974);
(c) F.J. Adrian, J. Chem. Phys. 61, 4875 (1974).
29. A.D. Trifunac and E.C. Avery, Chem. Phys. Letters 27, 141 (1974).
30. N.C. Verma and R.W. Fessenden, J. Chem. Phys. 65, 2139 (1976).
31. P.B. Ayscough, G. Lambert and A.J. Elliot, J. Chem. Soc. Faraday Trans. I 72, 1770 (1976).
32. G.T. Evans, P.D. Fleming and R.G. Lawler, J. Chem. Phys. 58, 2071 (1973).
33. J. B. Pedersen and J. H. Freed, J. Chem Phys. 58, 2746 (1973); J. Chem. Phys. 59, 2869 (1973).
34. These theories are reviewed in Reference 23b.
35.(a) J.I. Morris, R.C. Morrison, D.W. Smith and J.F. Garst, J. Am. Chem. Soc. 94, 2406 (1972);
(b) Reference 18;
(c) D.A. Hutchinson, H.M. Vyas, S.K. Wong and J.K.S. Wan, Mol. Phys. 29, 1767 (1975);
(d) G.T. Evans and R.G. Lawler, Mol. Phys. 30, 1085 (1975).
(e) J.A. den Hollander, Doctoral Dissertation, University of Leiden (1976), Chap. VII;
(f) F.S. Sarvarov, K.M. Salikhov and R.Z. Sagdeev, Chem. Phys. 16, 41 (1976).
36. For details see S. Chandrasekhar, Rev. Mod. Phys. 15, 1 (1943).
37. J.M. Deutsch, J. Chem. Phys. 56, 6076 (1972).
38. H.S. Carslaw and J.C. Jaeger, Conduction of Heat in Solids (Oxford, U. P., London, 1959), p. 382.

CHAPTER VI

EXPERIMENTAL OBSERVATION OF ELECTRON SPIN POLARIZATION IN REACTIONS

K. A. McLauchlan

Physical Chemistry Laboratory, University of Oxford, Oxford, England.

ABSTRACT. The different types of experiments in which Chemically Induced Dynamic Electron Polarization (CIDEP) has been observed are described. Two distinct types of spin polarization are recognised and reported results are interpreted in terms of various admixtures of these two types. Reasons why one or the other predominates in a given case are suggested.

1. INTRODUCTION

If at some instant of time a spin system is produced whose population distribution differs from that at thermal equilibrium a magnetic resonance spectrum with abnormal intensities can be observed for a time of the order of the spin-lattice relaxation time of the system. This may be ten seconds for nuclei but only microseconds for electrons. All the observations of the nuclear phenomenon CIDNP have been made furthermore in samples containing steady-state concentrations of radicals which are produced at random times. In such systems polarization can often be observed for long periods of time until reactant depletion becomes too great (although the maximum polarization is observed at a comparatively short time after the reaction is initiated, this being dependent on the precise relaxation time and kinetics of the system). A consequence is that CIDNP is easily observed with conventional spectrometers and from the start consistent results have been reported all of which support the origin of spin polarization being in pairs of free radicals. In high magnetic fields this has always entailed a spin-sorting mechanism, rather than a true spin polarization one, and the spectra are characterised by displaying equal amounts of intensity in emission and absorption (at least after correction

L. T. Muus et al. (eds.), Chemically Induced Magnetic Polarization, 107-118.

for possible differential relaxation effects).

CIDEP has proved more difficult. Experimental observations are still comparatively few and have been reported from diverse experiments ranging from observations in steady-state concentrations of radicals to pulse experiments which create radicals in nanoseconds, and with many time scales in between. Whilst the first observation of CIDEP showed the zero net polarization associated with radical pair effects [1] the next two were one of this type [2] and another in which a spectrum was entirely in emission [3]. After early attempts to rationalise this on radical-pair terms it is now acknowledged to involve an entirely separate polarization mechanism. CIDEP is consequently complicated by the occurrence, often together, of two different polarization mechanisms. The early literature failed to make this conclusion and contains a plethora of theories and of apparently curious results; the rationalisation of the field was largely due to Dobbs [4]. Here we shall examine the types of experiment that have been performed and describe the results obtained before attempting to explain them. An exhaustive summary of the literature is not planned but rather examples are chosen to illustrate the effects. Early reviews provide much of the background results [5,6].

2. EXPERIMENTAL

CIDEP is observed only in reacting systems and must be searched for using those methods appropriate to the study of transient free radicals. These are the steady-state methods involving flow and photolysis and radiolysis in situ, the pulse methods of flash photolysis and pulse radiolysis, and experiments involving chopping or modulating a photolysis source. In all of these ways CIDEP has been observed; for no apparent reason no-one has reported it (or probably looked for it) in systems involving the generation of free radicals electrochemically. Many spectra which display CIDEP effects appeared in the literature before the phenomenon was recognised [7,8,9].

In the flow methods reactants are mixed outside the micro - wave cavity of the spectrometer and the solution, now containing free radicals, is flowed within periods of about 0.5 ms into it. This period far exceeds the relaxation time of the radicals and any polarization observed originates in processes which are dissociated from their formation step : it arises in reactions between radicals which have diffused together. The spectra observed under favourable conditions display some lines in absorption, and some in emission, the net polarization being zero [10].

Continuous electron irradiation of samples within the cavity led to the first report of CIDEP in 1963 [1]: the hydrogen atom was observed with its low-field line in emission and its high-field one in absorption with zero net polarization. Similar spectra have been observed in other systems [11-15]. In these experiments high concentrations of radicals are produced and the polarization behaviour is dominated by radical pair effects.

Continuous photolysis in situ is a method of wide application to ESR studies and Livingston and Zeldes published spectra displaying CIDEP effects in 1970 [16,17]. More significant than was realised at that time they reported two different behaviours: many of their spectra exhibited a spectrum with all its lines in emission. This observation depended on radical lifetime and concentration and under some conditions the same radical exhibited the emission-absorption CIDEP behaviour. This was the first indication of two different polarization mechanisms occurring in one system and was observed at quite high concentrations of radicals which were however produced inside the cavity together as a pair; we now recognise that one of the types of polarization originated in the formation step whilst the other largely came from inter-radical effects.

Flash photolysis and pulse radiolysis appear ideal methods with which to study CIDEP for radicals can be produced by pulses of light or radiation which are over before the spectrum is examined, and measurements can be taken within the spin-lattice relaxation time of the radical. The first observation of purely emissive spectra was reported in such a study [3]. The techniques involved require considerable adaptation of conventional spectrometers and have been described in detail elsewhere for flash photolysis [18,19,20] and for pulse radiolysis [21-24]. Initially all the flash-photolysis results gave spectra which were completely in emission, although in general only one radical of the pair produced on photolysis could be observed. It should be realised that the two groups working in this field, in Oxford with P.W. Atkins and K.A. McLauchlan and in Canada with J.K.S. Wan and his colleagues, both used low-power nitrogen lasers in their experiments which produced very low radical concentrations (ca 10^{-6} M) in each pulse and were restricted to molecules which absorb at 337.1 nm. More recently a higher power nitrogen laser has yielded spectra in our laboratory which whilst being wholly in emission at short times following the flash exhibit intensity anomalies (unpublished work). Of particular significance were the observations [24,25] that when both radicals produced in photolysis could be observed the spectra of each were completely in emission. This emphasises the occurrence of a true polarization process un-related to the spin-sorting one which leads to emission-absorption behaviour.

The pulse radiolysis results have been entirely different and even at very short times after the radiation pulse spectra show only emission-absorption characteristics, with no evidence for a second polarization mechanism. The contrast with the flash-photolysis experiments is in the high concentrations of radicals produced.

The disadvantage of pulse methods is that the response time of the spectrometer must be short compared with the relaxation time of the radicals if quantitative information on polarization magnitudes or on relaxation times is required. This necessitates a large band-width for the spectrometer with a concomitant fall in signal-to-noise radio, a factor of considerable importance at the low concentrations attained in photolysis studies. However the occurrence of polarization can be detected at longer times in slower spectrometers but now the response is the convolution of the signal and the response function of the spectrometer. If the latter can be obtained accurately quantitative information results from a deconvolution procedure. This is an attractive method for it allows the ESR spectrometer to be run with normal 100 kHz field modulation and high sensitivity. It has a further advantage in that the requirements for a sharp initiation pulse are removed; a source which has an intensity-time profile which is negligible with respect now to a response time of the order of 100 μs (rather than 1 μs in pulse experiments) is all that is required. This allows the use of more intense sources and produces higher radical concentrations. This experiment has been performed very successfully both with the light chopped with a rotating sector [26-29] and with a square-wave modulated source [30,31]. In many of these experiments CIDEP has been observed which is consistent with the occurrence of two polarization mechanisms : that the emission-absorption type occurs is evidenced by a hyperfine-dependence of the polarization but the amounts of emission and absorption are not equal showing that there is some net spin polarization.

To summarise, whenever CIDEP is observed in radicals which exist in high concentrations the dominant behaviour is the emission-absorption one. In flow experiments this is associated with the recombination of radicals produced at random when they diffuse together; in radiolysis experiments the radicals may be created together but the polarization behaviour appears dominated by the encounter of randomly diffusing radicals (see below). Radiolysis experiments have not been reported at low radical concentrations but under these conditions photolysis may yield pure emission spectra, with both radicals in emission and with both showing no hyperfine dependence of polarization. Occasionally even under these conditions emission-absorption behaviour is observed [4,32] with no purely emissive contribution.

Observations of CIDEP in radicals produced photolytically in high concentration usually show some (or all) emission-absorption behaviour.

It should be emphasised that not all systems containing reacting molecules exhibit CIDEP : for it to be seen certain rate and/or magnetic property requirements must be met. Once a polarized radical exists it may react within its relaxation time to produce a secondary polarized species [18,27] but here we shall be concerned only with primary species.

3. POLARIZATION MECHANISMS

It is not our purpose here to develop polarization theory (which is discussed elsewhere in this volume) but rather to emphasise the chemical requirements for it to be observed and to describe it in just sufficient detail to explain some of the observations. Spin polarization can arise feasibly at two different times in a radicals' existence, at formation and at re-encounter. For a radical to be polarized at its instant of creation it must originate in a prior spin-polarized state of the molecule which reacts, with spin conservation, to produce it. Obviously this is possible from molecular states of higher multiplicity than singlet only, and it happens that all of the systems studied to date which show this initial polarization result from reaction of molecular triplet states. The triplet is formed by inter-system crossing from an upper state singlet produced by light absorption from the ground state; in the field of the spectrometer the triplet Zeeman levels are non-degenerate and inter-system crossing occurs preferentially (in the carbonyls and quinones studied in photolytic experiments) to the upper T_{+1} level to produce a spin-polarized triplet. This polarization is carried out to the radicals produced on reaction if the reaction rate competes with the very fast triplet spin-lattice relaxation process. Only diffusion-controlled reactions are sufficiently fast. Both radicals are produced in the same emissive state. Photochemical reactions proceed through excited states and those which do not involve singlets produce initial polarization in this way if the reaction criterion is met; large polarizations (up to 600 times the equilibrium population difference) can be obtained. We shall refer to this as the triplet mechanism (TM).

Initial polarization of the radicals in this way would not be expected in radiolysis which often involves homolytic breaking of bonds in ground state molecules, normally singlets. However consideration of symmetric scission in a molecule to yield two identical radicals does suggest that they should be formed in a spin-saturated condition and that polarizations of one equilibrium population difference should result; this has not been detected. This may imply that homolytic scission is not as important a source of radicals as electron capture.

Once a radical pair has been formed, either in an initially polarized state or not, further polarization can occur in a manner familiar in CIDNP. The two mechanisms normally occur sequentially at two different times (and are wrongly described as simultaneous) but even in the fastest pulse experiments they are both over long before measurements commence (in ca 1 μs) and we may observe the resultant effect of both.

In the radical-pair mechanism (RPM) we focus attention on the spin states of the radical pair either when formed with spin conservation from a molecular precursor or at subsequent encounter. Polarization arises from mixing the singlet (S) and triplet (T) levels as the radicals diffuse apart (as in CIDNP the initial radical separation is too fast to produce RP polarization and a pair formed geminately separate before re-encountering ca. 10^{-8}s later to be effective in polarization). The Zeeman levels $T_0, T_{\pm 1}$ of the radical pair triplet are non-degenerate in the field of the spectrometer and at high fields, including the 0.33 T of an X-band spectrometer, S-T_0 mixing predominates. It is induced by a perturbation which depends upon the difference in the Larmor precession frequencies of the unpaired electrons in the radical pair. This arises in general from differences both in the g-values and in the hyperfine coupling constants of the radicals. The constant phase relationship which exists between the electron spins in the pure S or T states created is perturbed and the states are interconverted. S-T_0 mixing simply determines the distribution of α and β spins in the pair and does not constitute a true polarization mechanism. However if the g-values of the radicals differ appreciably the spectrum of one of the radicals may show net emission although the other must then display net absorption. The importance of observing both radicals is obvious although it has rarely been possible and, since some radicals persist for very short times and some have very short relaxation times, the advantage of the pulse methods is apparent.

S-T_{-1} mixing, if it occurred, would yield both radicals with their spectra in net emission and it was suggested to rationalise some early results [25,26]. However the polarization observed showed no hyperfine dependence and it appears that, as in CIDNP in monoradicals at high fields, S-T_{-1} mixing is unimportant.

Radicals which diffuse together after having escaped from their respective original solvent cages do so with their unpaired electrons in random phase. Those that encounter as singlets may react immediately but triplet pairs evolve according to the RPM to produce further polarization and therefore mimic the behaviour of a triplet pair formed directly by reaction of a molecular triplet state. This effect is especially significant at high radical concentrations.

For pairs of radicals with quite similar g-values the CIDEP behaviour is normally that the low-field hyperfine lines in their spectra are in emission and the high-field ones in absorption. This is a consequence of the constancy of hyperfine coupling characteristics between similar radicals and of the origin of the radical pairs in states of similar multiplicity or in encounters in the different systems.

4. DISCUSSION

The model we take to discuss the observations is as shown overleaf.

We distinguish three different polarization regions and attempt to assess their relative importance; we note that only the last stage where RPM is effective at re-encounter depends upon the concentration of radicals. (The TM is of course dependent on the concentration of triplets and substrates, since these help to determine the rate of formation of radicals).

With this model is mind it is at first surprising that polarization solely due to the TM was observed at all: every time a pair of radicals is produced it may be polarized initially but then the RPM should cause hyperfine-dependent polarization. At very low radical concentrations in flash-photolysis experiments no RP effects are observed and we conclude that TM is so efficient a source of polarization that it swamps any geminate RP contribution. This is an important conclusion for the stoichiometry of the system is independent of the intensity of the light source: for every triplet produced leading to a radical pair the relative polarizations produced are constant. Thus when radical pair effects are obsèrved at higher light intensities both in flash and intermittent illumination experiments the RP polarization appears to originate in the re-encounter step. This is despite the fact that the radical concentration falls before it occurs (although not very quickly relates to the electron relaxation time) and implies that multiple re-encounters must be important. These arguments only hold when the TM operates and in its absence RP effects may be observable at the geminate stage.

The observation of a radical in emission in a continuous photolysis experiment [16] is suggestive that the TM was operative, particularily since the lines showed no hyperfine dependence of

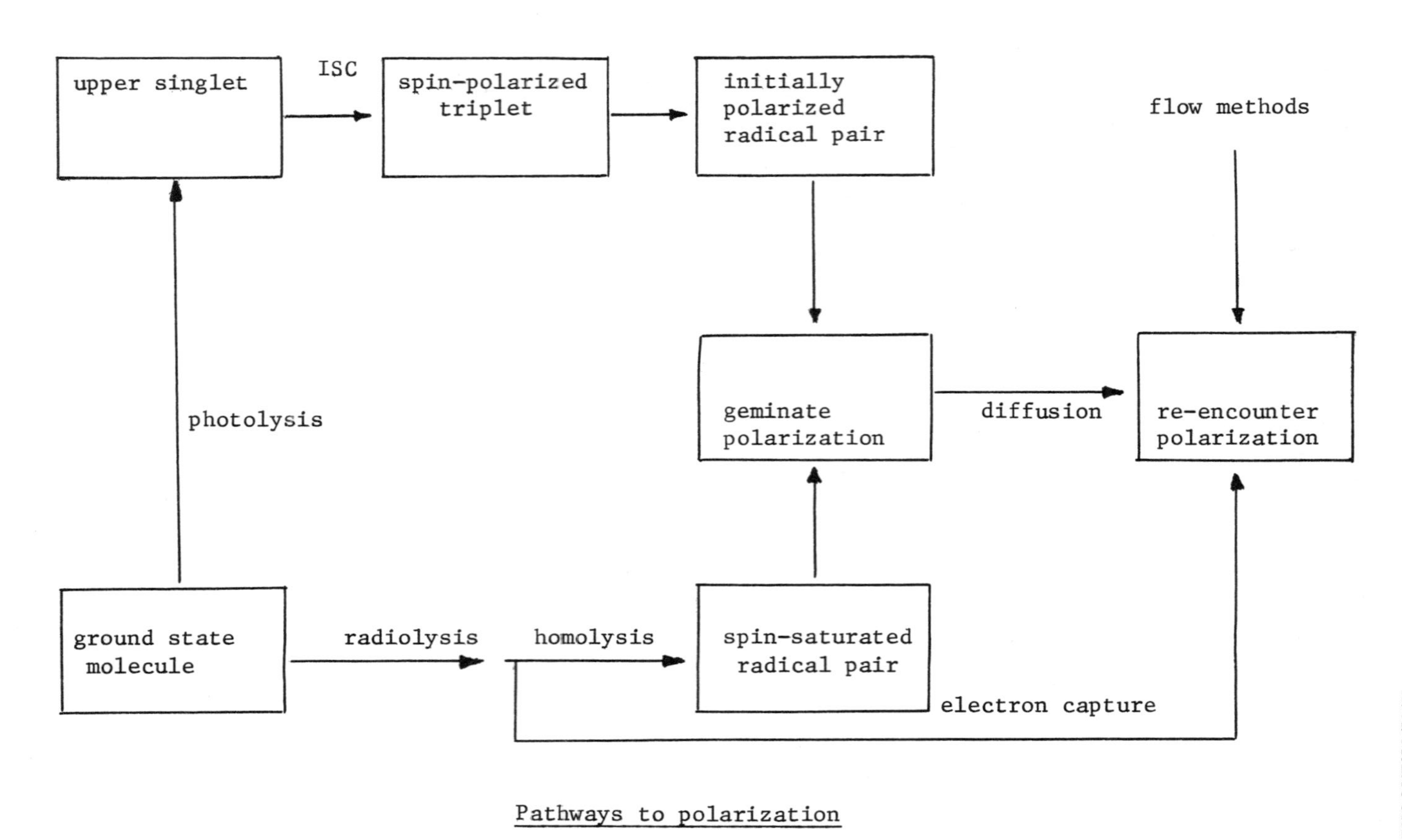

Pathways to polarization

intensity. The fact that CIDEP was observed only in systems in which the radical lifetime was short could have had several explanations: the reaction may have prevented build up of a normal thermal equilibrium signal which might have obscured the polarized one, the effective relaxation time may have become so short that polarization could not be observed or the reaction rate of the triplet itself may have been affected by the change in conditions. More recent analysis fully explains it [30].

Several reports have been made of the operation of both RP and TM mechanisms in photolysis systems [4,28,30,32-37]. Under the conditions of the experiments the photolysis of substituted benzoquinone, naphthoquinone and anthraquinone with 2,6-di-t-butyl peroxide showed the RPM in a minor role [28] whilst the photoreduction of pyruvic acid [33] showed it to be dominant. An elegant detailed analysis of the p-benzosemiquinone radical showed the RPM to contribute 7-23% of the polarization depending upon conditions; this study emphasised and analysed the influence of re-encounter polarization [30]. RP and triplet contributions to the polarization of the semidiones formed from pyruvic acid and biacetyl in alcoholic solvents have been separated in a kinetic analysis which gave their relative magnitudes independent of de-convolution techniques [37]. An investigation of the dependence of CIDEP on radical concentration and structure disclosed only RP effects [38]. Flash photolysis of pivalophenone in hydrogen-donating solvents gives triplet polarization but in other solvents bond-scission occurs to give benzoyl and t-butyl radicals which display RP effects [32]. At the concentrations studied this probably originates in geminate polarization, the RPM being particularly efficient and observable in this case because of the unusually high hyperfine coupling in the t-butyl radical. At high viscosities the normal emission-absorption RP pattern is reversed although the radicals still originate in a triplet precursor; this has not been explained convincingly. It is apparent that whether TM or RPM polarization, or a mixture of the two, is observed in photochemical experiments is a matter of experimental choice.

In radiolysis, whether continuous or pulsed, only RP effects are observed, and this would be expected (within a possible error of one Boltzmann population difference). CIDEP has been observed at periods after a radiation pulse which far exceed the electron relaxation times in the systems [39,40] which suggests that it arises on re-encounter. Even 0.2 μs after the pulse only normal emission-absorption RP patterns are observed [23]; assuming a singlet precursor or radicals produced randomly by electron capture, this again implies that re-encounter polarization is dominant. Radical pair polarization in radiolysis has been pursued to great effect both experimentally and theoretically by R.W. Fessenden [11,21,41-43] who has obtained an expression for the time-dependence of signal intensities, particularly for H-atoms, which allows rate constants to be determined. The importance of the g-factor difference in the RPM has been demonstrated clearly [40], although the precise nature of the radicals concerned has been questioned [4]. The RPM has been further confirmed in studies of the radiolysis of aqueous alcohols [44].

CIDEP has been observed from solvated electrons in ethereal solution in a photolysis experiment [45] and in aqueous solution by radiolysis [43]. In the latter study it was found that the presence of certain counter radicals caused the electron signal to be in emission and the counter radical to be wholly in enhanced absorption. This represents a case of the type discussed above where the g-value difference dominates and one radical is in net emission, the other in net absorption. The important mechanistic conclusion from the interpretation of these results using the RPM is that reaction between e^-_{aq} and the counter radical occurs preferentially into the triplet state of the product.

In flow experiments the only possible origin of spin polarization is in radical re-encounter unless mixing of reactants is incomplete before they enter the microwave cavity. However since ground-state molecules are concerned geminate polarization would evolve from singlet radical pairs and give opposite polarization effects from those of random encounter pairs. Once again inspection of the spectra shows that the latter dominate.

5. CONCLUSION

We have sought to report the different experiments in which CIDEP has been observed and to show that the disparate behaviour can be rationalised. The experimental tests of the different polarization mechanisms are described elsewhere in this volume and have been omitted here but it is these that have been the most significant in our understanding of CIDEP phenomena. The work of the Oxford Group is described quite fully later and it would be remiss not to acknowledge the outstanding contributions especially from the laboratories of R.W. Fessenden and J.K.S. Wan.

CIDEP is apparently well understood and there is little evidence at present for complicating factors. It might be that observed intensities should be corrected for different relaxation times of different hyperfine lines, which would especially affect interpretation of RP and mixed RP and triplet experiments, but the relaxation behaviour of electrons in radicals are much more dependent on spin-rotation and g-anisotropy effects than they are on hyperfine ones. Any corrections to observed polarizations are probably within experimental error. Electrochemical experiments involving the alternate production of reactive positive and negative ions might create the conditions for producing productive $S-T_{-1}$ mixing since Coulombic forces might hold the radicals sufficiently close for long enough for it to develop; otherwise $S-T_{-1}$ mixing seems unimportant.

As with CIDNP the future use of CIDEP is likely to be mainly in mechanistic studies and here triplet polarization is particularly attractive in providing a label for following a radical reaction pathway. However CIDEP does provide the means to obtain important physical information, which cannot be so easily obtained in any other way [18], on a wide variety of radicals. These physical experiments can only be performed to sufficient accuracy using pulse methods. Proton relaxation times are easily measured in many ways but CIDEP is likely to be particularly significant in providing accurate measurements of electron relaxation times in radicals and in precursor triplet molecules.

REFERENCES

1. R.W. Fessenden and R.H. Schuler, J.Chem.Phys., 2147, 39,1963.
2. B. Smaller, J.R. Remko and E.C. Avery, J.Chem.Phys., 5174 48. 1968.
3. P.W. Atkins, I.C. Buchanan, R.C. Gurd, K.A. McLauchlan and A.F. Simpson, Chem. Commun., 513, 1970.
4. A.J. Dobbs, Molec. Phys., 1073, 30, 1975.
5. P.W. Atkins and K.A. McLauchlan in "Chemically Induced Magnetic Polarization" ed. A.R. Lipley and G.L. Closs, John Wiley 1973.
6. J.K.S. Wan, S.K. Wong and D.A. Hutchinson, Accts.Chem.Res., 58, 7, 1974.
7. H. Fischer, Z. Naturforsch, 866, A19, 1964.
8. P.J. Krusic and J.K. Kochi, J.Amer.Chem.Soc., 7155, 90, 1968.
9. R. Livingston and H. Zeldes, J.Amer.Chem.Soc., 4333, 88, 1966.
10. H. Paul and H. Fischer, Z.Naturforsch, 443, A25, 1970.
11. K. Eiben and R.W. Fessenden, J.Phys.Chem., 1186, 75, 1971.
12. P. Neta, R.W. Fessenden and R.H. Schuler, J.Phys.Chem., 1654, 75, 1971.
13. N.C. Verma and R.W. Fessenden, J.Chem.Phys., 2501, 58, 1973.
14. R.W. Fessenden and K. Eiben, J.Phys.Chem., 1186, 75, 1971.
15. P.B. Ayscough and J.A. Elliott, J.Chem.Soc.Faraday 1, 791, 72, 1976.
16. R. Livingston and H. Zeldes, J.Chem.Phys., 1406, 53, 1970.
17. H. Zeldes and R. Livingston, J.Phys.Chem., 3336, 74, 1970.
18. K.A. McLauchlan. This volume.
19. P.W. Atkins, K.A. McLauchlan and A.F. Simpson, J.Phys.(E), 547, 3, 1970.

20. J.R. Bolton and J.T. Warden, Creat.Detect.Excited State, 63, 2, 1974.
21. R.W. Fessenden, Nucl.Sci.Abstracts., 24257, 38, 1973.
22. B. Smaller, E.C. Avery and J.R. Remko, J.Chem.Phys., 2414, 55, 1971.
23. A.D. Trifunac, K.W. Johnson, B.E. Clifft and R.H. Lowers, Chem.Phys.Letts., 566, 35, 1975.
24. S.K. Wong, D.A. Hutchinson and J.K.S. Wan, J.Amer.Chem.Soc., 622, 95, 1973.
25. P.W. Atkins, A.J. Dobbs and K.A. McLauchlan, Chem.Phys.Letts. 209, 22, 1973.
26. R. Livingston and H. Zeldes, J.Magn.Res., 331, 9, 1973.
27. R. Livingston and H. Zeldes, J.Chem.Phys., 4891, 59, 1973.
28. B.B. Adeleke and J.K.S. Wan, J.Chem.Soc.Faraday I, 1799, 72, 1976.
29. S.K. Wong and J.K.S. Wan, J.Amer.Chem.Soc., 7197, 94, 1972.
30. J.B. Pedersen, C.E.M. Hansen, H. Porbo and L.T. Muus, J.Chem.Phys., 2398, 63, 1975.
31. P.B. Ayscough, T.H. English and D.A. Tong, J.Phys.(E), 31, 9, 1976.
32. P.W. Atkins, A.J. Dobbs and K.A. McLauchlan, J.Chem.Soc. Faraday II, 1269, 71, 1975.
33. K.Y. Choo and J.K.S. Wan, J.Amer.Chem.Soc., 7127, 97, 1975.
34. H.M. Vyas and J.K.S. Wan, Canad.J.Chem., 979, 54, 1976.
35. H.M. Vyas and J.K.S. Wan, Int.J.Chem.Kinetics, 125, 6, 1974.
36. H.M. Vyas, S.K. Wong, B.B. Adeleke and J.K.S. Wan, J.Amer Chem.Soc., 1385, 97, 1975.
37. P.B. Ayscough, G. Lambert and A.J. Elliott, J.Chem.Soc. Faraday I, 1770, 1976.
38. P.B. Ayscough, T.H. English, G. Lambert and A.J. Elliott, Chem. Phys. Letts., 557, 34, 1975.
39. R.W. Fessenden, J.Chem.Phys., 2489, 58, 1973.
40. A.D. Trifunac and E.C. Avery, Chem.Phys.Letts., 141, 27, 1974.
41. R.W. Fessenden, J.Chem.Phys., 2489, 58, 1973.
42. N.C. Verma and R.W. Fessenden, J.Chem.Phys., 2501, 58, 1973.
43. R.W. Fessenden and N.C. Verma. J. Amer.Chem.Soc., 243, 98, 1976.
44. A.D. Trifunac and M.C. Thurnauer, J.Chem.Phys., 4889, 62, 1975.
45. S.H. Glarum and J.H. Marshall, J.Chem.Phys., 5555, 52, 1970.

CHAPTER VII

CHEMICALLY INDUCED ELECTRON POLARIZATION OF RADIOLYTICALLY PRODUCED RADICALS

Richard W. Fessenden

Radiation Laboratory*, University of Notre Dame,
Notre Dame, Indiana 46556

ABSTRACT. The study of chemically induced electron polarization of radicals produced by the radiolysis of aqueous solutions is described. The experimental arrangement necessary for pulse ESR experiments with submicrosecond time resolution is discussed. Direct detection of the ESR signal with no field modulation is preferred so that quantitative analysis is possible. The effect of the radiolysis pulse on the spectrometer is described. A brief review of the radiolytic formation of primary radicals in spurs is given to show that considerable reaction of the radicals occurs before complete separation of the pairs. Analysis of the observed time profiles of the ESR signals is by means of Bloch equations so modified to include both initial polarization and polarization by spin selective reaction. Data from direct observation of the hydrated electron and H atoms are given along with indirect data obtained by study of products of their reaction. Data for the products of OH reaction are also given. In each case it appears that the spin populations of the primary radicals can be transferred to the product. It can be concluded that e_{aq}^- initially has equal populations of the two spin states to within 0.2 of the Boltzmann value while OH appears to be relaxed to equilibrium in less than 1 nsec. Primary H atoms show a significant initial effect probably as a result of reactions before escape from the spur. Several unresolved aspects of the observations are also discussed.

---

*The research described herein was supported by the Division of Physical Research of the U.S. Energy Research and Development Administration. This is Radiation Laboratory Document No.-1738.

L. T. Muus et al. (eds.), Chemically Induced Magnetic Polarization, 119-150.

This discussion of chemically induced dynamic electron polarization (CIDEP) of radicals produced by high energy radiation will be divided into sections on the mechanism of radical production, special experimental details, the methods of analysis of the time profiles of ESR signals from pulse experiments including a consideration of the various effects involved, and a presentation of some typical results. Most of the ESR results to be discussed have come from the author's laboratory [1-4] with references to other work as appropriate. To provide a basis for the discussion to follow it should be pointed out that polarization or abnormal population of the electron-nuclear spin states of radicals is known to arise in two ways, namely, in the inital steps producing the radicals as from a photoproduced triplet state [5,6] or in their subsequent bimolecular reaction to form diamagnetic products. The radical pair process [7,8] which accounts for the spin-selection disappearance reaction is described in detail elsewhere in this volume. It is only necessary to note that this process involves the evolution of the electronic spin state for the pair of radicals either on the initial separation of the pair or during the interval between an encounter and a subsequent re-encounter. Normally a mixing of the singlet (S) and the m=0 triplet state (T_0) of the pair is important. The usual effect, a tendency toward emission for the signal at lower field arises because of preferential reaction to form singlet products. The significant quantity for pairs of chemically similar radicals is the hyperfine constant (or constants) while for unlike pairs the difference in the g factors is also important. The following section will examine in some detail the formation of radicals in radiolytic systems with particular attention to the pairwise behavior.

RADIOLYTIC PRODUCTION OF RADICALS

Because the specific experimental results to be presented all involve aqueous solutions, this discussion will be limited to water. Absorption of high energy radiation such as γ rays or fast electrons produces electrons of somewhat lower energy by the Compton effect which in turn produce electrons of lower energies and so on. A major fraction of chemical events are caused by electrons of less than 100 eV energy. This energy is deposited in small isolated regions called "spurs" which contain one or more pairs of radicals. The most important event is ionization of a water molecule to H_2O^+ and an electron. The latter species becomes solvated at some distance from the H_2O^+ and is then called the hydrated electron, e^-_{aq}. The H_2O^+ rapidly dissociates into H^+ + OH. The OH and e^-_{aq} then either diffuse together and react or diffuse apart. Other processes such as excitation of H_2O are possible. A smaller yield of H is produced, presumably together with OH. The H, OH, e^-_{aq}, and H^+ produced in the spur are called the primary species.

A detailed quantitative discussion of the evolution of the spur by diffusion and reaction has been given by Schwarz [9] among others. He considered data such as the yields of H_2 and H_2O_2, produced in part by reaction of the primary radicals in the spur, and their behavior as other reactants are added. Using a particular description of diffusion and experimental or estimated reaction rate constants and diffusion constants he has determined parameters such as the initial yields of the primary radicals and their spacial distribution. The results are probably dependent on details of the model but are indicative of the magnitudes involved. He finds that the average distance of OH from the center of the spur is 7.5 Å while that for e^-_{aq} is 23 Å as a result of the electron's high mobility before hydration. The initial yields (molecules/100 eV absorbed) are 0.62 for H, 4.78 for e^-_{aq}, and 5.7 for OH. At sufficiently long times after production of the radicals, no further reaction with spur partners occurs and the distribution in the soluton is uniform. This condition is reached after 100 nsec and the yields of e^-_{aq} and OH at that time are about 2.7 and 2.8, respectively. Direct photometric measurements of the decay of e^-_{aq} following pulse radiolysis with a 30 p sec pulse [10] confirm this drop and show that 10% of the initial e^-_{aq} decays between 0.1 and 1 n sec and that another 20% decays between 1 and 10 n sec. In all, e^-_{aq} corresponding to a yield of about 2.0 has reacted with other radicals within the spur. Most of this reaction (1.6) is with OH but Schwarz [9] calculates a yield (loss of e^-_{aq}) of 0.27 from $e^-_{aq} + e^-_{aq}$ and 0.13 from e^-_{aq} + H. In addition, H + OH occurs to a small extent and OH + OH produces a yield of about 0.7 for H_2O_2. Clearly a very significant extent of "back" reaction of geminate radicals occurs within the spurs. The calculations by Schwarz [9] are in reasonable agreement with direct experiments. It is important to point out, however, that the calculations do not attempt to consider the spin states of the radical pairs and that the rate constants were determined from data on homogeneous reactions with random encounters. The actual situation involves spins which are correlated in some way so that the quantitative behavior may be accordingly different.

Of the three primary radicals, only e^-_{aq} and H can be directly observed by ESR. Results on these two will be presented. All other radicals are secondary in that they result from reactions of the three primary radicals with various solutes. It is necessary, therefore, to consider the behavior of the electron or nuclear spins during the reaction. An abstraction reaction can be represented by

$$R\uparrow + H\downarrow\uparrow R' \rightarrow R\uparrow\downarrow H + \uparrow R' \qquad (1)$$

with the electrons in the product RH paired in an electronic singlet. Since all steps of the reaction occur at distances where there is a strong coupling between at least two of the electrons there should be no change in spin state and the product R' should have the same _z_ component of spin in the magnetic field as did R. If an ensemble of radicals R with particular populations in the two electron spin states react in a time shorter than their spin relaxation time, the product radicals should have the same populations as did the reactants. This concept of transmission of spin populations is also part of the triplet model of CIDEP in that the nonequilibrium populations of the triplet levels are transmitted to the two radical products. Demonstration of emission from reduced oxalic acid by Livingston and Zeldes [11] represents a clear example of transmission of an _inverted_ population in homogeneous reaction. If the radical reaction is an addition, as with H atoms adding to a double bond, it is also reasonable to suppose that no mechanism exists for changing the nuclear spin states and the population of the combined electron-nuclear spin states will be transmitted. For instance, if the low-field line of the H atom has an inverted population difference then after addition so should the corresponding transition (or transitions) in the product. Some of the results obtained with H atoms will be discussed using this concept.

Because the spin populations will be transmitted (an assumption as yet) it is possible to use the population distributions in the product radicals to investigate the state of the primary radicals at the time of reaction. The average time of reaction after the initial event producing the primary radicals is, of course, controlled by the concentration of reactant and its rate constant. Typical half lives range from greater than 1 μsec to less than 1 nsec. It is thus possible to have the reaction period be either longer or shorter than typical spin relaxation times or to be even faster than some of the spur reactions. In experimental terms, these principles represent a method for trapping and identifying the specific population distribution of spins in a selected radical species. The behavior of OH, which cannot be observed direct, has been investigated in this way as described below.

EXPERIMENTAL CONSIDERATIONS

The experimental sample arrangement used for the pulse experiments is essentially the same as for steady-state experiments [12]. A standard flat aqueous flow cell of 0.5 mm internal spacing and 8 mm width is used. For steady-state experiments a cell of high-purity silica is used to reduce the ESR signal caused by centers in the cell. For pulse experiments a pyrex cell can be used because the amplitude of any transient signal in the glass is small. The energy of the electron beam used in the work to be described was 2.8 MeV. The time-averaged intensities of the beam are similar for pulse and steady-state experiments with a total beam current of about 10-20 μA over a circle of about 5 mm diameter. Because the cell must be placed in the plane of low microwave electric field it is irradiated "edge-on". Thus the solution volume is defined by the thickness, 0.5 mm, the cell width, 8 mm (parallel to the beam direction) and the 5 mm beam diameter. The dose rate for a 10 μA beam corresponds to a production rate of 10^{-2}M sec^{-1} total radicals in water. A flow rate of 1 $cm^3 sec^{-1}$ changes the irradiated volume 50 times per second so that the accumulated dose is acceptable. Even so, solutes must be present to at least $5x10^{-4}$ M to prevent serious depletion problems. In the pulse experiment, a repetition rate of 100 sec^{-1} is used so that one volume element receives on the average only two pulses. A single pulse can produce 10^{-4} M of radicals although most pulse experiments to be reported used smaller currents corresponding more nearly to $3x10^{-5}$ M per pulse. At this concentration, a species with a rate constant for second-order disappearance of $1x10^9$ $M^{-1}sec^{-1}$ would decay with a first half life of 33 μsec. With a 100 sec^{-1} pulse repetition rate and a 1 μsec pulse, the current during the pulse is 100 mA. It should be noted that 2.8MV x 10μA = 28 watts so that water cooling of the cavity as well as a flowing sample solution is necessary to limit temperature rises.

A moderate degree of time resolution such as provided for the photolytic experiments by mechanically chopping the light beam can distinguish between polarization produced upon radical formation and that on subsequent reaction. However, more quantitative measurements require radical production periods and spectrometer time resolution to be shorter than electron spin relaxation times. Radical production periods are limited by the dose per pulse which can be delivered but pulse lengths of less than 1 μsec are practical. An ESR spectrometer response faster than 1 μsec places several constraints on its design. Several workers have chosen to use high field modulation frequencies to solve this problem [13, 14]. With field modulation, the signal from the microwave detector is amplified and synchronously rectified at the modulation frequency. At least one cycle of the AC signal must be so treated to produce a response. Thus a frequency of 2 MHZ has been used. A major disadvantage of field modulation is that the resultant derivative signals cannot readily be analyzed by reference to the Bloch equations in the way described below. At this point it is appropriate to ask the purpose of the field modulation. It is mainly used to convey signal information from the detector at a frequency where the signal-to-noise ratio is high. All dectector diodes exhibit noise which increases at lower frequencies (f) approximately as

f^{-1}. The frequency at which this excess noise begins is in the range 1-10 kHz with several modern microwave diodes. In pulse experiments one wants to observe the ESR signal after radical formation as it develops over times comparable to spin lattice relaxation times (∿1 μsec) and to follow it as the radicals react bimolecularly. Typical reaction periods are 100 μsec. Observation over such a reaction period corresponds to the same time domain as is encountered in an experiment with field modulation at 10 kHz and should, therefore, also give good signal-to-noise ratio. This approach with no field modulation allows wide band amplification of the signal from the microwave detector and has been applied in our studies [1,4]. Levanon [15] has also recognized the praticality of such an experiment.

The direct amplification and use of the detector output requires a different type of automatic frequency control (AFC) of the microwave frequency. Normally, the frequency of the klystron oscillator is modulated by a small amount at some frequency such as 10 or 70 kHz and the response of the cavity to this modulation is used to provide AFC. With wide-band amplification of the detector output, the response to the frequency modulation gives a large signal which interferes with observation of the ESR signal. To avoid this problem, an alternative form of AFC can be used which depends on the phase response of the microwave power reflected from the cavity as in a superheterodyne spectrometer [1].

A large pulse is produced at the output of the main detector when the cavity and sample are irradiated. This pulse is the result of changes in the cavity characteristics by the electron beam. Both the electrons in the beam itself and those formed by ionization of the gas in the cavity represent a conductivity and reduce the cavity Q [16]. This interference pulse is minimized by reducing the pressure to 1 torr or less and adding SF_6 to capture any residual electrons. Even then, the interference pulse is several hundred times the ESR signal. One other effect produced by the beam is a long-lived signal building slowly after the pulse and decaying over the period between pulses. This signal is comparable in size to the ESR signals but is not dependent on magnetic field. It is not present in the absence of the cell. This signal is believed to be a conductivity signal in the silica or glass of the cell. It is several times smaller with a pyrex than a fused silica cell so the pyrex cell is preferred.

The response time of the overall system depends mainly on two factors, the Q of the cavity and the response of the preamplifier. The time dependence of the energy absorption by the spins corresponds to amplitude modulation of the microwave power, thus producing side bands. The cavity acts as a filter and reduces the response to sidebands which are widely separated from the main signal or carrier. For a loaded cavity Q of 1700 (a typical value) sidebands separated from the carrier by 2.75 MHz will be reduced in amplitude by 0.707. A simple RC filter of time constant 0.06 μsec would also show this same reduction in amplitude so the effective time constants caused by the cavity must be of this magnitude. The observed decay of the preamplifier output caused by the interference during the pulse corresponds to 0.3 μsec so the time resolution must be determined mainly by the preamplifier. Because the pulse during radiolysis is so much larger than the ESR signal, it must decay for a number of time constants before observations are possible. In practice, valid data could be obtained starting about 0.7 μsec after the end of the beam pulse.

Two modes of operation of the spectrometer are possible, namely, display of a spectrum at a fixed time delay after the pulse and recording of the time profile of an ESR line at a fixed point in a spectrum. These alternatives are illustrated in Fig. 1. The four traces at the upper left and the data in

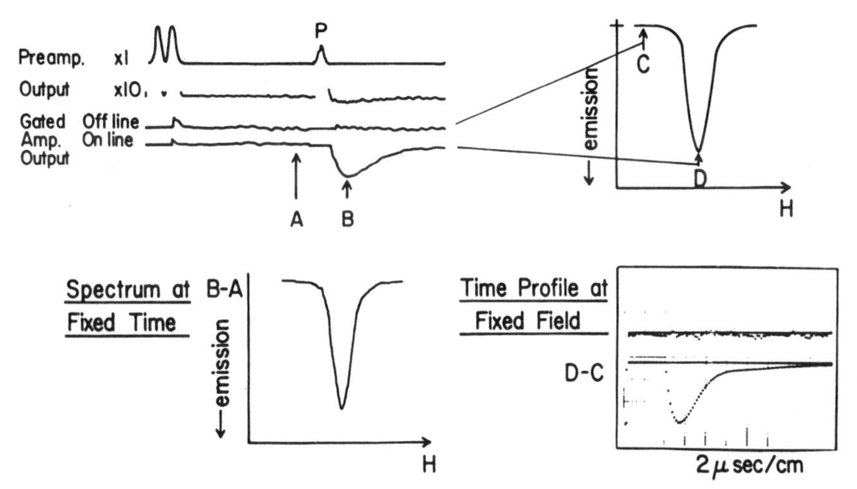

Fig. 1. Summary of the two modes of operation, see text. The spectrum at a fixed time after the pulse is obtained by sampling the signal at points before (A) and after (B) the radiolysis pulse with a boxcar integrator and displaying the difference of the two channels as the field is changed as at lower left. The time profile is obtained by accumulating data at fields C and D and subtracting the two curves to give the result at lower right. The four oscilloscope traces and the curves in the lower portion of the figure are data taken on the low-field H atom line in irradiated 0.1 M $HClO_4$ and so are drawn in the fashion appropriate for emission.

the lower portion represent actual data obtained for the low-field H atom line in irradiated 0.1 M $HClO_4$ in water. This line is the strongest we have observed on account of a large polarization and long relaxation time. The traces at the upper left represent the preamplifier output with pulses for the phase control circuit on the left and the interference by the irradiation pulse at P. At higher gain (x10) these pulses are off scale but the ESR signal can hardly be seen. After further amplification and gating to remove the interface, the ESR signal is more obvious as in the fourth line. Note that this oscillogram represents detection of the ESR signal in a "one shot" experiment. The third trace gives the signal ∿2G away from the ESR line. Use of the boxcar integrator mode to display the signal at a fixed time after the pulse involves sampling the signal at points A and B as indicated and averaging the amplitudes with a capacitor. The difference signal which represents the signal produced in response to the pulse is shown at lower left as a function of magnetic field. The time profile rather than the spectrum results when the signal is digitized at a number of times and a number of these traces averaged in a signal averager or minicomputer. If a similar number of traces is taken at a field off the resonance as at C (upper right) the difference in the stored data for points C and D will represent only the field-dependent (or ESR) signal. Data taken in this way is shown at lower right. Approximately 10,000 traces were accumulated. A base line at 50 times higher gain with both points C and D off the line is also given. For this very strong line the signal-to-noise ratio is about 500.

To make the time profile mode practical it is necessary to have control of the magnetic field so that corrections for drift of the microwave frequency and magnetic field can be made. This function is provided by the minicomputer which does the signal averaging. It is possible to maintain position in a spectrum to around ±5 mG for long periods. Detection of the center of a line is accomplished by putting fields C and D (see Fig. 1) on the sides of the line and adjusting the field for zero difference. With this approach the relative amplitudes of different lines in a spectrum can be accurately measured.

METHOD OF DATA ANALYSIS

The time-dependent curves giving the ESR response as a function of time are best analysed by means of modified Bloch equations [2,4,17]. The initial conditions at the time of radical formation are taken to be zero magnetization in the x and y directions (that is, the corresponding components of the spin for various radicals are uncorrelated) and some value in the z direction depending on the initial population difference for the upper and lower states of the transition. The magnetic moment of the sample then evolves under the influence of the static and microwave magnetic fields, the relaxation processes, and chemical reactions which destroy radicals. Except for the reaction process, this situation is the same as if the radicals were already present in the initial state and the microwave field were suddenly turned on. This latter case was examined by Torrey [18] for NMR and much of the analysis here is similar to that given by him.

Before proceeding further it is appropriate to consider how CIDEP will be introduced. An initial effect which populates the two spin states for the transition differently must, in the macroscopic scale, cause a nonzero z component of the sample magnetic moment as was mentioned above. Thus, this form of CIDEP is introduced mathemetically as an initial condition. To see how CIDEP produced by reaction can be included let us consider the populations n_1 and n_2 of the upper and lower spin states of a transition. The existence of CIDEP implies that the disappearance rates for the two populations are different. The differential equation for the difference in populations (n_2-n_1) is approximately

$$\frac{d(n_2-n_1)}{dt} = k_2 n_2 n_+ - k_1 n_1 n_- \qquad (A)$$

where k_1 and k_2 are allowed to be different and n_+ and n_- represent in each case all radicals in the system with spin opposite to that of n_1 and n_2 to allow for recombination through the singlet state only. Since the absolute magnitudes of n_+, n_-, n_1 and n_2 never depart greatly from the equilibrium values and so are close to the appropriate statistical fractions of all radicals, i.e., relaxation is effective in keeping the polarizations relatively small, this expression can be simplified to

$$\frac{d(n_2-n_1)}{dt} = (k_2-k_1)an^2 \qquad (B)$$

where n is the total number of radicals in the system and a is a numerical factor to take care of the fractions of n which the other quantities represent. Depending on the sign of (k_2-k_1) this equation represents formation of excess population either in the normal or inverted senses. The corresponding behavior for the macroscopic magnetization will be a production for the z component which is proportional to the rate of the second-order radical reaction, i.e. to the square of the radical concentration. Further analysis [1] shows that the quantity VT_c/T_1 should be independent of reaction rate for a given system. In this expression V is the enhancement factor or $(S-S_o)/S_o$ where

S and S_o are the signals corresponding to the perturbed and Boltzmann population differences, T_c is the first second-order half life, and T_1 is the spin relaxation time.

The form of the Bloch equations we have chosen to use has the x and y components transformed for the coordinate system rotating at the resonance frequency but with time unchanged. In this form the time derivatives of the components of the magnetization, $\dot{M}_x$ etc., are

$$\dot{M}_x = \Delta\omega M_y - (T_2^{-1} + \dot{R}/R)M_x$$
$$\dot{M}_y = -\Delta\omega M_x - (T_2^{-1} + \dot{R}/R)M_y - \omega_1 M_z \qquad (C)$$
$$\dot{M}_z = \omega_1 M_y - (T_1^{-1} + \dot{R}/R)M_z + M_o T_1^{-1} F(t) + PM_o[F(t)]^2$$

Here, T_1 and T_2 have their usual meaning, $\omega_1 = \gamma H_1$ with H_1 the rotating component of microwave magnetic field, γ the magnetogyric ratio and $\Delta\omega = \omega - \omega_o$ is the offset in angular frequency of the observing field from the center of resonance ω_o. The terms of the form M_x/T_1 etc. are the usual relaxation terms. In addition the terms $\dot{M} = M(\dot{R}/R)$ represent relaxation toward zero as a result of chemical reaction. This form was derived by assuming that the loss of magnetization follows the loss of radical concentrations, R(i.e., $\dot{M}/M = \dot{R}/R$). For a first-order decay $\dot{R}/R = -k$ where k is the rate constant. The term $M_o/T_1 F(t)$ concerns the equilibrium magnetization. For stable radicals $F(t)=1$ and this term represents a production driving the system toward M_o. (Usually the equation is written with $(M_z - M_o)/T_1$.) The function $F(t)$ is the time dependence of radical concentration to take into account the fact that the equilibrium magnetization must decrease as the radicals decay. The function is $F(T) = \exp(-kt)$ for a first-order radical decay and $F(T) = (1+t/T_c)^{-1}$ for second-order radical decay where t_c is the first half life. With this treatment M_o corresponds to the initial radical concentration only and does not represent the equilibrium magnetization at any other time. As discussed above, CIDEP by the radical pair process during radical decay represents a production term for either positive or negative M_z depending on the particular rate constants for the two levels. This term is made proportional to the square of the radical concentration by the factor $[F(t)]^2$. The coefficient, P, corresponds to the value at the initial radical concentration and is specific to that concentration. It will have a different value at different initial concentrations. The quantity P can be replaced by V/T_1 where V is the enhancement factor as normally used for relatively slow radical decay. It should be noted that in the earlier paper [1] the production term was not proportional to the square of the radical concentration as intended. These equations are essentially as given by Pedersen [19] except for the relaxation by reaction for M_x and M_y.

The method of solution of the differential equations depends on the particular approximations which may be valid. Pedersen [19] has considered a number of specific cases but does not give an analytic solution allowing for both initial magnetization and that produced by reaction. Such an equation was derived by considering exponential decay for F(t) in which case an analytic solution is possible [4]. Except for the effect of reaction on the relaxation time this solution is that for no decay multiplied by exp(-kt) in appropriate plates. In the limit $T_1^{-1} >> k$ it is possible to replace the terms exp(-kt) by the second-order decay function. The resulting equation is

$$S(t) = \frac{M_o T_c}{t+T_1}\left(1+\frac{VT_c}{t+T_c}\right)\left(\frac{\omega_1 T_1}{1+(\delta^2+1)(\omega_1 T_1)^2}\right) \times$$

$$\left[1-\exp(-t/T_1)\cos(\delta^2+1)^{1/2}\omega_1 t - \frac{\exp(-t/T_1)}{\omega_1 T_1(\delta^2+1)^{1/2}} \sin(\delta^2+1)^{1/2}\omega_1 t\right]$$

$$+ f\,\frac{M_o T_c}{t+T_c}\,\frac{\exp(-t/T_1)}{(\delta^2+1)^{1/2}} \sin(\delta^2+1)^{1/2}\omega_1 t \qquad \text{(D)}$$

where $\delta = \Delta\omega/\omega_1$ is the reduced offset from the center of the line. The initial value of M_z was taken to be fM_o so that f is the fraction of the equilibrium Boltzmann value. Equality of T_1 and T_2 (which was found experimentally for $\dot{C}H_2CO_2^-$) has also been assumed. Although this expression is rather complex, a number of features are evident. There are transient terms containing $\exp(-t/T_1)$ which are damped out over several relaxation periods. If ω_1(i.e., H_1) is small then the terms with sin $\omega_1 t$ are zero and the signal will grow as $1-\exp(-t/T_1)$. If ω_1 is large and T_1 is long then the terms in sin $\omega_1 t$ and cos $\omega_1 t$ are expected to lead to oscillatory behavior. Such behavior has been seen in radiolysis experiments [2] but is clearer in photolytic experiments [20] where a large initial magnetization is produced. After all the transient terms have decayed the form is that derived on kinetic grounds alone [1] namely

$$S(t) \propto \left(1+\frac{VT_c}{t+T_c}\right)\frac{T_c}{t+T_c} \qquad \text{(E)}$$

where V is the enhancement factor at t=0. Note that in this approximation, if V>>1, the signal should be proportional to the square of the radical concentration. The long term saturation behavior is also included through the term $\omega_1 T_1/1+(\delta^2+1)(\omega_1 T_1)^2$. The parameters which must be determined in fitting a curve based on Eq. D to data are T_1, T_c, H_1, f, V and a scaling factor. (At the center of a line, δ=0). These parameters affect different portions of the curve differently so the set obtained is relatively unique. The fitting of data taken at several different power levels further restricts the range of some of the variables. The microwave field, H_1, can be determined from measurements of cavity Q and power level or from the transient oscillations as described below and is not a free parameter.

If no approximations are appropriate, the differential equations can be solved numerically from the initial condition. The analytic equation has been verified in this way. When using such an approach it is convenient also to treat the radical concentration by means of a differential equation rather than to put in analytic forms for F(t) and $\dot{R}/R$. The differential equation

$$\dot{R}=-k_1[R]-k_2[R]^2 \quad \text{(F)}$$

for the radical disappearance by both first- and second-order reactions has been used in treating H atom [21] and e^-_{aq} [4] signals in the presence of reactant. The initial condition $[R]_o=1$ is used and this equation integrated numerically along with the Bloch equations. The value of R at a given time is inserted for F(t) and the value of $\dot{R}/R$ also determined from this equation for the relaxation terms.

In a number of cases to be discussed the rate of rise of the signal is such that T_1 must be greater than 4 μsec. Such a value suggests a very narrow line (<15 mG) if $T_1=T_2$. The observed width was greater than this (∿70 mG) and must be produced by field inhomogeneity. Although Pedersen [19] has suggested a method to treat this problem by taking $T_2 \ll T_1$ the conditions under which this approach is valid are rather restrictive (low power, but $\omega_1 T_1 \gg 1$) and often not met in practice. The effect of field inhomogeneity can be included by evaluating Eq. (D) for various values of δ and then summing the results with a suitable line shape function as weighting factor. In the same way, sets of differential equations can be numerically integrated for a number of offsets from the center of the line. In previous work only the analytic function was used in this way because the multiple integration was too time consuming with the computer available. With more computing speed, which we now have, it will be routine to include the effects of field inhomogeneity in this way. The effect of including field inhomogeneity will be illustrated by data on bromomaleic acid.

TYPICAL RESULTS

A complete understanding of the results of time-resolved ESR experiments on radiolytically produced radicals clearly requires an understanding of the primary radicals themselves. Direct observations on the two, e^-_{aq} and H, which can be studied in this way, will be discussed. The behavior of the primary radicals is also of concern in studies of secondary radicals since the initial condition of magnetization of the secondary radicals will depend on the polarization of the primary radical upon reaction. In the discussion to follow it will become evident that the inverse process, the study of the magnetization of primary radicals by means of the initial conditions for the products of reaction will in fact be more practical. The hydrated electron, e^-_{aq}, will be discussed first.

The hydrated electron, e^-_{aq}

The ESR signals of e^-_{aq} cannot be detected in steady state experiments because of its great reactivity toward impurities and radiolysis products such as H_2O_2. It can be detected in pulse experiments. Early experiments by Avery _et al._ [22] reported the detection of e^-_{aq} but did not give quantitative data on its life time. Experimentally, we have found that it is much easier to detect e^-_{aq} when the OH radicals are converted to some other radical by a solute which does not react with e^-_{aq}. Suitable reactants are CH_3OH and SO_3^{2-} which produce $\dot{C}H_2OH$ and $\dot{S}O_3^-$, respectively. These radicals also do not seem to react rapidly with e^-_{aq}. Acid conditions which allow $e^-_{aq}+H^+$ also must be avoided. Curves showing the growth and decay of the ESR signal of e^-_{aq} in a solution of 5mM Na_2SO_3 at pH [4] 12 are given in Fig. 2. Results for two different initial concentrations as measured by the beam current are shown. The disappearance reaction is not well defined. In part, $e^-_{aq}+ e^-_{aq}$ must be involved but relatively little difference is seen between the data for the two beam currents. Examination of similar solutions by optical absorption spectrophotometry at a dose about tenfold lower showed a half life of 35-40 μsec so that reaction with impurities is not very important. In part, decay by $e^-_{aq}+H_2O_2$ will remove some e^-_{aq}. Most probably several reactions contribute. For purposes of analysis, a version of Eq. D with first-order decay was used. The value of H_1 (13 mG) was determined, in this case, from cavity Q and power level measurements.

For these curves, the initial magnetization f was taken to be zero (see below). The other parameters determined by fitting the calculations to the data are for the lower curve T_1=T_2=4 μsec, decay half life=12 μsec and for the upper curve T_1=T_2=2.5 μsec, half life=10 μsec. The two sets of experimental data are plotted with the same scaling factor so that the upper calculated curve should have a five fold higher scaling factor based on the beam currents. In fact, the curve given needed only a factor of two increase in scaling factor. Data to be given

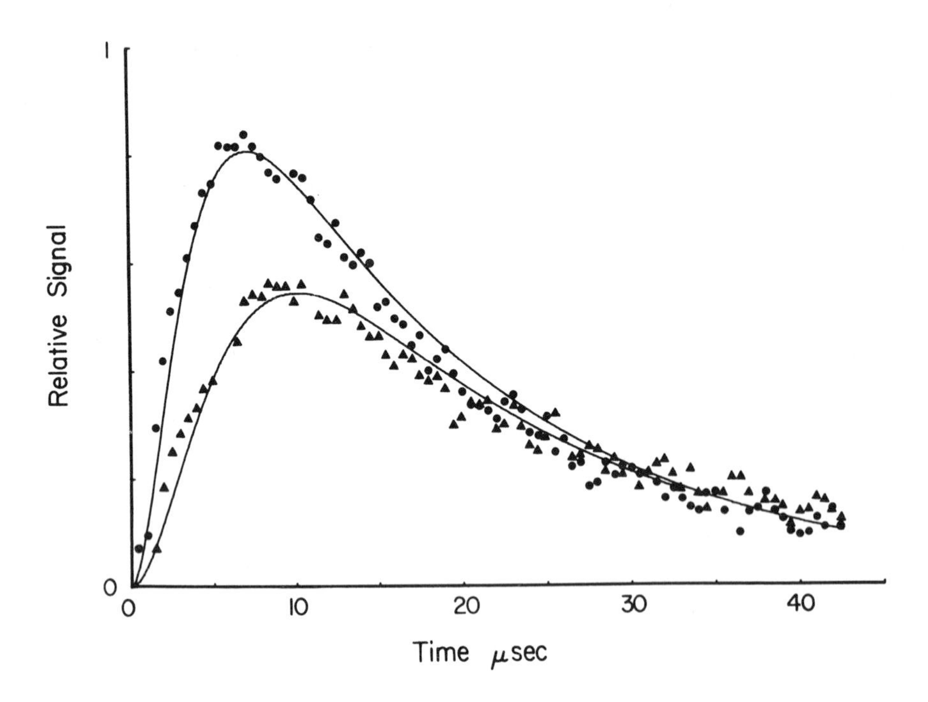

Fig. 2. ESR time profile of e^-_{aq} for beam currents of 0.2 (upper) and 0.04 μA (lower) and a power level corresponding to H_1 = 13mG. The calculated curves used T_1 = 2.5 and 4.0 μsec respectively.

below shows that the radical concentration is not strictly proportional to the measured current, accounting for much of this effect. The figure shows that an excellent fit of calculation to data is possible. The change in apparent relaxation times with radical concentration is peculiar and will be discussed in a later section.

Curves like the upper one were also recorded for constant beam current (0.2μA) but different microwave power levels corresponding to H_1 of 7, 13, and 23 mG [4]. The value of T_1 was determined by fitting curves to the data for the lowest power level with f=0 assumed. The same parameters also fit the data for the two higher power levels. If f had not been close to zero a much faster rise would have been found for the highest power level. By this approach it was estimated that f must be <0.3.

The rapid rise and decay of the signal makes this determination less sensitive than one might wish. The concept of transfer of spin populations suggests that the initial

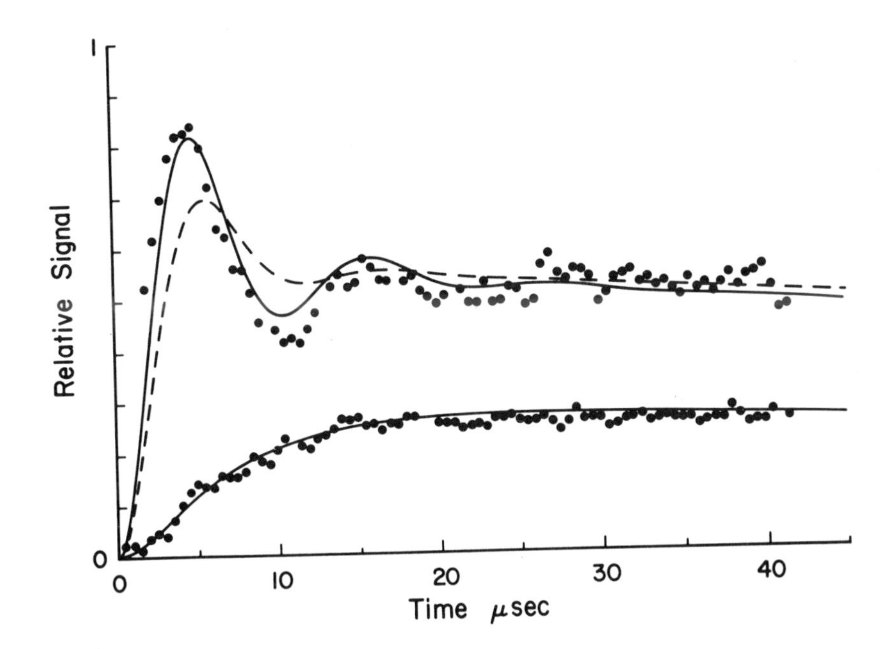

Fig. 3. Time dependence of the high-field line of $^{-}O_2CCH{=}\dot{C}CO_2^-$ in a solution of 2 mM bromomaleate at pH 9 (borate buffer). Upper curve was taken at a power level of -5 dB while the lower was at -25 dB. The calculated solid curves took field inhomogeneity into account (see text) and used H_1 = 31 and 3 mG respectively. The dashed curve was calculated from Eq. D without consideration of field inhomogeneity.

magnetization for e_{aq}^- can also be determined from the behavior of a product. Because the directly measured value of T_1 for e_{aq}^- is at least several μsec the reaction must be over in about 0.1 μsec or less. One such result is shown in Fig. 3 for $O_2CCH{=}\dot{C}CO_2^-$ produced from bromomaleic acid by

$$e_{aq}^- + {}^-O_2CCH{=}CBrCO_2^- \rightarrow {}^-O_2CCH{=}\dot{C}CO_2^- + Br^- \qquad (2)$$

(Although the <u>cis-trans</u> configuration of the radical is not certain all evidence suggests that the final geometry is obtained very rapidly.) The radical has a single hyperfine constant of 49.79 G [23]. Data is shown for the high-field line at two power levels with otherwise identical conditions. The chemical decay is slow because of the double charge of the radical and a value of 0.3 was determined from the ratio of intensities of the high and low field lines at about 30 μsec. Attempts to calculate curves to fit the data using only Eq. D were unsuccessful in that the value of $T_1=T_2=4$ μsec which gave a reasonable fit to the lower power curve gave a much quicker damping of the oscillation at the higher power level as shown by the dashed line. (At the lower power level this calculation essentially overlaps the solid curve.) Introduction of a Gaussian field inhomogeneity line-shape function with a width of 63 mG full-width at half height allowed a much better fit. The low-power curve was fit with $T_1=T_2=7.5$ μsec, f=0 and a value of H_1 (3 mG) determined from the period of the oscillation at high power and the known power ratio. These same parameters with H_1=31 mG gave the solid curve as shown for the higher power. Although there is some departure, this fit also is good. Since we are particularly concerned with the value of f an attempt was made to fit the data with $f \neq 0$. A value of $f<0$ would cause an initial emission for a short time and so is unacceptable. (Even f=-0.2 would do so.) The lower power curve can be fit with f=0.2 and a longer T_1 to compensate for the otherwise more rapid rise. With $T_1=T_2=12$ μsec the fit was acceptable but not as good as is shown for f=0. However, at the higher power level the long term height was only slightly above the lower power curve as a result of strong saturation. Similar results were obtained for $[{}^-O_2CCH{=}CHCO_2^-]^-$ produced by electron attachment to fumarate ion [4]. In this case curves calculated for f=±0.3 were far from the observed curve. Clearly, such experiments are a sensitive test of the value of f for the product radicals and by inference for the magnetization of e_{aq}^- upon reaction. In the case of bromomaleate it can also be concluded that the initial electron adduct containing bromine must loose Br^- before it relaxes.

Radicals produced from OH

The OH radical cannot be detected in solution because of its short relaxation time. It is useful, therefore, to examine the behavior of a reaction product. Data [4] for the time dependence of one of the central second-order components of $\dot{C}H_2CO_2^-$ obtained by

$$OH + CH_3CO_2^- \rightarrow \dot{C}H_2CO_2^- + H_2O \qquad (3)$$

is shown in Fig. 4 along with similar data obtained when the same radical is formed by

$$e_{aq}^- + ClCH_2CO_2^- \rightarrow \dot{C}H_2CO_2^- + Cl^- \qquad (4)$$

Clearly there is a marked difference in behavior. We have already discussed radicals produced from e_{aq}^- and so probably f=0

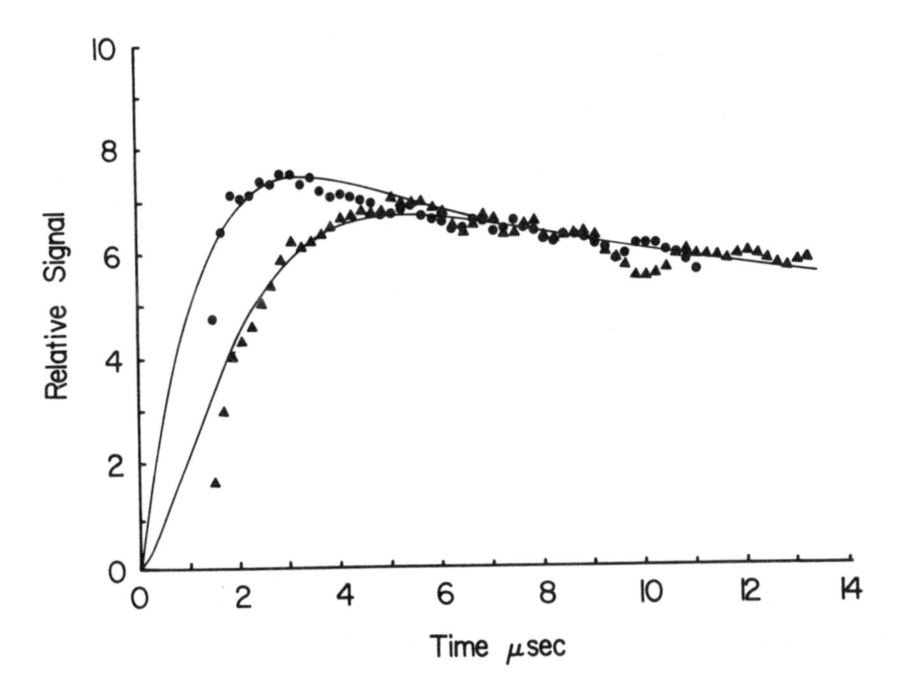

Fig. 4. Time dependence of the "center line" (higher-field second-order component) of $\dot{C}H_2CO_2^-$ formed from OH in N_2O saturated, 50 mM NaO_2CCH_3 at pH 10.5 (●) and from e_{aq}^- in 40 mM $ClCH_2CO_2^-$ at pH 10.5 (▲). The calculated curves correspond to f = 1 (upper) and f = 0 (lower).

for $\dot{C}H_2CO_2^-$ formed by this route. The calculated curves here are easily obtained because the natural T_1 and T_2 produce line widths closer to the observed values. Thus Eq. D was used as given. The relaxation times for the outer lines of this radical have been measured to be 1.4 μsec [4] and that for the central line should be similar. With $CH_2ClCO_2^-$, OH is converted to $\dot{C}HClCO_2^-$ so that a polarization of the central line of $\dot{C}H_2CO_2^-$ could be produced in the cross reaction because of the differing g factors. We assume this effect is small and take V=0. The value of H_1 is known, so the only parameters are a scaling factor and the chemical decay determined from the long term behavior. The calculated curve fits very well with f=0. A 1 μsec recovery period after the end of the beam pulse puts the first data points at about 1.5 μsec. The reaction is over very rapidly since the rate constant is $1x10^9 M^{-1}sec^{-1}$ and the concentration 40 mM.

The upper curve in Fig. 4 was obtained with an N_2O saturated solution. In that case about half the radicals are produced from primary OH and half from OH produced by

$$e^-_{aq} + N_2O \rightarrow N_2 + O^- \qquad (5)$$

$$O^- + H_2O \rightarrow OH + OH^-. \qquad (6)$$

These reactions are over in about 10 nsec. For this portion, one is concerned with the relaxation of O^- as well as of OH. As shown in Fig. 4 a calculation with f=1 works excellently showing that relaxation of OH (or O^-) occurs before reaction. The half life for the OH reaction is 0.2 μsec.

Essentially identical results were obtained [4] with $CH_3\dot{C}(O^-)CO_2^-$ produced by

$$OH + CH_3CHOHCO_2^- + OH^- \rightarrow CH_3\dot{C}(O^-)CO_2^- + 2H_2O \qquad (7)$$

or $$e^-_{aq} + CH_3COCO_2^- \rightarrow CH_3\dot{C}(O^-)CO_2^- \qquad (8)$$

A value of f=0 fits the data for e^-_{aq} reaction and f=1 fits the OH data. In this case the OH rate constant is about $3\times10^8 M^{-1}sec^{-1}$ and experiments were possible at 1 M concentration of lactate. An identical curve as for the lower concentration with f=1 was obtained. The half life for pseudo first-order reaction of OH is 2 nsec so that nearly complete relaxation occurs in less than 1 nsec. An experiment with a solute which reacts even more rapidly with OH is possible so that this limit could be pushed to even shorter times.

CIDEP as a result of radical-radical reactions was studied in detail for $\dot{C}H_2CO_2^-$ formed in N_2O saturated solutions of $CH_3CO_2^-$. In steady-state experiments this radical shows a value of V of about 0.2 under typical conditions. Previously, [1] data for the steady-state and pulse experiments were compared and it was shown that a common value of VT_c/T_1 fit both experiments. More complete curves of the time dependence of the three lines was taken in a later work [4]. That data is shown in Fig. 5. Again, the values of H_1=31 mG, T_1=1.4 μsec and f=1 were already fixed. The value of second-order half life was determined for the central line with V=0 and it was found that V=±2 with the <u>same</u> scaling factor would also fit the curves for the high-and low-field lines. The initial positive-going signal for the low-field line reflects the initial magnetization obtained from OH and the change-over to an emission signal is the result of the second-order reaction. This curve demonstrates most dramatically the source of the inversion of the low field line. The data taken for $\dot{C}H_2CO_2^-$ produced from e^-_{aq} are similar except that the low-field line immediately starts negative. As for the central line, the high field line in this latter case also grows

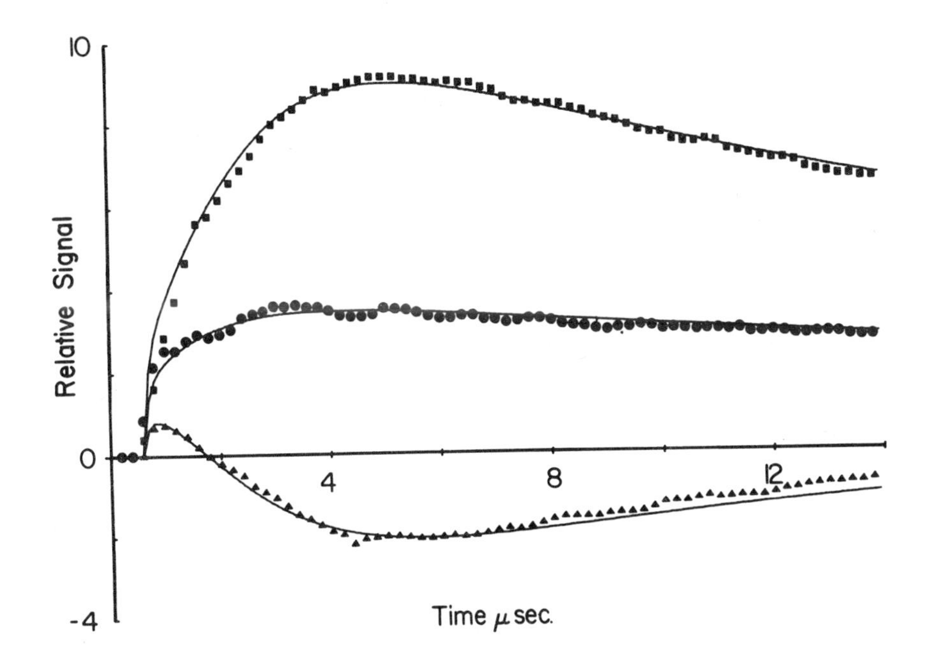

Fig. 5. Data for $\dot{C}H_2CO_2^-$ formed from OH and $CH_3CO_2^-$ as in Fig. 5 but showing all lines. The calculated curves used the parameters as in Fig. 4 except for f = 1 and V = -2, 0, and 2. Note the initial positive going signal for the low-field line at short times.

somewhat more slowly. Trifunac _et al_. [24], more qualitatively, have also shown different time dependences for $\dot{C}H_2CO_2^-$ produced in the two ways.

The excellent fit of the calculation to the data led us to take data for a number of initial radical concentrations to see how V varied [4]. The data is given in Table I. The radical half life was determined from the decay of the central line and the value of V from the amplitudes of the high- and low-field lines. As expected, the value of V increased with beam current while the half life decreased. However, the range of half lives was considerably less than implied by the beam currents. We believe the radical half lives measure the radical concentration and so conclude that the beam current is not necessarily a good measure of the concentration or production rate. Probably, a change in beam focus occurs so that the collected current does not reflect the actual dose in the aqueous sample. The product

Table I

Measured Values of Signal Enhancement Factor, V for $\dot{C}H_2CO_2^-$ [a]

Beam Current [b] i (μA)	Radical Half Life T_c (μsec) [c]	$V^{c,d}$	i x T_c	V x T_c
0.06	55	1.05	3.30	57.7
0.17	26	2.15	4.42	55.9
0.21	25	2.35	5.25	58.7
0.44	15.5	3.60	6.82	55.8

[a] The sample solution contained 0.1 M sodium acetate at ∿ pH 11.5 and was saturated with N_2O.

[b] Average beam current (μA) collected from cell; pulse repetition rate 100 Hz.

[c] Values were determined by use of Eq. D and refer to the radical concentration at t = 0.

[d] Value of V determined from high- and low-field lines. The amplitudes of the central lines were consistent with this value.

VxT_c is very constant as expected. Unfortunately, the experiment for T_c=15.5 μsec was at the highest beam current practical at the time. Trifunac [25] has obtained data at still higher radical concentration (half lives ∿10 down to 2 μsec) which seem to show no further variation of V with increasing dose after a value V∿5 has been reached. There are some detailed problems with his conclusion since V was determined at a fixed time after the pulse (3 μsec) and on the basis of the amplitudes measured at that time. As the half life becomes shorter a significant decay has occured by the time the ESR signal reaches its peak or before measurement is made. A detailed fit of the data such as described here is preferable. However, as described below, this observation may reflect a new process which occurs at high concentration, namely, Heisenberg spin exchange.

Hydrogen atoms

Hydrogen atoms can be studied either in neutral or basic solution at a yield of about 0.5 molecules/100 eV absorbed or in acid solution at a yield of about 3.5 as a result of the conversion of e_{aq}^-.

$$e_{aq}^- + H^+ \rightarrow \dot{H} \qquad (9)$$

In acid solution only the yield of 0.5 is primary; the remainder should have properties reflecting their origin from e_{aq}^-. In steady-state experiments on acid solutions [26] it was found that the amplitudes of the H atom ESR signals were approximately equal but of opposite sense with the low-field line in emission. The amplitudes of the signals varied with the first power of the irradiating beam current rather than the square root as might be expected for a steady-state experiment with second-order disappearance. This behavior could result from an initial effect on radical formation and the kinetic behavior was analyzed on the basis of this assumption [26]. Alternatively with a polarization produced on reaction, the signal will be proportional to the square of the concentration which is itself proportional to the square root of the beam current. Pulse experiments are necessary to select the correct explanation. A study of the time profile of the H atom signals only resolves the question upon quantitative analysis. Experiments were performed at several beam currents and at several power levels [2]. The data could best be fit by calculation only with the assumption that polarization occurred on reaction. At low beam current the signal reached a maximum at 10-15 μsec and decayed to half that level at about 40 μsec. The slow rise implies a long T_1. At the power level used, decay could not correspond to saturation of an initial population difference as such an effect would occur more quickly. Thus a continuing source of polarization is necessary.

To investigate the behavior of the H atom signals in more detail new data has been obtained [21]. The mathematical treatment involved integration of the Bloch equations since the half life approaches the relaxation time. To take field inhomogeneity into account we, in this case, used Pedersen's suggestion [19] and set T_2=1.5 μsec. Very low power was used (H_1=4 mG) and it was found that the rise of the curve could be fit with T_1=20 μsec. In this determination a small initial effect corresponding to f=-0.5 (for the low-field line) was, in fact, included as will be described below. For the low-field ESR line the value of P must be sufficient to keep the line in emisson throughout the decay as observed. A value of P=-0.9 (i.e., V=-18 with T_1=20 μsec) seems to fit the data well. The shape of the curves is not strongly dependent on P in this range. The remaining parameters are a second-order half-life and the scaling factor. Figure 6 shows data in the lower curve for the low-field ESR line of H in 0.1 M $HClO_4$ [21]. The parameters are half life=4 μsec and scaling factor -4 (negative because of the inversion). As previously reported, the addition of <u>tert</u>-butyl alcohol (2 mM) which reacts with OH but not the H atoms increases the intensity as is shown in the upper curve. In this case the curve could be fit with P=-1, the same scaling factor, and 7.8 μsec

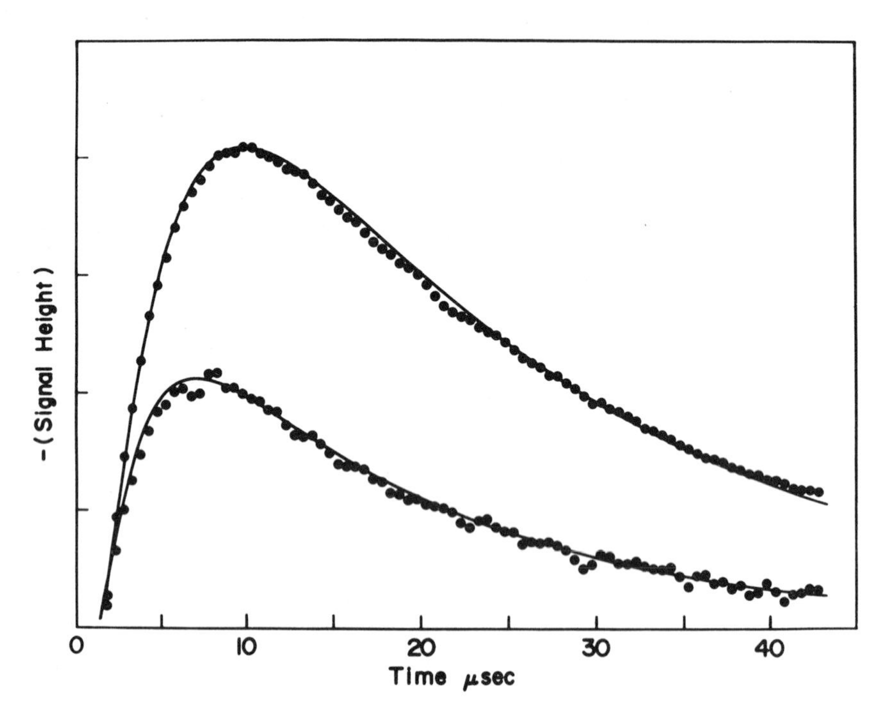

Fig. 6. Time profile of the low-field ESR line of H atom in pulse irradiated 0.1 M $HClO_4$ both with (upper curve) and without (lower curve) 2mM tert-butyl alcohol. The upper curve was taken with a beam current of 0.06 μA, the lower with 0.07 μA. A power setting of -25 dB (H_1=4 mG) was used for both. Note that the upward direction represents emission. The calculated curves used the parameters described in the text with a second-order half life of 7.8 μsec (upper) and 4 μsec (lower).

half life. The increase in half life shows that H+OH is faster than $H+\dot{C}H_2C(CH_3)_2OH$. In fact, with rate constants of $2k=1.5\times10^{10}$ M^{-1} sec^{-1} for H+H [27] and 2.0×10^{10} $M^{-1}sec^{-1}$ for H+OH [28] and yields of 3.6 for H and 2.8 for OH the half life should decrease by just a factor of two if $H+\dot{C}H_2C(CH_3)_2OH$ were very much slower than H+H. This interpretation seems the best one to explain this effect. The fact that the values of P are similar shows that H+OH does not produce polarization. This conclusion supports the observation that the e^-_{aq}+OH reaction in the spur also does not produce polarization. The rapid relaxation of OH probably accounts for these facts.

The steady-state experiments [12] showed a difference in amplitude of the two H atom lines with the low-field line more intense. The earlier pulse experiments [2] failed to find such an effect. However, more recent work [21] also detected such a difference with the low-field line again more intense by about 30%. Because of the improved means of setting the field on the center of the line the more recent data is probably correct. Several explanations are possible. However, the observation by Trifunac and Nelson [29] of a net negative polarization for $\dot{C}H_2CO_2H$ in the presence of H atoms suggests very strongly that mixing of singlet states with triplet $T_{\pm 1}$ states is responsible.

The behavior of the primary H atoms can be investigated either by direct observation or by looking at a product radical. The latter approach has led to extensive studies of the reaction

$$H + {}^{-}O_2CC \equiv CCO_2^- \rightarrow {}^{-}O_2CCH = \dot{C}CO_2^-. \tag{10}$$

The product radical is the same as that formed by reaction of e_{aq}^- with bromomaleic acid so its relaxation time is known. The time profiles of the two lines are shown in Fig. 7. Clearly an initial effect is present as seen most clearly by the emission for the low-field line. At longer times a polarization from the disappearance reaction is also evident from the approximate 2:1 ratio of signal intensities. The detailed discussion of this radical is complicated somewhat by the fact that the reactions

$$e_{aq}^- + {}^{-}O_2CC \equiv CCO_2^- \rightarrow [{}^{-}O_2CC \equiv CCO_2^-]^- \tag{11}$$

$$[{}^{-}O_2CC \equiv CCO_2^-]^- + H_2O \rightarrow {}^{-}O_2CCH = \dot{C}CO_2^- + OH^- \tag{12}$$

also occur. Thus the ESR signals observed represent the sum of the two behaviors. On the basis of previous discussions, the product formed from e_{aq}^- should show no initial polarization and certainly no different behavior for the two lines as is found here. The initial effect for ${}^{-}O_2CCH{=}\dot{C}CO_2^-$ is attributed to differing population differences among the H atom spin states before addition. No information is available at this time on the time scale of the protonation reaction. However, no inconsistencies have arisen from the assumption that this reaction occurs in less than 1 μsec.

The data of Fig. 7 can be fit very accurately as is shown. The relaxation time obtained from bromomaleic acid (7.5 μsec) was used as well as measured H_1 (9 mG here). The chemical decay is quite slow and was determined from data taken over longer times. The value of V was determined from the difference in the amplitude of the two lines at 30-40 μsec. (It was assumed that

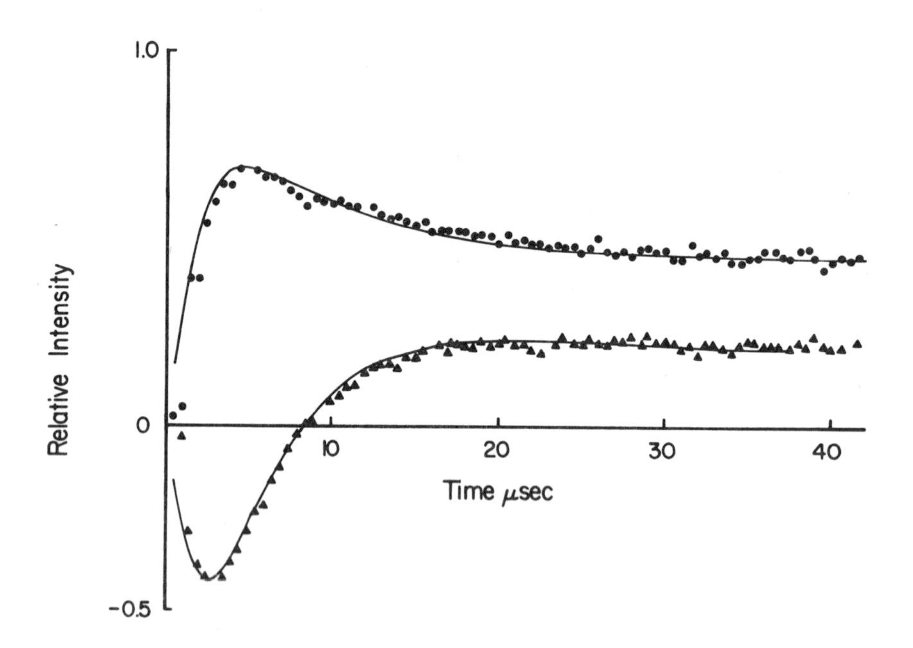

Fig. 7. Time dependence of the low-field (▲) and high-field (●) lines of $^{-}O_2CCH{=}\dot{C}CO_2^-$ in a solution of 3.3 mM $^{-}O_2CC{\equiv}CCO_2^-$ at pH 11.2. The calculated curves took field inhomogeneity into account and used T_1 = 7.5 μsec, H_1 = 9 mG, V = ±0.425, f = ±1.2 and T_c = 300 μsec.

no net polarization was caused by the presence of the OH reaction product, $^{-}O_2CCO\dot{C}HCO_2^-$.) The calculated curves used V=±0.425, f=±1.2 and T_c=300 μsec. Although equality of the magnitudes of f for the two lines was not imposed it was found that the best fit occurred when such was the case. The equality was better than 10%. Table II gives data for other concentrations of reactant. It is seen that for 3.3 and 10 mM the same f values are obtained at two beam currents so that the polarization cannot arise from homogeneous radical-radical reactions before addition. Larger values of f were found at lower concentrations. Pronounced changes in V and T_c with radical concentration (beam current) also were found. The decrease in first half life clearly is the result of an increased ionic strength since the disappearance reaction is hindered by repulsion between the doubly-charged radicals. High ionic strength will decrease this effect. The increase of V with concentration is approximately in inverse proportion to the decrease in T_c.

Table II

Parameters[a] for $^{-}O_2CCH{=}\dot{C}CO_2^-$ Produced from $^{-}O_2CC{\equiv}CCO_2^-$

Concentration mM	Beam Current μA	f^b	v^c	$T_c{}^d$ μsec
3.3	0.11	± 1.15	0.28	800
3.3	0.32	± 1.2	0.42	300
10	0.14	± 1.1	0.28	550
10	0.31	± 0.8	0.6	225
20	0.24	± 0.7	0.75	180
30	0.23	± 0.6	0.9	160
50	0.23	± 0.5	1.3	110

[a]Determined as described in text.

[b]The positive value corresponds to the high-field line, the negative to the low-field line. In each case the same magnitude fit both lines well.

[c]Determined from the difference in intensity of the two lines after any initial transients had decayed.

[d]Approximate values only.

A given value of f for the radical $^{-}O_2CCH{=}\dot{C}CO_2^-$ implies a value about 7 times higher for the H atoms since their yield is only 0.5 out of 3.5 total for the radical. On this basis the time profile of the ESR signals of H atoms in basic solution should be significantly different than for H atoms in acid solution where most come from e_{aq}^-. The f value in acid should, in fact, be similar to that for $^{-}O_2CCH{=}\dot{C}CO_2^-$ since similar proportions come by the direct and secondary routes. A detailed study of the signal profiles in basic solution has not yet been made but is quite practical.

The time scale of the reaction of H with acetylene-dicarboxylate is not know with precision. We have attempted to study the H atom ESR signal in the presence of $1\text{-}2\times10^{-4}$M compound and found a weak signal with a decay period of a few μsec. This experiment is complicated somewhat by the possibility of solute depletion. Nevertheless, the rate constant is clearly over $1\times10^9 M^{-1}sec^{-1}$. The reaction half lives for the concentrations given in Table II thus extend from 200 nsec at 3.3 mM to 14 nsec at 50 mM if 1×10^9 is taken as the rate constant. It is possible that the change in f with concentration represents interference with the slower spur reactions.

The explanation given here for the initial effect for $^{-}O_2CCH=\dot{C}CO_2^-$ is similar in some aspects to that offered by Wong and Wan [30] for the observations on cyclopentyl radical by Smaller et at. [31]. Wong and Wan suggested that the emission observed for cyclopentyl radical (Fig. 4 of Ref. 31) was the result of an inverted population difference of H atoms which added to cyclopentene produced in the radiolysis. The equation discussed here is appropriate to describe the curve for cyclopentyl and an attempt was made to do so [4]. In particular it should be possible to distinguish between an initial effect and one produced during the disappearance reaction. Careful study of the published data shows that the change from emission to absorption at 15 μsec after the pulse occurs rather suddenly. Attempts to fit the curve with f=0 and a negative value of V were not successful in that the time of zero crossing and the longer term decay could not be matched simultaneously. In contrast, a calculation with H_1=6 mG, f=-10, V=0, $T_1=T_2$=4 μsec and T_c=75 μsec matched almost exactly. The curve presented by Smaller et al. [31] represents the time profile at the peak of a first derivative signal and so is not strictly covered by Eq. D. However, the values of the parameters seem reasonable. At this point, an initial effect does seem appropriate to explain those results. The source of the polarization for the H atoms is not necessarily as Wong and Wan suggested.

It was pointed out above that the ESR signal for the low-field H atom line is about 30% more intense than the high-field line in both steady-state and pulse studies. In addition, transient H atoms in ice show a marked effect in the same sense [32]. A further experiment on this point was carried out by Trifunac and Nelson [29] who studied solutions of 0.33 M sodium acetate in sulfuric acid. They recorded the spectrum of $\dot{C}H_2CO_2H$ at 2 μsec after the pulse for several pH values. At higher pH (3.7) where e_{aq}^- is not rapidly converted to H atoms (the fate of e_{aq}^- is not clear under these conditions), the spectrum of $\dot{C}H_2CO_2H$ is essentially as given here for $\dot{C}H_2CO_2^-$ with an unpolarized central line and a low-field in emission with a slightly smaller amplitude than for the high-field line. At pH 1.2 where all e_{aq}^- are rapidly converted to H the spectrum of $\dot{C}H_2CO_2H$ shows a central line in emission and a larger amplitude (in emission) for the low-field line. Clearly, the whole spectrum has shifted toward emission. This fact, together with the greater emission for the low field H atom line seen in separate experiments, has been suggested by Trifunac and Nelson [29] to be the result of $S-T_{\pm 1}$ mixing. It is hard to counter this suggestion.

Other effects

In the discussion to this point, we have considered single radical systems or ignored the presence of other radicals. It is to be expected that the reaction of two radicals with different g factors should lead to net absorption for the radical with lower g factor and net emission for that with higher g factor. Although no quantitative studies similar to those discussed above have been carried out some qualitative results have been presented. Experiments [33] were carried out by studying radicals such as $\dot{C}H_2OH$ in solutions containing halogen compounds such as CCl_3CH_2OH or $CHBr_3$. The spectra of $\dot{C}H_2OH$ taken at 2 μsec after the pulse showed significant shifts toward increased emission. These shifts were taken to be the result of partial reaction of $\dot{C}H_2OH$ with the halogen-containing radicals which should have higher g factors. Unfortunately, changes in reaction rates with the changed conditions also could contribute. A more quantitative study is needed.

Although the behavior of e^-_{aq} is "normal" in the presence of $\dot{S}O_3^-$ and $\dot{C}H_2OH$ radicals it is definitely abnormal in the presence of certain other radicals. In an attempt to find an effect of g factor, solutions of carbonate were irradiated [3]. The e^-_{aq} does not react with carbonate but OH is converted to $\dot{C}O_3^-$ with a g factor of 2.01 The ESR signal of e^-_{aq} in such an experiment shows an emission which decays toward zero with a half life of about 3-4 μsec. At no time does the signal cross over to absorption. Experiments were then carried out with many of the compounds which react with OH but do not react with e^-_{aq}. Several classes of behavior were found. Four other reactants for OH gave radicals which also caused emission from e^-_{aq}, namely, phosphite (radical $\dot{P}O_3^{2-}$), hypophosphite ($H\dot{P}O_2^-$), phenol at pH 11 (phenoxyl), and hydroquinone at pH 11 (benzosemiquinone ion). Data for the hydroquinone system is shown in Fig. 8. For this system it was also possible to record the signal of the other radical, the benzosemiquinone ion. For reference, the signal of the semiquinone is also given for N_2O solution where the radical is produced with a yield of about 6.0. Other classes of behavior found for the signal of e^-_{aq} were a larger absorption signal when ethylene, formate, or tert-butyl alcohol rather than methanol was used as reactant for OH and a weaker e^-_{aq} signal when Br^- or ferrocyanide was used. Only the behavior in the cases showing emission will be discussed further.

The signal of the benzosemiquinone ion as given in Fig. 8 shows enhanced absorption for the period of the electron decay. The polarizations are attributed to the cross reaction with the g factor difference being the driving force. The decrease of the semiquinone signal must be, in part, the result of decay by

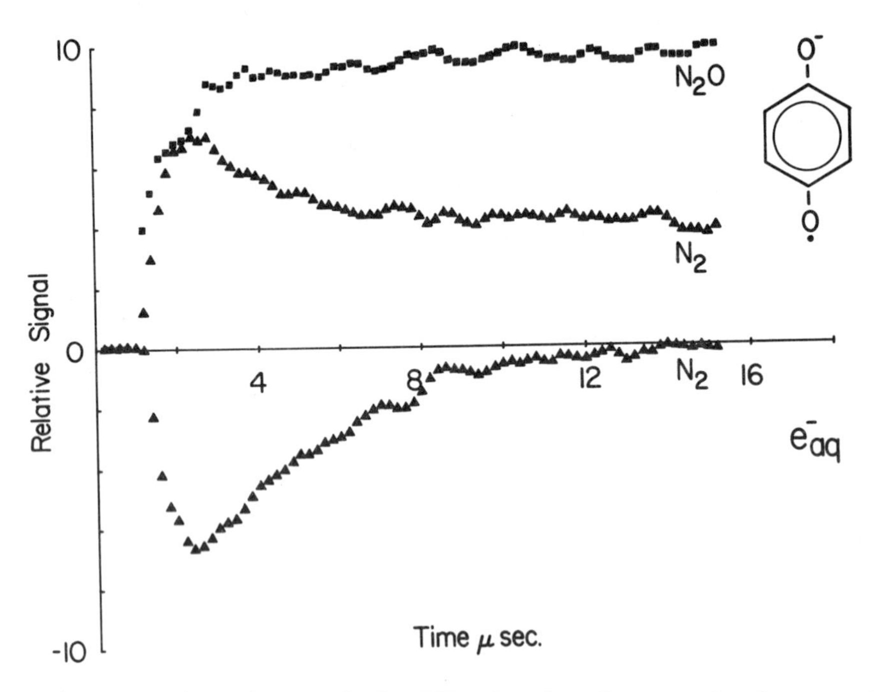

Fig. 8. Time dependence of the ESR signals of e^-_{aq} and p-benzosemiquinone ion in a solution of 2 mM hydroquinone (pH 11.9). The lower trace gives the signal for e^-_{aq} in N_2 saturated solution while the center trace shows the central line of the semiquinone ion for the same conditions. The upper trace gives the signal for the semiquinone ion in N_2O saturated solution where the yield is roughly twice as great. Most of the decay of the central trace is relaxation of an enhanced absorption and not chemical decay since the long term level is nearly 50% that in the N_2O solution.

this cross reaction but all of the decay cannot be explained in this way because the signal at about 2 μsec is too large for the relative yields (2.8 without N_2O, 6.0 with N_2O). In addition, the signals at longer time are reasonably close to the yield ratio showing that only a small fraction has decayed.

The emission effect for e^-_{aq} seems clearly to be the result of cross reaction but the sense of the effect is anomalous. The g factor of the semiquinone is 2.0045 vs 2.0003 for e^-_{aq} so that absorption for the latter rather than emission should have been found. To rescue the radical pair theory it is necessary to say that preferential reaction of the pair e^-_{aq}-semiquinone occurs to give triplet product. It is to be remembered that the species

e^-_{aq} is rather special so that the usual ideas which are derived from considerations of covalent bonds may not apply. It is possible to offer a rationalization for reaction to give triplet state. Henglein [34] has proposed that e^-_{aq} may react by tunneling so that one must have a matching of the energy levels of e^-_{aq} which the electron will leave and the unoccupied levels of the acceptor which the electron is entering. The center of the occupied levels of e^-_{aq} has been placed at -1.7 eV with respect to the unsolvated electron. The ground state levels which the electron can enter when it reacts with radicals such as $\dot{C}O_3^-$ or phenoxyl will be much below this value because of the high energy for photoionization of CO_3^{2-} or phenoxide ion. Thus there is room for an electronically excited state which may match that of e^-_{aq} better than the ground state. If this state is a triplet then preferential reaction to enter it may occur. Differences between various radicals arise because of their differing excited states. Potentially, one might be able to find luminescence associated with the formation of such excited states but none has yet been reported.

The lack of a further increase of enhancement factor above 5 for $\dot{C}H_2CO_2^-$ as found by Trifunac [25] has already been attributed to spin relaxation by Heisenberg spin exchange. Several other effects also suggest such this process. Among these are the linear dependence on dose of the emission signal from e^-_{aq} in the system with $\dot{C}O_3^-$ [3] and the change with dose of the apparent relaxation time of e^-_{aq} in the system with $\dot{S}O_3^-$ [4]. In the first two cases the qualitative behavior suggests an effect which cancels further expected increases in signal. Such an effect can come from relaxation by spin exchange when this relaxation mechanism becomes comparable to the intrinsic one for the radical. The data for e^-_{aq} in the system with $\dot{S}O_3^-$ show a decrease from 4 to 2.5 μsec for $T_2(=T_1)$ at radical concentrations [R] of about $1.5x10^{-5}M$. Thus $k[R]=(2.5x10^{-6})^{-1} -(4.0x10^{-6})^{-1}$ $=0.15x10^6 sec^{-1}$ and $k \sim 1x10^{10} M^{-1}sec^{-1}$. Typical rate constants for radicals such as $\cdot ON(SO_3^-)_2$ or di-tert-butyl-nitroxide in water are $2\text{-}4x10^9 M^{-1}sec^{-1}$ [35] or perhaps twice the values one might expect for reaction of, for example, carbon centered radicals of similar size. On this basis the rate estimated for e^-_{aq} is quite reasonable since $e^-_{aq}+e^-_{aq}$ has a rate constant (2k) of $1x10^{10}$ M^{-1} sec^{-1} [36]. In the case of the data for $\dot{C}H_2CO_2^-$, the observed behavior suggests a pseudo first-order rate for relaxation of about $3x10^5 sec^{-1}$ at a concentration where the half life is 5-10 μsec or $k(R)=1\text{-}2x10^5 sec^{-1}$. Again, the spin exchange rate constant appears to be similar to the reaction rate constant. A better test of this idea could be made with a radical of longer natural relaxation time so the effect would begin at lower radical concentration. Such a test is planned.

SUMMARY

At this point it is appropriate to summarize the results obtained so far. The mathematical treatment of the time profiles of the ESR signals can generally be handled by the modified Bloch equations. Fitting of calculated curves to data has worked well and quantitative parameters have been obtained which probably represent physically meaningful quantities. With this experience it is clear that a number of other quantitative measurements will be possible. All the radicals with hyperfine splitting which have been investigated show the normal multiplet CIDEP with the low-field lines weak or in emission depending on the rates of the radical-radical reactions. It has proved possible to determine initial population differences in suitable cases to within 0.2 of the Boltzmann value and by this means to demonstrate, quantitatively, transfer of populations from one radical to another. The population difference for e^-_{aq} has been found to be zero within an experimental error of about 0.2 of the Boltzmann value both by direct observation or, more sensitively, by the initial state for product radicals. This fact shows that the very significant extent of reaction of e^-_{aq} in the spurs does not lead to polarization even though much of that reaction is with OH which has a much higher g factor. In contrast, the H atoms seem to have an initial multiplet type polarization with population differences about 3 times the Boltzmann value. It is possible this effect results from spur reactions and so is not strictly an initial effect. The product radicals formed from OH all show initial populations equal to the Boltzmann value, presumably as the result of rapid relaxation before reaction. In addition to the straight-forward behavior observed in most systems several instances of unexpected behavior have been found. Anomalous emission from e^-_{aq} is observed when that species reacts with certain radicals such as CO_3^-, phenoxyl, or benzosemiquinone ion. This effect may be the result of preferential reaction to give triplet rather than singlet product. In addition, several effects have been observed which may be the result of relaxation by Heisenberg spin exchange. More work is needed in these latter problems.

ACKNOWLEDGEMENTS

The author is pleased to acknowledge the collaboration of Dr. Naresh C. Verma in much of the work covered here.

REFERENCES

1. R.W. Fessenden, J. Chem. Phys., 58, 2489 (1973).

2. N.C. Verma and R.W. Fessenden, J. Chem. Phys., 58, 2501 (1973).
3. R.W. Fessenden and N.C. Verma, J. Am. Chem. Soc., 98, 243 (1976).
4. N.C. Verma and R.W. Fessenden, J. Chem. Phys., 65, 2139 (1976).
5. S.K. Wong, D.A. Hutchinson, and J.K.S. Wan, J. Chem. Phys., 58, 985 (1973).
6. P.W. Atkins and K.A. McLauchlan, in "Chemically Induced Magnetic Polarization," edited by A.R. Lepley and G.L. Closs (Interscience, New York, 1973).
7. F.J. Adrian, J. Chem. Phys., 54, 3918 (1971); 57, 5107 (1972).
8. J.B. Pedersen and J.H. Freed, J. Chem. Phys., 58, 2746 (1973).
9. H.A. Schwarz, J. Phys. Chem., 73, 1928 (1969).
10. C.D. Jonah, M.S. Matheson, J.R. Muller, and E.J. Hart, J. Phys. Chem., 1267 (1976).
11. R. Livingston and H. Zeldes, J. Chem. Phys., 59, 4891 (1973).
12. K. Eiben and R.W. Fessenden, J. Phys. Chem., 75, 1186 (1971).
13. B. Smaller, E.C. Avery, and J.R. Remko, J. Chem. Phys., 55 2414 (1971).
14. P.W. Atkins, R.C. Gurd, and A.F. Simpson, J. Physics E: Sci. Instruments 1970, 547.
15. H. Levanon and S.I. Weissman, J. Am. Chem. Soc., 93, 4309 (1971); H. Levanon, Varian Instrument Applications, 5, No. 4, 6 (1971).
16. R.W. Fessenden and J.M. Warman, Adv. Chem. Ser., 82, 222 (1968).
17. P.W. Atkins, K.A. McLauchlan, and P.W. Percival, Mol. Phys., 25, 281 (1973).
18. H.C. Torrey, Phys. Rev., 76, 1059 (1949).
19. J.B. Pedersen, J. Chem. Phys., 59, 2656 (1973).
20. P.W. Atkins, A.J. Dobbs, and K.A. McLauchlan, Chem. Phys. Lett., 25, 105 (1974).
21. N.C. Verma and R.W. Fessenden, Discussions of the Faraday Society, No. 63, in press.
22. E.C. Avery, J.R. Remko, and B. Smaller, J. Chem. Phys., 49, 951 (1968).
23. P. Neta and R.W. Fessenden, J. Phys. Chem., 76, 1957 (1972).
24. A.D. Trifunac, K.W. Johnson, B.E. Clift, and R.H. Lowers, Chem. Phys. Lett., 35, 566 (1975).
25. A.D. Trifunac, J. Am. Chem. Soc., 98, 5202 (1976).
26. P. Neta, R.W. Fessenden, and R.H. Schuler, J. Phys. Chem., 75, 1654 (1971).
27. P. Pagsberg, G. Christensen, J. Rabani, G. Nilsson, J. Fenger, and S.O. Nielson, J. Phys. Chem., 73, 1029 (1969).

28. M. Anbar, Farhataziz, and A.B. Ross, Natl. Stand. Ref. Data Ser., Natl. Bur. Stand., 51 (1975).
29. A.D. Trifunac and D.J. Nelson, J. Am. Chem. Soc., 99 289 (1977).
30. S.K. Wong and J.K.S. Wan, Chem. Phys., 4, 289 (1974).
31. B. Smaller, J.R. Remko, and E.C. Avery, J. Chem. Phys., 48, 5194 (1968).
32. H. Shiraishi, H. Kadoi, Y. Katsumura, Y. Tabata, and K.Oshima, J. Phys. Chem., 78, 1336 (1974).
33. A.D. Trifunac and E.C. Avery, Chem. Phys. Lett., 27, 141 (1974); A.D. Trifunac and M.C. Thurnauer, J. Chem. Phys., 62, 4889 (1975).
34. A. Henglein, Ber. Bunsenges. Phys. Chem., 78, 1078 (1974); 79, 129 (1975); M. Grätzel, A. Henglein, and E. Janata ibid, 79, 475 (1975).
35. M.P. Eastman, G.V. Bruno, and J.H. Freed, J. Chem. Phys., 52, 2511 (1970).
36. M.S. Matheson and J. Rabani, J. Phys. Chem., 69, 1324 (1965).

CHAPTER VIII

FLASH-PHOTOLYSIS ELECTRON SPIN RESONANCE

K. A. McLauchlan

Physical Chemistry Laboratory, South Parks Road,
Oxford, England.

ABSTRACT. Free radicals produced by flash-photolysis can be identified from their electron spin resonance spectra at times exceeding a microsecond after the flash. If they are produced by reaction from triplet precursors and observed within their spin-lattice relaxation time the spectra may be completely in emission. This results from a true spin-polarization process in contrast to the spin-sorting characteristic of radical-pair processes at high magnetic fields. The time-evolution of the signal at a single magnetic field value yields measurements of the degree of polarization and of the rates of the relaxation and kinetic processes associated with the radical; the electron spin-lattice relaxation time in the triplet precursor can be deduced.

The spin-polarization can be transmitted from the primary radical by sufficiently fast reaction to secondary and successive radicals thus disclosing radical reaction pathways to other radicals. This provides complementary mechanistic information to CIDNP which shows pathways to diamagnetic products. Secondary polarization can be used to produce radicals with non-equilibrium spin populations in great variety for relaxation and kinetic studies.

1. INTRODUCTION

Although flash-photolysis is a powerful method for studying free radicals in solution, the normal detection technique, ultra-violet absorption spectroscopy, suffers from the low resolution associated with broad spectral lines; radicals are detected but only some specific chromophore they contain can be identified.

L. T. Muus et al. (eds.), Chemically Induced Magnetic Polarization, 151-167.

Electron spin resonance (ESR) spectra of the radicals on the other hand are of high resolution, often allow positive identification, and can distinguish different radicals containing identical chromophores. It was attractive to investigate the possibility of using flash-photolysis with ESR detection of the transient free radical.

The experiment to be described consists in applying a pulse of light to a sample held inside the microwave cavity of an ESR spectrometer within an external magnetic field. The radicals, produced in 2-3 μs, persist for about 10 ms, which times have significance in the design of a suitable spectrometer. The spin-states of the radicals can be defined at the instant of their creation but their energies within the applied field cannot. This is a consequence of the Uncertainty Principle and implies that there is a delay before the ESR transitions can be observed. In particular the width of the transition varies with time and ESR lines of 0.1 mT width are attained only after about 1 μs [1]; lines of at least this sharpness are required to resolve the hyperfine structure. Uncertainty broadening is exacerbated in ESR spectroscopy by the small size of the quantum. If the identity of the radical is desired the spectrometer should be designed to display the complete spectrum 1 μs after the flash if the inherent time-resolution of the technique is to be exploited fully; polarization measurements could be obtained at shorter times but with a concomitant fall in the sensitivity of the apparatus, and have yet to be performed in a photolysis experiment.

Usually in ESR the spectrum is scanned by varying the current applied to the windings of an electromagnet but induction prevents scanning even one line in about 1 μs. Use of a separate sweep coil allows shorter scan times but these are still too long for this experiment. Fourier methods are difficult at microwave frequencies in systems with very short relaxation times and so a novel sampling technique was devised to display the spectrum on the microsecond timescale [2]. This no longer involves scanning; rather the field is stepped manually between magnetic field values separated by small increments. At each field value a flash occurs and the resulting signal is sampled after 1 μs and subsequently about 1000 times thereafter to obtain its time-dependence. Approximately one thousand field values are investigated and a similar number of decay curves obtained; the spectrum at any time after the flash is simply reconstructed by taking a cross-section through these curves arranged in their correct field order. The radical concentration produced in each flash (which occurs in a period much less than 1 μs) is small and in practice signal-averaging of multiple flashes is employed at each field value. This mode of operation (mode 1) yields the full time-dependence of the system and the maximum of information.

In practice the radical present rarely changes after 1 μs and an alternative technique (mode 2) is more efficient in identifying the radical, although the time-behaviour is lost. Now the light source is pulsed repetitively (50 p.p.s.) and the magnetic field is swept continuously but slowly (typically 5 mT in 2 hours; it would be better stepped). The signal at a selected time after each flash is sampled and stored in a specific address of a digital data store, which is associated with a specific field value, and after one complete field sweep the entire spectrum is obtained at this time. The high pulse repetition rate and slow sweep result from the need to signal-average: the results of up to 5000 pulses are stored in each address before the field and the store advance. This unusual technique is better than its obvious alternative, the signal-averaging of multiple field-scans with one flash per address, since it is neither reliant on the precise start not the precise extent of the field sweep being reproducible. Having identified the radical the polarization and relaxation behaviour of each line can be obtained from mode 1 experiments at a limited number of field points.

2. EXPERIMENTAL

A further problem of operating an ESR spectrometer on the microsecond timescale is its response. Normal spectrometers which employ 100 KHz field modulation have response times which are much too long (ca 40 μs). Originally it was hoped to overcome this by using a broad-band (ca 6 MHz) super-heterodyne spectrometer but its sensitivity proved too low for photolysis experiments and its bandwidth was subsequently narrowed by use of 2 MHz modulation; the measured response function of the existing spectrometer is exponential with a characteristic response time of 1.1 μs. This yields reasonably undistorted (but see below) mode 1 decay curves but has the disadvantage of introducing 2 MHz sidebands across the spectrum. It becomes convenient to separate the two different types of information required and to use 2 MHz modulation when polarization and relaxation studies are performed and 100 KHz modulation when the radical's identity and decay kinetics are investigated.

The basic spectrometer is a Decca X-band one equipped with an irradiation cavity using 100 KHz modulation coils; it proved possible to wind further 2 MHz coils within these and to obtain up to 2 mT modulation field at the sample thanks to the cavity walls consisting of thin layers of evaporated gold supported on a dielectric. The magnet system is a Varian Fieldial II with which a specific field value is maintained by use of a temperature-compensated Hall effect probe. The entire experiment is controlled from a small dedicated digital device (BIOMAC 1000) which besides acting as a data store initiates a pulse sequence to control the experiment. The basic device has a time resolution of 5 μs between addresses, with a sampling time of 300 ns; for the 2 MHz measurements a fast buffer store (Datalab 905) is used which has a resolution of 0.2 μs per address. In mode 1 the pulse which initiates the sweep of the data store is fed also via a pulse delay and amplifier unit to the flash photolysis source, a nitrogen gas laser (Lambda Physik M2000). This was chosen to give sharp pulses (ca 3ns) of u.v. radiation (337.1 nm) at high repetition rates (up to 50 p.p.s.), with a flash energy of 20 mJ/pulse (this is 20 times the output of the laser originally used [2], and produces 20 times the radical concentration). The extreme sharpness of the pulse implies that it has decayed to zero before measurements commence. The flash creates radicals in the sample held within the cavity and the ESR response is obtained from the spectrometer held at resonance.

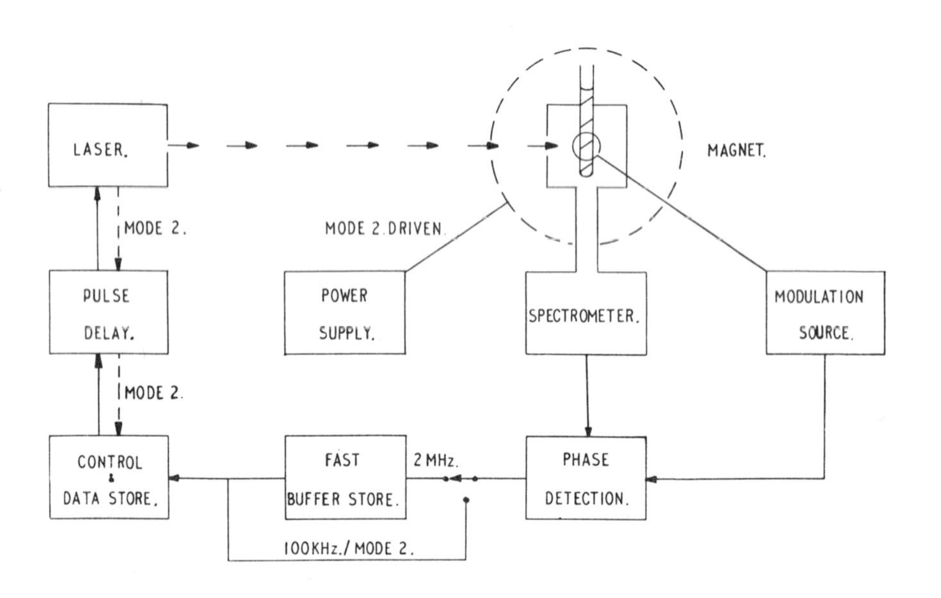

The spectrometer as used in the mode 1 experiments.

The signal, delayed by the response time, is sampled at a series of subsequent times by the buffer store but because of the pulse delay unit it first appears some way along the address register, which allows a base-line to be established. Once the 1000 addresses of the buffer store have been accessed their contents are read into the slower backing store after which the buffer store is re-set and a further pulse generated to initiate the procedure once more. In this way the results of typically 2^{14} flashes are averaged in each experiment. Mode 1 information obtained at 100 KHz is useful in obtaining chemical kinetic information, the rate processes usually being quite slow (~ 1 - 10 ms), for here the extra sensitivity associated with the narrower band-width gives better signals.

In mode 2 the experiment is controlled by the internal pulse system of the laser run at its maximum repetition rate, and the buffer store is disconnected. A pulse synchronised with the flash is fed via a second pulse delay unit to an external advance input of the data store. On arrival it initiates the signal sampling process (which therefore occurs after a time defined by the delay unit and can be adjusted at will) and provides the means to advance the data store between addresses. By interrupting the internal logic of this device a pre-set number of these pulses are consumed in a counter before this in turn generates a pulse to advance the address; in this way the results of several flashes are averaged in each address. It would be simple to use this pulse to step the magnetic field in synchrony with the addresses, but this has not yet proved necessary. The present mode does however limit the number of flashes which can be averaged since all the samples are taken before the field has changed perceptibly.

The sample can be maintained at temperatures between 130-380 K by holding the sample in a nitrogen gas stream inside a Dewar vessel constructed from thin-walled silica. In polarization and relaxation experiments the concentration of the sample is adjusted for optimum signal (it is usually 0.5 - 1.0 M) but kinetic studies are performed at low concentration to obtain uniform radical concentration throughout the irradiated sample; this is essential practice in other than first-order reaction processes.

3. RESULTS

Many of the important polarization experiments were performed using only a weak (1mJ/pulse) laser source and the bulk of the results reported here were obtained in this way; use of the more intense new source will be specifically noted as it occurs. This is important since the radical concentrations produced by the new laser are such that both pure polarization and spin-sorting effects are visible in the spectra whereas the old laser normally showed only true spin polarization phenomena. This dependence of polarization behaviour on concentration can be exploited to separate the two effects quantitatively.

A typical early mode 2 result is shown in the figure. It shows the spectra of a pair of free radicals observed 100 μs after the flash when a solution of anthraquinone in di-t-butyl phenol is photolysed [3]. The spectra show exactly the hyperfine structures and intensities expected for the corresponding semi-quinone and phenoxyl radicals, without distortion. The sharper lines are from the semiquinone and the broader, which are partially obscured, are from the phenoxyl. The phases of all the lines of both radicals are the same and all, by comparison with spectra observed from stable radicals, correspond to emission.

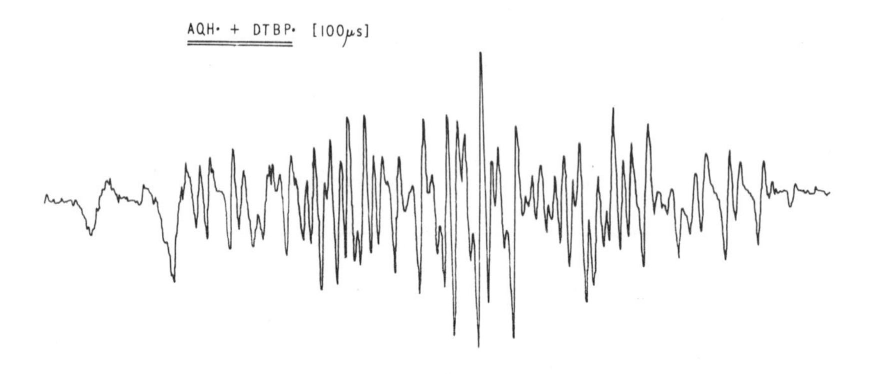

A mode 2 spectrum 100 μs after the flash.

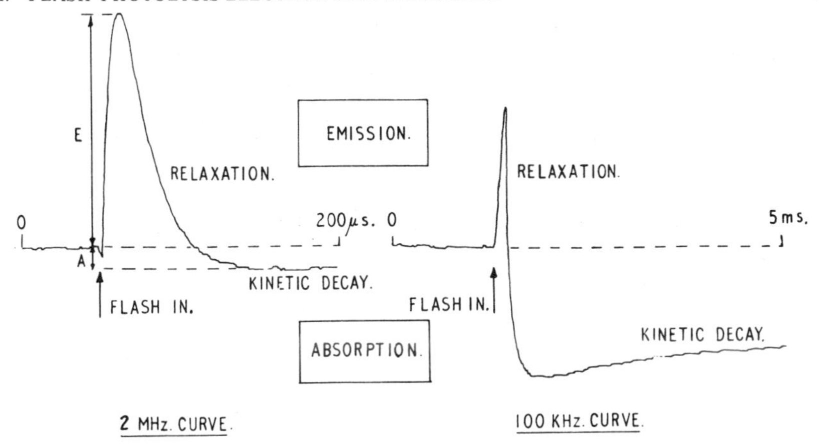

Mode 1 decay curve using 2 MHz and 100 KHz modulation.

This behaviour is entirely inconsistent with the polarization which would result if S-T_0 mixing in a radical pair process was important for in this case the total absorption and emission intensities (corrected for relaxation) would be the same. Rather both radicals are observed with the same polarization and no hyperfine-dependence of polarization is apparent.

The emissive nature of the spectra is demonstrated very clearly in the mode 1 decay curve obtained at a specific field value. The figure shows two curves obtained from the duroquinone anion produced by photolysis of a solution in isopropanol in the presence of triethylamine.

The first curve above is the response obtained using the 20 mJ/pulse laser with the 2 MHz system and the second, at longer time scale, with the 100 KHz system. Each is basically similar; following a short base line the flash occurs and the signal rises with the response time of the spectrometer to a maximum followed by a rapid fall due to spin-lattice relaxation which establishes thermal equilibrium with the surroundings. Since the signal is observed whilst relaxation proceeds this is driven by the microwave field and a true measure of the relaxation time can only be obtained by making observations at different field strengths and extrapolating to zero (see below). The curve crosses the axis, corresponding to a change in phase of the ESR signal, and the subsequent absorption signal decays by chemical processes at a much slower rate; on the timescale of the first trace it is barely discernible. The emission (E) and (hypothetical) absorption (A) signals immediately following the flash can be evaluated by extrapolation and the polarization ratio is defined as

$$\gamma = (E + A)\, A^{-1}$$

This value is dependent on microwave field strength also for the more relaxation that occurs the lower the value; it too must be extrapolated to zero field. A further correction is required for the finite response of the spectrometer; for radicals with relaxation times of ca. 5-20 μs the measured E value is 0.74 of the true one.

The occurrence of polarization is apparent also on the 100 KHz trace but its magnitude cannot be measured accurately; deconvolution procedures give γ-values which are highly dependent on the (complex) response function of the spectrometer. This trace shows clearly the effect of second-order kinetics on the chemical decay of the radical.

4. INTERPRETATION

4.1 Polarization ratios

Elsewhere in this volume the origin of electron spin polarization in radicals produced on reaction from triplet precursors has been discussed by P.W. Atkins. It suffices here to emphasise that it reflects an earlier spin polarization attained in the triplet by preferential inter-system crossing from an excited singlet produced on irradiation to the non-degenerate Zeeman levels of the triplet in the field of the spectrometer. In the laboratory frame of reference the upper level is the more highly populated in aromatic ketones and quinones, as shown schematically in the figure. This polarization

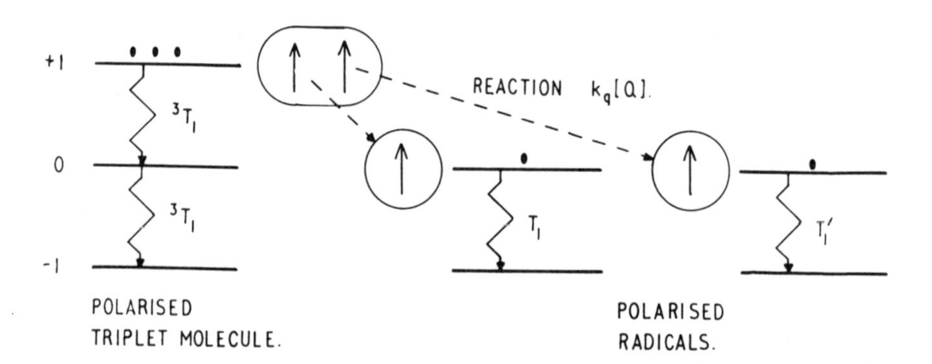

Triplet relaxation competes with chemical reaction in the production of polarized radicals.

results in spin polarization in the radicals produced on reaction of the triplet, since reactions conserve spin (we can think of the parallel pair of electrons in the upper Zeeman level simply being dragged apart by the reaction to end up on different radicals), provided that the reaction rate is sufficient to compete with the very fast (ca 1 ns) spin-lattice relaxation process which tends to remove the population disequilibrium in the triplet [5]. Relaxation in the radical is much slower, as shown in the decay curve above. Only diffusion controlled reactions - electron, proton and H-atom transfers - are sufficiently fast and the basic requirement to observe "triplet" polarization is apparent. Interestingly, the conclusion of this process yields a radical pair in which, under suitable magnetic conditions, S-T_o polarization can develop at a later time which is still less than the 1 μs observation time of the spectrometer. The two processes are resolved in time but the spectrum observed reflects both their occurrences besides that of any relaxation which preceeds the measurement. The occurrence of both is apparent at the radical concentrations produced by the 20 mJ/pulse laser but at low concentrations of radicals without unusually high hyperfine coupling the triplet mechanism predominates.

This mechanism has been tested in several ways. Firstly its requirement for the two radicals both to show emission spectra has been verified above. Furthermore it has recently been shown that the polarization ratios of each are the same [6] (see below). Following a suggestion of F.J. Adrian [7] direct confirmation has come from the observed dependence of polarization ratios on the plane of polarization of the photolysis light pulse [8]; this especially probes the inter-system crossing step in the precursor. The experimental result was later confirmed [9], although the magnitude reported was different; polarization ratios are notoriously difficult to measure and the difference may not be significant. The bulk of the evidence for the polarization mechanism, and the greatest use to the chemist, has resulted from experimental test of a relationship obtained by simple kinetic analysis of the competing rate processes shown in the figure. This yields

$$4(3\gamma - 1)^{-1} = \gamma_o^{-1} + (\gamma_o\, {}^3T_1 k_q [Q])^{-1}$$

where γ_o is the initial polarization in the triplet and [Q] is the concentration of the substrate with which the radical reacts (see Atkins loc. cit.). Thus a plot of $4(3\gamma-1)^{-1}$ versus $[Q]^{-1}$ should give a straight line with a slope $\gamma_o{}^3T_1k_q$ whilst γ_o can be obtained most accurately from a plot of $4[Q](3\gamma-1)^{-1}$ against [Q]. This relationship has been tested in a large number of observations in our laboratory, although few have been published. From the two plots the product $k_q{}^3T_1$ is obtained from which the triplet relaxation time in solution can be deduced if the quenching rate constant is determined independently, either by direct experiment or by a Stern-Volmer analysis of Mode 1 decay curves obtained at different quencher concentrations. This method has been used on the triplet state of duroquinone in various solutions and values of 3-17 ns obtained [10]. The possibility exists of using these values as internal time standards for determining quenching rate constants. Exploitation of this aspect of CIDEP is in its infancy.

Radical polarization ratios are difficult to measure if they become large and usually conditions are chosen to enhance the measurement accuracy by restricting them to around 10. Values of up to 150 ± 20 have been observed in the duroquinone anion (unpublished work).

The kinetic analysis was performed assuming that the rates of reaction of the triplet molecules are independent of the Zeeman level which they populate. This is of relevance for if reaction occurs from different levels at different rates this implies a third spin polarization process for the radicals. Such differing rates occur in solids but experiments to investigate this effect in solution are not yet conclusive [11].

4.2 Relaxation times

The relaxation behaviour of the non-equilibrium spin distribution attained in the polarization process and under the influence of a microwave field (of strength B_1) is analysed by reference to the appropriate Bloch equations. These are solved under the initial magnetisation conditions that the longitudinal component differs from its thermal equilibrium value and the transverse components are zero. The general solution of the equations is cumbersome and we consider two particular cases of relevance to the work in our laboratory, that in which $T_1 \gg T_2$ (as in viscous liquids) and that with $T_1 = T_2$ (as in low-viscosity liquids). In each case the longitudinal magnetisation decays in an exponential manner [12].

A feature of the derivative of an absorption ESR line is that the signal is zero at resonance; any experiment must be performed off-resonance by an amount $\delta\omega$ Hz. This is not so for the derivative of a dispersion mode signal which is maximum at resonance, and both experimentally and analytically it is advantageous to work with it.

For $T_1 \gg T_2$, the effective relaxation times T_{1eff} for absorption and dispersion curves are given by

$$(T_{1eff}^{abs})^{-1} = T_1^{-1} + b^2T_2(1 + \delta\omega^2T_2^2)^{-1}$$

and

$$(T_{1eff}^{disp})^{-1} = T_1^{-1} + b^2T_2 \ ,$$

where $b = g\mu_B B_1 h^{-1}$. These equations have been verified experimentally and provide a linear extrapolation to yield the true value of T_1. Results on the benzophenone ketyl radical in paraffin solutions yield identical values from both equations of around 10^{-4}s in these high-viscosity liquids. From the slope of the plots T_2 can be evaluated if b is known. This is normally very difficult to calculate or measure but the transient response following a photolysis flash can yield it directly.

If a system is chosen for which $T_1 = T_2$, solution of the Bloch equations shows that the observed signal should show a damped oscillation (the Torrey wiggles) whose amplitude decays with the true relaxation time of the radical. In this case T_1 is not dependent on the presence of the microwave field. For an observation at resonance on the dispersion mode signal the frequency of oscillation is a direct measure of the microwave field in the cavity [13]. Wiggles occur only if $b > \frac{1}{2}(T_1^{-1} - T_2^{-1})$, in practice if $T_1 \sim T_2$, and they can be used to detect the onset of exchange contributions to T_2.

Most spectra are obtained near, but rarely at, the $T_1 = T_2$ limit. In general T_{1eff} is dependent upon B_1 and an empirical extrapolation is used to zero power to determine the true T_1; values down to 2 μs can be measured with reasonable accuracy.

This method of measuring spin-lattice relaxation times has some important advantages over more conventional methods. The radicals are observed under transient conditions and have not to be produced in steady-state concentrations; this greatly extends the variety of species which can be investigated. The radical concentration produced in each flash (10^{-4} - 10^{-5} M) is very low and the observations are free from exchange effects. The hyperfine structure is resolved and allows systematic investigation of the dependence of relaxation on the chemical nature of the radicals [14]; it also allows investigation of the hyperfine-dependence of relaxation times. To extend the usefulness of the method considerable effort has been made to increase the variety of radicals available for study by use of secondary polarization techniques (see below).

4.3 Chemical studies

Although identification and kinetic study of transient free radicals forms an important part of our research programme, this is largely independent of polarization phenomena and will be omitted here (but see section 6). The exception is that the increased magnitude of polarized signals often allows radicals to be detected at lower concentrations than can be employed in conventional experiments.

5. SECONDARY POLARIZATION

If a spin-polarized primary radical reacts within its relaxation time it may produce a secondary polarized radical. This is akin to CIDNP observations where the primary nuclear polarization results in polarized product either directly or, through memory and substitution effects, indirectly via intermediate radicals or diamagnetic molecules. Its importance is two-fold: it provides a method with which radical reaction pathways to other radicals can be followed (the polarization simply acts as a label) and it provides a wide variety of species for relaxation and kinetic studies. The polarization transmitted to the secondary radical depends upon the reaction rate and the T_1 of the primary radical.

Its observation depends upon careful choice of reaction conditions but it can easily be shown to occur. If a solution of benzophenone in methanol is photolysed the hydroxymethyl radical is observed but its spectrum is not polarized; evidently the rate of reaction of the triplet is very much less than its relaxation rate. In the presence of triethylamine the spectra of both the benzophenone ketyl radical and hydroxymethyl are polarized but no radical is observed from triethylamine [15]. This results since the triplet abstracts hydrogen very rapidly from the amine to produce initially-polarized radicals:

$$Ph_2CO^T + NEt_3 \rightarrow Ph_2\dot{C}^*OH + Me - \dot{C}^*H - NEt_2 \qquad (1)$$

The amine-derived radical subsequently reacts with methanol, within its relaxation time, to give polarized hydroxymethyl. This reaction process occurs before the radical can be observed in our spectrometer.

In higher alcohol solutions it has been demonstrated that the amine-derived radical reacts with ground-state benzophenone to form the ketyl by a secondary route [16] so that the quantum yield for ketyl radical formation is two and the secondary route also contributes to the observed polarization ratio. The reaction

$$\mathrm{Me\text{-}\dot{C}H^{*}\text{-}NEt_2 + Ph_2CO \rightarrow CH_2{=}\ CH - NEt_2 + Ph_2\dot{C}^{*}OH} \qquad (2)$$

is sufficiently fast for polarization to be transferred, as is shown by the dependence of the polarization of the ketyl on benzophenone concentration. Addition of carbonyl compounds which react faster than benzophenone with the primary radical causes the ketyl quantum yield to fall to one but also yields carboxyl radicals in a spin-polarized state. By this means a great variety of transient radical anions has been produced for study [14,16].

For example, with biacetyl:

$$\mathrm{Me - \dot{C}H^{*}\text{-}NEt_2 + MeCOCOMe \xrightarrow{k_1[B]} CH_2 = CH\text{-}NEt_2 + \begin{matrix} Me\dot{C}(OH)COMe \\ \Downarrow\Uparrow\ NEt_3 \\ Me\dot{C}O\overset{-*}{C}OMe + H\dot{N}Et_3^{+} \end{matrix}} \qquad (3)$$

The neutral radical is not observed in our experiments; the anion exists in two isomers whose equilibrium ratio depends upon the concentration of biacetyl [6].

The polarization carried over onto the secondary radical depends upon a simple competition between the rate of reaction (3) and the relaxation time of the primary species, $T_1(p)$. For an observed polarization ratio in the secondary radical γ_2 and an unobserved primary polarization γ_1,

$$(\gamma_2-2)^{-1} = (\gamma_1-2)^{-1}\{k_1[B]T_1(p)\}^{-1} + (\gamma_1-2)^{-1} ,$$

where B represents biacetyl. One significance of this expression is that a plot of $(\gamma_2-2)^{-1}$ versus $[B]^{-1}$ allows the polarization ratio of the primary radical to be assessed from measurements of that in the secondary one; with the primary ketyl radical directly observed, the initial polarization of both primary radicals is obtained for the first time [6]. In 2-octanol solution the plot gave $\gamma_1 = 33 \pm 5$ for the amine radical, and direct observation of the ketyl gave $\gamma_1 = 30 \pm 4$ (both uncorrected for finite instrument response); the equality of these values accords with triplet theory of spin polarization. From the plot the product $k_1T_1(p)$ is also obtained and this promises to be a useful technique for measuring relative reaction rates of different substrates with the primary radical; independent knowledge of one rate constant would serve to calibrate the series absolutely.

The description above appears deceptively simple for reaction conditions must be chosen with care to yield meaningful data. Concentrations are chosen so that > 99% of the light absorbed is absorbed by benzophenone, and not the carbonyl compound directly; energy transfer is minimised by using concentration ratios of > 10:1 and with biacetyl has been shown not to be a significant polarization route. The triethylamine concentration is adjusted so as to make direct reaction of the alcohol solvent with either triplet benzophenone or primary radical insignificant. If these conditions are satisfied a complete reaction scheme can be written and analysed to yield an expression for the signal from the secondary radical as a function of time; this involves incorporation of the spectrometer response function, r(t) which at 2 MHz is a simple exponential with a response time t_s. We perform this analysis for the short-time behaviour of the signal for a reaction scheme consisting of reactions (1) and (3) above, the decay of the biacetyl radical anion at rate k_2 s^{-1} (assumed pseudo first-order; this is not an important assumption in the interpretation of polarization and relaxation data) and the relaxation of the secondary species with relaxation time $T_1(s)$.

The calculation is based on the rate equations for the populations of the α and β spin states of the secondary radical S. These include terms in the rate of formation with spin conservation from primary radical P, in the transition probabilities for relaxation between states W and in the rate of reaction of the secondary radical:

$$\frac{dS_\alpha}{dt} = k_1 P_\alpha - S_\alpha W_{\alpha\beta} + S_\beta W_{\beta\alpha} - k_2 S_\alpha$$

$$\frac{dS_\beta}{dt} = k_1 P_\beta + S_\alpha W_{\alpha\beta} - S_\beta W_{\beta\alpha} - k_2 S_\beta ,$$

where

$$W_{\alpha\beta} + W_{\beta\alpha} = T_1(S)^{-1} .$$

The rate equation for the difference in population of the spin states, S(t), is obtained by subtraction:

$$\frac{dS(t)}{dt} = k_1 P(t) - 2S_\alpha W_{\alpha\beta} + 2S_\beta W_{\beta\alpha} - k_2 S(t) .$$

The solution of this equation is straightforward [17] and yields an expression for S(t) which represents the input signal to the spectrometer. The output, $\phi(t)$, is related to it by a proportionality factor (taken as unity) and by a convolution integral with the spectrometer response function:

$$\phi(t) = \int_0^\infty S(t-u)r(u)du = \int_{-\infty}^{t} S(u)r(t-u)du \; .$$

Evaluation of this integral gives a formidable expression:

$$\phi(t) = \frac{1}{t_s} P_e(0) \left[\frac{k_1}{k_1-k_2} \left\{ \frac{1}{t_s^{-1}-k_2} e^{-k_2 t} - \frac{1}{t_s^{-1}-k_1} e^{-k_1 t} \right\} \right.$$

$$+ \frac{k_1}{(t_s^{-1}-k_2)(t_s^{-1}-k_1)} e^{-t/t_s}$$

$$+ \frac{(\gamma_1 - 2)\, k_1}{k_1-k_2+T_1(p)^{-1}-T_1(s)^{-1}} \left\{ \frac{1}{t_s^{-1}-k_2-T_1(s)^{-1}} e^{-(k_2+T_1(s)^{-1})t} \right.$$

$$\left. - \frac{1}{t_s^{-1}-k_1-T_1(p)^{-1}} e^{-(k_1+T_1(p)^{-1})t} \right\}$$

$$\left. + \frac{(\gamma_1 - 2)\, k_1}{(t_s^{-1}-k_2-T_1(s)^{-1})(t_s^{-1}-k_1-T_1(p)^{-1})} e^{-t/t_s} \right] ,$$

where $P_e(0)$ is the equilibrium signal of P at time zero.

At very short times the first bracket is small: the first term is dominated by t_s^{-1} and the second by k_1 (k_1 must be large for secondary polarization to occur at all). The effect of the two terms in e^{-t/t_s} dies away very rapidly as t becomes greater than t_s. Thus it is the second term which dominates throughout our measurement region and with $k_1 \gg k_2$ (if $k_2 \sim k_1$ tertiary polarization can be obtained; we choose our conditions to make $k_1 \gg k_2$), and $T_1(s)^{-1} \gg k_2$ this complex expression reduces to

$$\phi(t) = \text{const. } e^{-t/T_1(s)}$$

It is consequently possible to extract the relaxation time of the secondary radical from its decay curve, as would be expected intuitively. This is allowing study of the relaxation behaviour of a wide range of radicals of diverse chemical type [14].

Secondary polarization studies are much in their infancy with little work yet done on alternative sensitisation sources or on diverse reactive primary radicals. From the relaxation point of view they remove a major restriction of our original experiments: that the radicals' precursor must absorb at 337 nm, the laser wavelength. From the chemical point of view they provide a unique method for following the reactions of radicals in solution [6] and of measuring their rates.

6. RELEVANCE TO MECHANISTIC STUDIES

The reaction pathways of free radicals in solution are often disclosed with extraordinary clarity by nuclear polarization, CIDNP, studies but only rarely can these be quantified. It is common to discover that the observation of CIDNP phenomena give little or no indication as to the importance of a given reaction route. ESR techniques, and in particular flash-photolysis methods, when used in conjunction with NMR observations provide a particularly powerful combination for studying free radical reaction routes in solution. The ESR technique gives positive identification of the radicals and the rate constants for their decay and it helps to establish a precise reaction pathway which may allow quantitative analysis of time-dependent CIDNP signals. This interplay of the two techniques is exemplified in our laboratory by detailed studies of the photolysis of benzaldehyde [18,19] and of pivalophenone [20,21] whilst its mechanistic power has been shown in an examination of the photosensitized decomposition of oxy-and thio-acids [22].

Not all radicals can be detected using the flash-photolysis method, particularly those with broad lines, those with very short lifetimes or those formed in low quantum yield. In these cases recourse has been made to spin-trapping techniques.

7. ACKNOWLEDGEMENTS

The original work was performed in happy collaboration with Dr P.W. Atkins and we are both indebted to our co-workers, particularly Drs A.J. Dobbs and P.W. Percival; the secondary polarization studies have further increased my indebtedness to Dr R.C. Sealy and Mr J.M. Wittmann.

REFERENCES

1. P.W. Atkins, K.A. McLauchlan and A.F. Simpson, Molecular Spectroscopy, Institute of Petroleum, London, 177, 1968.
2. P.W. Atkins, K.A. McLauchlan and A.F. Simpson, J. Phys.(E), 547, 3, 1970.
3. A.J. Dobbs and K.A. McLauchlan, unpublished results.
4. S.K. Wong and J.K.S. Wan, J. Amer. Chem. Soc., 7197, 94, 1972.
5. S.K. Wong, D.A. Hutchinson and J.K.S. Wan, J. Chem. Phys., 985, 58, 1973.
6. R.C. Sealy, K.A. McLauchlan and J.M. Wittmann, J. Chem. Soc. Faraday II, in the press.
7. F.J. Adrian, J. Chem. Phys., 4875, 61, 1974.
8. A.J. Dobbs and K.A. McLauchlan, Chem. Phys. Letts., 257, 30, 1975.
9. B.B. Adeleke, K.Y. Choo and J.K.S. Wan, J. Chem. Phys., 3822, 62, 1975.
10. P.W. Atkins, A.J. Dobbs and K.A. McLauchlan, Chem. Phys. Letts. 616, 29, 1974.
11. A.J. Dobbs, C.G. Joslin and K.A. McLauchlan, to be published.
12. P.W. Atkins, K.A. McLauchlan and P.W. Percival, Molec. Phys., 281, 25, 1973.
13. P.W. Atkins, A.J. Dobbs and K.A. McLauchlan, Chem. Phys. Letts., 105, 25, 1974.
14. K.A. McLauchlan, R.C. Sealy and J.M. Wittmann, in preparation.
15. K.A. McLauchlan, and R.C. Sealy, unpublished results.
16. K.A. McLauchlan and R.C. Sealy, Chem. Phys. Letts., 310, 39, 1976.
17. K.A. McLauchlan, R.C. Sealy and J.M. Wittmann, in preparation.
18. P.W. Atkins, J.M. Frimston, P.G. Frith, R.C. Gurd and K.A. McLauchlan, J. Chem. Soc.,Faraday II, 1542, 69, 1973.
19. P.G. Frith and K.A. McLauchlan, J. Chem. Soc. Faraday II, 1984, 71, 1975.
20. P.W. Atkins, A.J. Dobbs and K.A. McLauchlan, J. Chem. Soc. Faraday II, 1269, 71, 1975.
21. P.G. Frith and K.A. McLauchlan, J. Chem. Soc. Faraday II, 87, 72, 1976.
22. P.R. Bowers, K.A. McLauchlan and R.C. Sealy, J. Chem. Soc. Perkin II, 915, 1976.

CHAPTER IX

TIME-DEPENDENCE OF ESR INTENSITIES

J. Boiden Pedersen

Department of Physics, Odense University,
DK-5230 Odense M, Denmark

1. INTRODUCTION

The quantitative information we have about many aspects of CIDEP is obtained by the time-resolved ESR technique[1-3]. The radicals under investigation are generated by a pulsed light[1] or electron source[2], or alternatively by a modulated source[3], e.g. a rotating sector or an electronically modulated light source. The signal intensity is recorded as a function of time from the instant of time when radical formation starts. Rapid response ESR spectrometers[1,2,4] with a time resolution in the microsecond domain have been developed and this makes it possible, at least in principle, to determine T_1, ω_1, chemical decay constants, and various polarization terms from the time dependence of the intensity. Such a determination requires a very careful curve fitting and the purpose of this chapter is to provide the necessary theory[5]. The approach is based on the Bloch equations, which have been modified to include chemical reactions and polarization production. The time resolution or response time of the spectrometer is a lower limit to the time constants that can be determined[3,6]. Whenever the response time is comparable to or larger than the smallest time constant (e.g T_1) then one must include the effect of the spectrometer response on the intensity curve and this is discussed at the end of the chapter.

2. DERIVATION OF THE FUNDAMENTAL EQUATION

In the following the radical under investigation is denoted A and other radicals with which A may react are denoted B. To specify that a given quantity depends on the nuclear state of A, we use

L. T. Muus et al. (eds.), Chemically Induced Magnetic Polarization, 169-180.

a subscript a, but this subscript is often omitted for typographical reasons. The absorption signal for the hyperfine line a is proportional to $M_y(t)_a=M_y(t)$ the y-component, in the rotating frame, of the macroscopic magnetization of the sample. The dispersion signal is proportional to the x-component of the magnetization. The macroscopic magnetization $\vec{M}$ in the rotating frame of a system with no chemical reactions is assumed to satisfy the Bloch equations, which we write as

$$d\vec{M}/dt=\bar{\bar{L}}\vec{M}(t)+T_1^{-1}\vec{M}_{eq} \tag{1a}$$

where

$$\bar{\bar{L}} = \begin{bmatrix} -T_2^{-1} & \Delta\omega & 0 \\ -\Delta\omega & -T_2^{-1} & -\omega_1 \\ 0 & \omega_1 & -T_1^{-1} \end{bmatrix} \tag{1b}$$

and $\vec{M}_{eq}$ is the equilibrium value of the magnetization in the presence of a static magnetic field in the z-direction, i.e.

$$\vec{M}_{eq} = nP_{eq} \begin{bmatrix} 0 \\ 0 \\ 1 \end{bmatrix} \tag{2}$$

where n is the number of radicals in the sample and P_{eq} is the equilibrium polarization (population difference between the two electron spin eigenstates) in the presence of the static field. $P_{eq} \simeq 0.7\times 10^{-3}$ for X-band experiments and is almost independent of the nuclear state. The other quantities in Eq.(1) have their usual meanings.

The following generalization of the Bloch eq.'s to include chemical reactions are based on Ref.(5) and Ref.(7). When the number of radicals in the sample changes with time, we assume that the z-component of the magnetization at any instant of time relaxes towards $n(t)P_{eq}$. This is obtained by changing n in Eq.(2) to n(t), the instantaneous value of n. The disappearance of radicals will cause a lifetimebroadening and[8] this effect may be included in Eq.(1) by addition of the term

$$-T_c^{-1}(t)\vec{M}(t)=-K(t)\vec{M}(t)=\left[n(t)^{-1}dn/dt\right]_{disap}\vec{M}(t) \tag{3}$$

where $T_c(t)$ is the instantaneous lifetime of the radical defined as the reciprocal of the instantaneous decay constant $K(t)=[n^{-1}dn/dt]$ The subscript disap. is to remind one that only processes causing a disappearance of radicals give a contribution to K, i.e. radical source terms give no contribution to this term.

A radical source term will, however, give a contribution to the z-component of Eq.(1) equal to[5,7]

$$\text{initial pol. contribution}=k_o(t)_a P_a(I)=x_A^{-1} k_o(t) P_a(I) \qquad (4)$$

where $k_o(t)_a$ is the rate of formation of the radical in state a which is assumed to equal the total rate of formation of A divided by x_A, the number of nuclear states coupled to the unpaired electron, i.e. the number of hyperfine lines of A. $P_a(I)$ is the initial polarization of the radical in nuclear state a, i.e. the polarization of the radical at the instant of formation. There are at least three possible mechanisms that can give initial polarization. If the radicals are produced photolytically then they may have initial polarization due to the triplet mechanism (TM)[9]. The TM-polarization is independent of the nuclear state, i.e. all the hyperfine lines are equally polarized, and the polarization may be quite large $\lesssim$ o.48. The radical pair mechanism (RPM)[7,10] may give initial polarization whenever the radicals are produced in pairs. If the initial radical pair has an excess singlet or triplet character then no recombination reaction is necessary for the RPM to be operative. Polarization by the RPM is generally less than o.o1. It is interesting to note that if the TM is operative then the RPM is also operative although it may of course give a negligible contribution. The two mechanisms discussed so far can give initial polarization of primary radicals. Secondary radicals may also have initial polarization and this polarization is assumed to be equal to the polarization of the primary radical at the instant of reaction, i.e. one assumes that the spin state does not change in the reaction (primary radical → secondary radical). This transfer of polarization may be used to estimate the average lifetime of the primary radical, e.g. if we let a prime and a double prime denote primary and secondary radical respectively and assume that no polarization production occur during the lifetime T_c' of the primary radical then approximately

$$P(I)''=P_{eq}'+(P(I)'-P_{eq}')\exp(-T_c'/T_1'). \qquad (5)$$

This effect has been observed[6,8,1].

A spin selective reaction of radical A with radical B (B may be equal to A) will result in a polarization production due to the RPM[7,10]. This radical termination reaction is sometimes called recombination reaction and the resulting polarization is called recombination polarization. The recombination polarization gives a contribution to the z-component of Eq.(1) equal to

$$\text{recomb. pol. contribution}=\sum_b k_2 n_a(t) n_b(t) P_{ab}^{\infty} \qquad (6a)$$

where $k_2 n_a n_b$ gives the number of collisions per second of A and B radicals in the specified nuclear states and P^{∞}_{ab} is the polarization produced per collision. P^{∞}_{ab} depends both on the nuclear states as indicated and on the probability of reaction per collision F_{ab}. However, P^{∞}_{ab}/F_{ab} is independent of F_{ab}[7] and F_{ab} is approximately constant and equal to F for radicals produced independently of each other, and this is the situation discussed here. To a very good approximation Eq.(6a) can then be rewritten to

$$\text{recomb. pol. contribution} = k_2 F P_a^{*} n_A(t) x_A^{-1} n_B(t) \tag{6b}$$

where P_a^{*} is defined as

$$P_a^{*} = x_B^{-1} \sum_b P^{\infty}_{ab}/F \simeq x_B^{-1} \sum_b P^{\infty}_{ab}/F_{ab} \tag{7}$$

and we have assumed that

$$n_a(t) \simeq n_A(t)/x_A \text{ and } n_b(t) \simeq n_B(t)/x_B,$$

which can be justified by noting that $(n_a(t)-n_{a'}(t))/n_a(t)$ is just the CIDNP polarization which typically is less than 10^{-2}. Note that k_2F is the second order rate constant to appear in the rate equation for $n(t)=n_A(t)$, which for example may look like

$$dn/dt = k_o(t) - k_1 n(t) - k_2 F\, n_B(t)\, n(t). \tag{8}$$

Unreactive collisions between radicals will induce spin exchange and this will cause a polarization quenching, which can be included in the description[11]. We have chosen to neglect spin exchange in the present work and this may be justified by the observation that recombination polarization requires short lived radicals, i.e. few unreactive collisions.

It is convenient to redefine $\vec{M}$ as $x_A\vec{M}$ since the final general rate equation for $\vec{M}(t)$ can, by addition of Eq.'s (1), (3), (4), and (6b), be written as

$$d\vec{M}/dt = \bar{\bar{L}}\vec{M}(t) - T_c^{-1}(t)\vec{M}(t) + \vec{F}(t) \tag{9}$$

where

$$\vec{F}(t) = \begin{bmatrix} o \\ o \\ f_a(t) \end{bmatrix} \tag{10a}$$

and

$$f_a(t) = P_{eq} T_1^{-1} n(t) + P_a(I) k_o(t) + P_a^{*} k_2 F n_B(t) n(t) \tag{10b}$$

3. GENERAL SOLUTION FOR A SLOW RADICAL DECAY

We define slow radical decay as the situation for which $\max(T_c^{-1}(t)) \ll T_1^{-1}$, c.f. Eq.(3), and note that the term $T_c^{-1}(t)\vec{M}(t)$ in Eq.(9) can be neglected for this situation. For a first order reaction $T_c^{-1}=k_1$ and the condition is then $k_1T_1 \ll 1$. By Laplace transformation of Eq.(9) we obtain

$$\vec{M}(s) = \bar{\bar{L}}(s)^{-1} \left[\vec{M}(o)+\vec{F}(s)\right] \tag{11}$$

where

$$\bar{\bar{L}}(s) = -\bar{\bar{L}}+s\bar{\bar{1}} \tag{12}$$

and $\vec{M}(o)$ is the initial magnetization. By an inverse Laplace transformation of Eq.(11) we obtain the general, although formal, result

$$\vec{M}(t) = \bar{\bar{L}}^{-1}(t)\vec{M}(o)+\int_o^t\bar{\bar{L}}^{-1}(t')\vec{F}(t-t')dt' . \tag{13}$$

It is convenient to rename some of the matrix elements as only a few are to be used. These are

$$g_y(t) = L^{-1}(t)_{yz} = -L^{-1}(t)_{zy} \tag{14a}$$

and

$$g_z'(t) = L^{-1}(t)_{yy} . \tag{14b}$$

The following definition will also prove useful

$$G_y(t) = T_1^{-1} \int_o^t g_y(t')dt' . \tag{15}$$

The absorption signal can then be written as

$$M_y(t)=y_og_z'(t)+m_og_y(t)+\int_o^t g_y(t')f(t-t')dt' . \tag{16}$$

In order to simplify Eq.(16) further the initial polarization term in Eq.(1ob) is separated from the other terms of f(t) as nothing has been assumed for the time dependence of $k_o(t)$. The time dependence of the other terms of f(t) comes from n(t) and this time variation is much slower than $\exp(-t/T_1)$ due to our assumption of a slow radical decay. Since, as we shall see, $g_y(t)$ decays essentially as $\exp(-t/T_1)$ we may take n(t) to be constant in the integration range and approximate Eq.(16) to

$$\begin{aligned}M_y(t)=&y_og_z'(t)+m_og_y(t)+P_a(I)\int_o^t g_y(t-t')k_o(t')dt'\\&+\left[P_{eq}n(t)+P_a^*T_1k_2Fn_B(t)n(t)\right]G_y(t) .\end{aligned} \tag{17}$$

Equation (17) is the natural starting point for a description of the time dependence of the absorption signal for a system with a slow radical decay, i.e. the lifetime of the radical is much longer than the spin lattice time T_1. No further simplifications are possible before the experimental conditions are specified such that the functions g(t), and $k_o(t)$ are known.

4. THE FUNCTIONAL FORM OF $g_y(t)$ AND $g_z'(t)$

The functions $g_y(t)$ and $g_z'(t)$ are defined by Eq's(14) as elements of $\bar{\bar{L}}^{-1}(t)$, which in turn represents the general solution to the homogeneous part of the Bloch equations Eq.(1). Unfortunately it is not possible to write an expression for $L^{-1}(t)_{ij}$ in the general case but the structure of the time dependence is given by

$$L^{-1}(t)_{ij}=A\ \exp(at)+B\ \exp(bt)+C\ \exp(ct)$$

where a, b and c are the eigenvalues of $\bar{\bar{L}}$. In fact, the general solution may be obtained numerically in this form by application of the orthonormal transformation that diagonalizes $\bar{\bar{L}}$, i.e. by solving the eigenvalue problem for $\bar{\bar{L}}$. Another method, although in principle identical, is to invert the matrix $\bar{\bar{L}}(s)$ directly by use of the adjoint matrix and then perform an inverse Laplace transformation. This latter method was used by Torrey[13] who first considered time dependent solutions to the Bloch eq's. and gave solutions for $g_y(t)$ but not for $g_z'(t)$, which, however, may be obtained similarly. The following useful expressions are given without proof

On resonance, i.e. $\Delta\omega$=o

$$g_y(t) = \exp\ (-(T_1^{-1}+T_2^{-1})t/2)\sin([1-\omega_1^{-2}(T_1^{-1}-T_2^{-1})^2/4]^{\frac{1}{2}}\omega_1 t) \times [1-\omega_1^{-2}(T_1^{-1}-T_2^{-1})^2/4]^{-\frac{1}{2}} \tag{18}$$

On resonance and $T_1=T_2$

$$g_y(t) = \exp(-t/T_1)\ \sin\ (\omega_1 t) \tag{19a}$$

$$g_z'(t) = \exp(-t/T_1)\ \cos\ (\omega_1 t) \tag{19b}$$

$$G_y(t) = \omega_1 T_1/(\omega_1^2 T_1^2+1)[1-\exp(-t/T_1)(\cos\omega_1 t+\sin\omega_1 t/\omega_1 T_1)] \tag{19c}$$

Off resonance and $T_1=T_2$

$$g_y(t)=(1+\delta^2)^{-\frac{1}{2}}\exp(-t/T_1)\sin[\omega_1(1+\delta^2)^{\frac{1}{2}}t] \tag{20a}$$

$$G_y(t) = x/(1+x^2)\left[1-\exp(-t/T_1)\cos(xt/T_1)\right.$$

$$\left. - \exp(-t/T_1)\sin(xt/T_1)/x\right] \tag{20b}$$

where the dimensionless parameters x and δ are defined as

$$\delta = (\omega-\omega_o)/\omega_1 \tag{20c}$$

$$x = \omega_1 T_1 (1+\delta^2)^{\frac{1}{2}} = T_1\omega_1\left[1+(\omega-\omega_o)^2/\omega_1{}^2\right]^{\frac{1}{2}}. \tag{20d}$$

5. PULSE EXPERIMENTS. (Slow radical decay)

In a pulse experiment the radicals are produced by a pulsed source (light or electron beam), i.e. the source is applied for a short time T_p, and the absorption signal is recorded after the termination of the pulse. The time between consecutive pulses is sufficiently long for all radicals to disappear before application of the next pulse.

The initial magnetization, i.e. m_o and y_o, is zero if we define t=o as the instant of time when the pulse starts and $M_y(t)$ is given simply by Eq.(17) and this initial condition. The pulse time T_p is often much less than T_1 and the initial polarization production term in Eq.(17) can then be simplified for $T_p \ll T_1$ and $t > T_p$ since the integral

$$\int_o^t g_y(t-t')k_o(t')dt' \simeq n_o g_y(t) \tag{21}$$

becomes independent of the form of $k_o(t')$, n_o is defined as the total number of radical A formed during the pulse

$$n_o = \int_o^{T_p} k_o(t)dt. \tag{22}$$

Inserting Eq.(21) into Eq.(17) gives the following simple expression for $M_y(t)$, which is valid for $t > T_p$ and $T_p \ll T_1$

$$M_y(t)/(n_o P_{eq}) = g_y(t)P(I)/P_{eq} + G_y(t)n(t)/n_o\left[1+Vn_B(t)/n_B(o)\right] \tag{23}$$

The enhancement V is defined as

$$V = P^* T_1 K_2/P_{eq} = P^* T_1/(P_{eq}T_{c2}) \tag{24}$$

and is generally smaller than one since $T_1K_2 \ll 1$ for a slow radical decay. K_2 is the second order decay constant

$$K_2 = k_2F\ n_B(o) = {}'T_{c2}{}^{-1} \tag{25}$$

and T_{c2} the lifetime for a pure second order reaction.

One observes that the absorption signal goes from zero to $P(I)/P_{eq}$ in a time smaller than T_1, then decays to $1+V\simeq 1$ during a time of order T_1, and finally decays to zero with rate constant K_2. The initial rise of the signal depends on ω_1, e.g. by Eq.(19a) we see that the initial slope of $M_y(t)$ equals ω_1. The decay from $P(I)/P_{eq}$ to 1 is close to an exponential decay, $\exp(-t/T_1)$, for low observing power, i.e. $\omega_1 T_1 \ll 1$, and $T_1=T_2$, but a determination of T_1 will require an exact curve fitting.

The sine and cosine terms in Eq's.(18)-(2o) may result in wiggles damping out in time. These wiggles, with period $\simeq \omega_1^{-1}$, are the transient nutations of the magnetization, which are damped by relaxation. Therefore one requires $\omega_1 T_1 \gtrsim 1$ for this effect to be observable. These wiggles have been used for experimental determination of the microwawe field in the cavity[1,2,8,12].

If the duration of the pulse T_p is not short compared with T_1 then Eq.(23) is not strictly applicable and one must use Eq.(17). However, other simplifications arise if one can assume that the pulse has a square wave form

$$k_o(t) = \begin{cases} k_o & \text{for } o \le t \le T_p \\ o & \text{otherwise} \end{cases} \tag{26}$$

This assumption requires that the initial rise and final decay of the pulse is fast compared to T_1 and that the top level is relatively constant. With $k_o(t)$ given by Eq.(26) one can simplify Eq.(17) to an expression identical to Eq.(23) but with the first term replaced by

$$k_o T_1 P(I)/(n_o P_{eq})\ G_y(t) \qquad \text{for } t<T_p \tag{27a}$$

or by using $n_o=k_o T_p$, c.f. Eq.(22),

$$T_1 P(I)/(T_p P_{eq})\ G_y(t) \qquad \text{for } t<T_p \tag{27b}$$

and

$$T_1 P(I)/(T_p P_{eq})(G_y(t)-G_y(t-T_p)) \qquad \text{for } t \ge T_p \tag{27c}$$

6. TIME RESOLVED STEADY STATE EXPERIMENTS

In a time resolved steady state experiments[3,6,14] the light source is modulated periodically[3] and the periods are sufficiently long for a steady state to be reached in any period.

We assume for simplicity that the modulation has form of a square wave

$$k_o(t) = \begin{cases} k_o & \text{in the light period} \\ \varepsilon^2 k_o & \text{in the dark period,} \end{cases} \tag{28}$$

i.e. the intensity of the lamp alternates between two constant levels called light and dark. The relative intensity of the two levels are given by the adjustable parameter ε^2 ($o \leq \varepsilon < 1$). We assume again slow radical decay and consequently the absorption signal is given by Eq.(17). The begining of a period is taken as t=o. In general m_o and y_o will be different from zero since they equal the steady state values,of M_z and M_y, in the preceding period. The steady state values of M_z and M_y, denoted by ∞, are for $\Delta\omega$=o related by Eq.(1)

$$M_y(\infty) = -\omega_1 T_2 \, M_z(\infty) \tag{29}$$

The steady state value of M_y may also be found as the $t \to \infty$ limit of Eq.(17)

$$M_y(\infty) = \left[P_{eq} T_1^{-1} \, n(\infty) + P_a^* \, k_2 F n_B(\infty) n(\infty) + k_o T_1 P(I)\right] G_y(\infty) \tag{3o}$$

where we have used Eq.(27a) for obtaining the last term.

By the use of Eq.'s (27a) and (29) we may write Eq.(17) as

$$M_y(t) = M_y^o \left[g_z'(t) - g_y(t)/(\omega_1 T_2)\right] + T_1 f(t) G_y(t)/G_y(\infty) \tag{31}$$

where $M_y(t)$ has been divided by $G_y(\infty)$, M_y^o is the steady state value of M_y (divided by $G_y(\infty)$) in the preceding period,and f(t) is defined in Eq.(lob). This equation is independent of the kinetics of the system.

If we assume that $n_B(t)=n(t)$ and that n(t) follows the rate equation

$$dn/dt = k_o(t) - k_2 F n(t)^2 \tag{32}$$

then the steady state value n_o and the decay constant K_2 in the light period are given by

$$k_2 F n_o^2 = k_o \tag{33a}$$

$$K_2 = n_o k_2 F \tag{33b}$$

By application of Eq.'s (3o), (31) and (33) and division by $n_o P_{eq}$, we obtain for the light period

$$M_y^o = \varepsilon^2 V(I)+\varepsilon+\varepsilon^2 V \quad (34a)$$

$$T_1 f(t) = V(I)+\ell(t)+V(R)\ell(t)^2 \quad (34b)$$

and for the dark period

$$M_y^o = V(I) + 1 + V(R) \quad (34c)$$

$$T_1 f(t) = \varepsilon^2 V(I)+d(t)+V(R)d(t)^2. \quad (34d)$$

The enhancement factors V(I) and V(R) are defined as

$$V(I) = P(I)T_1 K_2/P_{eq} \quad (35a)$$

$$V(R) = P^* T_1 K_2/P_{eq} \quad (35b)$$

and $\ell(t)$ and d(t) equal $n(t)/n_o$ in the light and dark period respectively.

The time dependence depends strongly on the magnitude of V(I)[5]. Systems having both V(I) and V(R) different from zero have been observed[3,6].

In this kind of experiment the rise and decay time of the light is usually longer than T_1 and the square wave approximation, Eq.(28), is not strictly applicable. However, for such a case the signal is simply proportional to f(t), Eq.(1ob), with the actual time dependence of the light source included in $k_o(t)$. If the response time of the spectrometer is longer than the rise and decay time of the light then the square wave approximation can be used[3,6].

7. FAST RADICAL DECAY

The T_c-term of Eq.(9) becomes important when the radical decay is comparable to or faster than the spin lattice relaxation. This term is unpleasant due to the time dependence of T_c.

We shall only consider the case where the duration of the pulse $T_p \ll T_c$ in which case the initial production term (the $k_o(t)$-term) may be omitted from f(t) and instead placed in the initial condition, c.f. Eq.(23),

$$m_o = P(I)n_o \quad (36)$$

where n_o is defined in Eq.(22). By application of the transformation

$$M'_y(t) = M_y(t)/n(t) \tag{37a}$$

one finds that M'_y satisfies a rate equation identical to Eq.(9) but without the T_c-term, and with f(t) replaced by

$$f'(t) = P_{eq} T_1^{-1} + k_2 F P_a^* n_B(t). \tag{37b}$$

The general solution to this rate equation is obtained as before with the result

$$M'_y(t) = M_y(t)/n(t) = P(I)g_y(t)$$
$$+ P_{eq} G_y(t) + \int_o^t g_y(t')k_2 F P_a^* n_B(t-t')dt'. \tag{38}$$

One should note that when solving the rate equation for n(t) the production term should be treated as an initial condition.

8. EFFECT OF SPECTROMETER RESPONSE

Assuming linear response the output signal I(t) from the spectrometer is related to the input signal $M_y(t)$ by the relation[3,6]

$$I(t) = \int_{-\infty}^t M_y(t-t')r(t')dt' \tag{39}$$

where r(t), the spectrometer response function, is the response to a δ-function input. If r(t) decays to zero during a time t_s and $T_1 \ll t_s \ll T_c$ then the equations for a slow radical decay can be used with $g_y(t)$ replaced by r(t), i.e. all information about T_1 is lost. If $t_s \ll T_1$ then spectrometer response need not be included. In other cases one must perform the convolution, e.g. by a numerical integration.

9. FIELD DERIVATIVE SIGNALS AND DISPERSION SIGNALS

The time dependence of the field derivative of the absorption signal is obtained by differentiating the expression for the absorption signal with respect to ω (the angular frequency of the rf. field). Only the functions g(t) defined in Eq.(14) depend on ω. Consequently all the expressions derived above for the absorption signal can be used for the derivatived absorption signal simply by replacing all functions g(t) by their derivative with respect to ω, evaluated at the frequency of observation. This procedure requires that the ω-dependence of the function $g_y(t)$ is known. Unfortunately, only for very special cases as $T_1=T_2$, Eq.(2o), and strong observing power[5] can this dependence be given analytically. For other cases one must use numerical techniques, cf. Sect. 4. The on-resonance value of the field derivative of the dispersion signal is different from zero and is related to the ($\Delta\omega$=o) - absorption signal by the following relation[15] which follows from Eq.(1)

$$(\partial M_x(t)/\partial\omega)_{\Delta\omega=o} = -\exp(-t/T_2)\int_o^t dtM_y(t)\exp(t/T_2) \quad . \qquad (4o)$$

1o. COMMENTS

The general feature of the presented solutions is that the time dependence due to the kinetics is contained in n(t) and is completely separated from the relaxtion behaviour which is included in the function g(t). The general solutions, e.g. Eqs. (17), (23) and (38), are independent of the actual form of n(t) and g(t). The problem is therefore essentially reduced to that of solving the appropiate kinetic rate equation. The time dependence of the radical producing source is contained in the term $k_o(t)P(I)$. We note that this term may also describe the production of secondary radicals by interpreting the term as the rate of production of z-magnetization at time t, associated with the formation of the secondary radical.

REFERENCES

1) See Chapter VI by K.A. McLauchlan and references therein.
2) See Chapter VII by R.W. Fessenden and references therein.
3) See Chapter XII by L.T. Muus and references therein.
4) A.D. Trifunac, K.W. Johnson, B.E. Clifft and R.H. Lowers, Chem.Phys.Lett., 35, 566 (1975).
5) J.B. Pedersen, J.Chem.Phys., 59, 2656 (1973).
6) J.B. Pedersen, C.E.M. Hansen, H. Parbo and L.T. Muus, J.Chem.Phys., 63, 2398 (1975).
7) J.B. Pedersen and J.H. Freed, J.Chem.Phys., 58, 2746 (1973).
8) N.C. Verma and R.W. Fessenden, J.Chem.Phys., 65, 2139 (1976).
9) See Chapter XI by P.W. Atkins and references therein. J.B. Pedersen and J.H. Freed, J.Chem.Phys., 62, 17o6 (1975).
1o) See Chapter V by F.J. Adrian and references therein.
11) J.B. Pedersen and J.H. Freed, J.Chem.Phys., 59, 2869 (1973).
12) P.W.Atkins, A.J. Dobbs and K.A. McLauchlan, Chem.Phys.Lett., 25, 1o5 (1974).
13) H.C. Torrey, Phys. Rev., 76, 1o59, (1949).
14) R. Livingston and H. Zeldes, J.Chem.Phys., 59, 4891 (1973).
15) P.W. Atkins, K.A. McLauchlan and P.W. Percival, Mol.Phys., 25 281 (1973).

CHAPTER X

PHOTOCHEMICAL PROCESSES

P.W. Atkins

Physical Chemistry Laboratory, South Parks Road,
Oxford, England.

ABSTRACT. A simple account of the description of some basic photochemical processes is presented. The contents are (1) a general outline of nonradiative processes, (2) the relation of irreversible decay to the strength of the perturbation and the density of final states using the Bixon-Jortner model, (3) an outline of the Born-Oppenheimer approximation and its role in internal conversion, and (4) an account of the group theoretical analysis of intersystem crossing processes.

1. GENERAL OUTLINE

The detailed history of a photochemical process can be very complicated, but the main features can be discerned, and modern work has made a brave attempt to describe them theoretically. In this Chapter we examine these features from a simple point of view, and lay the foundations for more detailed work on aspects of them.

A photoexcited molecule has open to it a variety of reactive channels [1,2]. One is radiative decay, where the molecule ejects a photon into the electromagnetic field. If this radiative decay occurs at the frequency of the exciting illumination, as it may in rarefied media where the radiative lifetime is significantly shorter than the time between molecular collisions or some other relaxation process, the process is known as resonance fluorescence. Many molecules decay non-radiatively into some lower state before fluorescing, and the actual fluorescent state is normally that predicted by Kasha's rule, that the emitting level of a given multiplicity is the lowest of that multiplicity. There are

L. T. Muus et al. (eds.), Chemically Induced Magnetic Polarization, 181-190.

exceptions, azulene and ferrocene for example, and the rule appears to be incomplete from the point of view of group theory; but it is a very helpful rule of thumb. The radiative lifetime, τ^*, can be identified as the inverse of the Einstein coefficient of spontaneous emission, A, and related through that to the transition dipole moment between the excited and the ground electronic states:

$$P_e(t) = P_e(0)\exp(-t/\tau^*) \qquad (1.1)$$

$$\tau^* = 1/A, \quad A = (8\pi^2/3\varepsilon_o c^3 h)\nu^3|\mu_{eg}|^2 \qquad (1.2)$$

The processes of more interest to us are the non-radiative changes of state, and we shall pay some attention to their irreversible nature. Two classes of non-radiative process are of prime importance. First there are the internal conversions, the name specifically implying a change of state without change of multiplicity (e.g. singlet-singlet interconversions). Then there are the intersystem crossings, this name being reserved for state conversion involving a change of multiplicity (e.g. singlet-triplet interconversions). The intersystem crossings (ISC) play a central role in spin polarization processes because they are a source of spin angular momentum.

2. INTERCONVERSION AND IRREVERSIBILITY.

Elementary time-dependent perturbation theory can be used to answer the following basic question. If two states ψ_1 and ψ_2 are separated in energy by $\hbar\omega$, and a constant perturbation of strength $\hbar V$ mixes them, what is the probability of finding the system in ψ_2 if at t = 0 it was known to be in ψ_1? The exact solution is

$$P_2(t) = \{4V^2/(\omega^2 + 4V^2)\}\sin^2\{\tfrac{1}{2}(\omega^2 + 4V^2)^{\frac{1}{2}}t\}. \qquad (2.1)$$

The features of this expression conform entirely to what intuition leads one to expect: if the perturbation V is absent P_2 is zero for all time; as the separation ω increases, the maximum value of the probability decreases; the strongest mixing occurs when the levels are degenerate (ω=0, P_{max}=1). Moreover, the probability oscillates between the two states with a frequency $(\omega^2+4V^2)^{\frac{1}{2}}$. The last point is crucial: in this two-level system there is no irreversibility. Such periodic behaviour can be expected to survive even when there are several levels interacting, and it was this behaviour that led to early speculations that irreversible internal conversions and intersystem crossings could not occur in isolated molecules and that nonradiative behaviour is governed by environmental influences.

Irreversible behaviour, however, does emerge when a state is in perturbative contact with a continuum, and virtually irreversible behaviour emerges when the acceptor states, although not truly continuous, have a high density. Quasi-irreversibility is turned into true irreversibility if the state remains in the quasi-continuum long enough for a further perturbation to put it into contact with a true continuum (e.g. a mobile environment).

The emergence of irreversibility as a consequence of a high density of acceptor states can be demonstrated by the following argument [3].

Consider some state ψ populated by some photochemical process. Although ψ may be an eigenstate of a model Hamiltonian, it is not an eigenstate of the true molecular Hamiltonian, and there exists a perturbation, $H' = H(\text{true}) - H(\text{model})$, which can mix ψ with other states. Suppose there is an array of states ϕ_n with an energy similar to that of ψ and which H' can couple into ψ. For simplicity it is supposed that the states ϕ_n form a uniform ladder, Fig. 1, with spacing ε. It is also supposed that H' has equal matrix elements with all ϕ_n, and we set $(\phi_n, H'\psi) \equiv \hbar V$ for all n.

Let the combination

$$\Psi = a\psi + \sum_n b_n\phi_n \tag{2.2}$$

be an eigenstate of the true Hamiltonian, and its energy be $\mathcal{E}$. Substitution of this expression into $H\Psi = \mathcal{E}\Psi$ and making use of the relations $H(\text{model})\psi = E\psi$, $H(\text{model})\phi_n = E_n\phi_n$, gives

$$aE\psi + aH'\psi + \sum_n b_nE_n\phi_n + \sum_n b_nH'\phi_n = \mathcal{E}a\psi + \mathcal{E}\sum_n b_n\phi_n. \tag{2.3}$$

Assuming orthogonality of ψ and all ϕ_n, asserting that H' has no diagonal elements, and invoking the equivalence of all non-vanishing matrix elements of H', leads to the coupled equations

$$\begin{aligned} Va + (E_n - \mathcal{E})b_n &= 0 \\ (E - \mathcal{E})a + V\sum_n b_n &= 0 \end{aligned} \tag{2.4}$$

From this it follows that

$$b_n = -Va/(E_n - \mathcal{E}) \tag{2.5}$$

and
$$\left[(E - \mathcal{E}) - V^2\sum_n\left\{\frac{1}{E_n - \mathcal{E}}\right\}\right]a = 0. \tag{2.6}$$

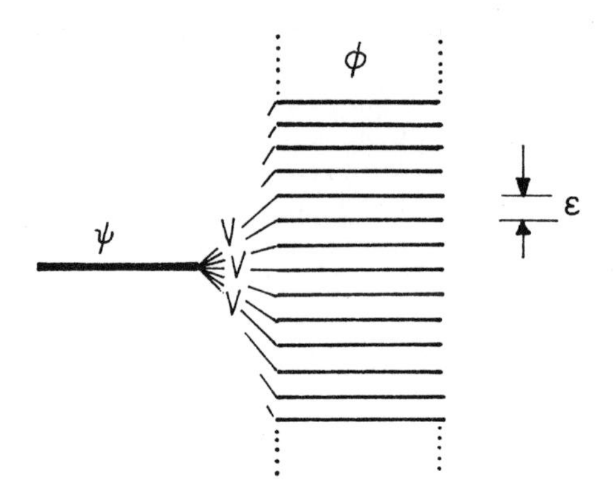

Fig. 1. Model for a nonradiative transition.

Since the quasi-continuum eigenvalues form a uniform ladder they may be expressed as

$$\mathcal{E} - E_n = \varepsilon\gamma - n\varepsilon \tag{2.7}$$

where $\varepsilon\gamma$ is the separation of energies of the true state ψ and the state assigned n=0 in the receptor array. Since

$$\sum_{n=-\infty}^{\infty}\left(\frac{1}{\gamma-n}\right) = -\pi\cot(\pi\gamma) \tag{2.8}$$

and the density of states is $\rho = 1/\varepsilon$, eqn (2.6) becomes

$$E - \mathcal{E} - \pi\rho V^2 \cot\pi\gamma = 0 \tag{2.9}$$

as an equation for the determination of $\mathcal{E}$.

Our concern, however, is with the probability of finding the system in the quasicontinuum. Since Ψ is normalized to unity,

$$a^2 + \sum_n b_n^2 = 1 . \tag{2.10}$$

Expressing b_n in terms of a using eqn (2.5) gives

$$a^2 + a^2V^2\sum_n(E_n - \mathcal{E}_n)^{-2} = 1 . \tag{2.11}$$

Since $$\sum_{n=-\infty}^{\infty} (\gamma-n)^{-2} = \pi^2 \operatorname{cosec}^2 \pi\gamma \tag{2.12}$$

it follows that

$$\begin{aligned} a^2 &= \{1 + \pi^2\rho^2V^2 \operatorname{cosec}^2\pi\gamma\}^{-1} \\ &= \{1 + \pi^2\rho^2V^2 + \pi^2\rho^2V^2\cot^2\pi\gamma\}^{-1} \\ &= \{1 + \pi^2\rho^2V^2 + \pi^2\rho^2V^2[(E-\xi)/\pi\rho V^2]^2\}^{-1} \\ &= V^2/\{(E-\xi)^2 + V^2 + (\pi V^2\rho)^2\}. \end{aligned} \tag{2.13}$$

If the density of states is very large (in the sense $\pi V\rho \gg 1$), the last term in the denominator is much larger than the second, and so the probability of the system being in ψ is given by a Lorentzian function. The width of a transition can be identified with the square root of the $V^2 + (\pi V^2\rho)^2 \sim (\pi V^2\rho)^2$ term in the denominator, and hence the rate of the conversion in a time-dependent formulation is

$$1/\tau \sim 2V^2\rho\pi/\hbar. \tag{2.14}$$

This demonstrates that when there is a dense manifold with which the initial state is in contact, there is a virtually irreversible conversion into the dense set with a time dependence governed both by the strength of the perturbation and the density of receptor states.

3. INTERNAL CONVERSION

In this section we consider briefly one of the principal driving forces for internal conversion. This is the perturbation that arises from a failure of the Born-Oppenheimer approximation. In order to find the perturbation we shall run through the sequence of converting an accurate Hamiltonian into a Born-Oppenheimer Hamiltonian, and then identifying the discarded term as the perturbation. The physical basis of the approximation, it will be recalled, is the neglect of the mobility of the nuclear framework relative to the mobility of the electrons.

The true Hamiltonian for a molecule is

$$H = T(q) + V(q,q) + V(q,Q) + V(Q,Q) + T(Q), \tag{3.1}$$

where q and Q signify collective electronic and nuclear coordinates respectively, T denotes kinetic energy, and V denotes potential energy. The electronic part of the Hamiltonian, $T(q)+V(q,q)+V(q,Q)$, where the Qs are regarded as parameters, has eigenfunctions $\phi_n(q,Q)$, which still depend parametrically on the Q. The true eigenfunctions of the system are linear combinations of the complete set $\{\phi_n\}$, with coefficients depending on the nuclear configuration Q:

$$\Psi_i = \sum_n \chi_n^{(i)}(Q)\phi_n(q,Q) \tag{3.2}$$

The coefficients are obtained by solving $H\Psi_i = E_i\Psi_i$ using the fact that $\{H-V(Q,Q) - T(Q)\}\phi_n = E_n(Q)\phi_n$, which together imply that

$$\sum_n\{E_n(Q) + V(Q,Q) - E_i\}\phi_n(q,Q)\chi_n^{(i)}(Q) + \sum_n T(Q)\phi_n(q,Q)\chi_n^{(i)}(Q)=0 \tag{3.3}$$

Since $T(Q)\phi_n(q,Q)\chi_n^{(i)}(Q) = -(\hbar^2/2M)(\partial^2/\partial Q^2)\phi_n\chi_n^{(i)}$

$$= \chi_n^{(i)}T(Q)\phi_n + \phi_n T(+)\chi_n^{(i)} - (\hbar^2/M)\left(\frac{\partial\phi_n}{\partial Q}\right)\left(\frac{\partial\chi_n^{(i)}}{\partial Q}\right) \tag{3.4}$$

it follows from eqn (3.3), by multiplying by ϕ_k^* and integrating over q, that

$$\{E_k(Q)+V(Q,Q)-E_i\}\chi_k^{(i)}(Q)+\sum_n\int dq\phi_k^*(q,Q)T(Q)\phi_n(q,Q)\chi_n^{(i)}(Q) = 0$$

$$\{E_k(Q)+V(Q,Q)-E_i\}\chi_k^{(i)}(Q)+\sum_n\chi_n^{(i)}\int dq\phi_k^*(q,Q)T(Q)\phi_n(q,Q)$$
$$+ T(Q)\chi_k^{(i)}(Q) - (\hbar^2/M)\sum_n\int dq\phi_k^*(q,Q)(\partial\phi_n/\partial Q)(\partial\chi_n^{(i)}/\partial Q)= 0$$

$$\{E_k(Q)+V(Q,Q)-E_i + T_k(Q) + T(Q)\}\chi_k^{(i)}(Q)$$
$$+ \sum_n{}'\{T_{kn}(Q) - (\hbar^2/M)\int dq\phi_k^*(q,Q)(\partial\phi_n/\partial Q)(\partial/\partial Q)\}\chi_n^{(i)}(Q)= 0. \tag{3.5}$$

The <u>adiabatic Born-Oppenheimer approximation</u> consists of neglecting all off-diagonal matrix elements. Of the last equation the only term to survive is then

$$\{T(Q) + [E_i - E_k(Q) + T_{kk}(Q) + V(Q,Q)]\}\chi_k^{(i)}(Q) = 0, \tag{3.6}$$

which can be interpreted as the Schrödinger equation for the motion of the nuclei in some effective molecular potential. The approximation has therefore achieved separation of electronic and nuclear motion.

The perturbation responsible for internal conversion can now be identified with the off-diagonal parts of the full expression. In particular, by multiplying by $\chi_m^{(j)}(Q)^*$ and integrating over Q, the off-diagonal matrix elements are of the form

$$H^{(1)}_{jm,ik} = \sum_n{}' \int dQ\chi_m^{(j)*} T_{kn}(Q)\chi_n^{(i)} - (\hbar^2/M)\sum_n{}' \int dQdq\chi_m^{(j)*}\phi_k^* \times (\partial\phi_n/\partial Q)(\partial\chi_n^{(i)}/\partial Q) \quad (3.7)$$

These matrix elements give rise to a mixing between the vibronic states $\phi_k\chi_k^{(i)}$ and $\phi_m\chi_m^{(j)}$. Now since

$$[T(q)+V(q,q)+V(q,Q)]\phi_n(q,Q) = E_n(Q)\phi_n(q,Q) \quad (3.8)$$

it is quite easy to show (by multiplying by $\phi_k^*\partial/\partial Q$, and integrating over q at constant Q) that

$$\int dq\phi_k^*(\partial\phi_n/\partial Q) = \left(\frac{\int dq\phi_k^*[\partial V(q,Q)/\partial Q]\phi_n}{E_k(Q)-E_n(Q)}\right). \quad (3.9)$$

Hence the second term in $H^{(1)}_{jm,ik}$ is large for vibronic states that lie close together in energy. Since the first term can be expressed in terms of $(\partial/\partial Q)\{C.S\}(\partial/\partial Q)$, where {C.S.} indicates a complete set, a similar remark applies to that too. Hence we find strongest couplings between energetically similar states.

The other quantity that is necessary to the discussion is some knowledge of the density of receptor states. A variety of standard expressions are available. There is the Whitten-Rabinovitch semi-classical parametrized equation [4]

$$\rho(E) = \left\{\frac{\lambda(E + \alpha E_o)^{n-1}}{(n-1)!\ \langle\nu\rangle^n}\right\}\left\{1 - \left(\frac{n-1}{n}\right)\beta f\right\} \quad (3.10)$$

where E_o is the zero-point energy, n the number of normal modes, $\langle\nu\rangle$ the mean vibrational frequency of the modes, and

$$\beta = \langle\nu^2\rangle/\langle\nu\rangle^2 \qquad \lambda^{-1} = \prod_i \nu_i/\langle\nu\rangle$$

$$\alpha = 1 - \left(\frac{n-1}{n}\right)\beta g \qquad \eta = E/E_o$$

$$g = (5.00\eta + 2.73\eta^{\frac{1}{2}} + 3.51)^{-1}$$

$$f = dg/d\eta = -(5.00 + 1.37\ \eta^{-\frac{1}{2}})g^2.$$

A more complex expression applies in the case of rovibronic states. Although this expression is quite good at high excitations, the expression due to Haarhoff [5] is normally more accurate for lower energies:

$$\rho(E) = \left\{\frac{(2/\pi n)^{\frac{1}{2}}(1 - \frac{1}{12n})\lambda}{h\langle\nu\rangle(1 + \eta)}\right\}\left\{(1+\tfrac{1}{2}\eta)(+\frac{2}{\eta})^{\frac{1}{2}\eta}\right\}^n\left\{1 - \left(\frac{1}{1+\eta}\right)^2\right\}^{\gamma(n)}$$

where $\gamma(n) = \{(n-1)(n+2)\beta - n^2\}/6n$.

The density of vibronic states is a rapidly increasing function of the energy in molecules of moderate size. For example, in anthracene the $^1B_{2u} \rightarrow {}^3B_{2u}$ transition releases 12000 cm^{-1}, and this corresponds to $\rho \sim 10^{10}$ vibrational levels per wavenumber in the $^3B_{2u}$ receptor state. Even in naphthalene the $^1B_{2u} \rightarrow {}^1B_{1u}$ internal conversion releases 3400 cm^{-1}, corresponding to $\rho \sim 10^3$ per wavenumber in $^1B_{1u}$. Such densities easily account for the irreversible nature of the nonradiative transitions, and estimates of internal conversion rates of the order of 10^{12} s^{-1} can be made.

4. INTERSYSTEM CROSSING

The perturbation responsible for intersystem crossing is the spin orbit coupling. Why this is successful can be illustrated by considering a two-electron system in which the coupling has the form

$$H_{so} = \zeta\vec{\ell}_1\cdot\vec{s}_1 + \zeta\vec{\ell}_2\cdot\vec{s}_2$$

$$\equiv \tfrac{1}{2}\zeta(\vec{\ell}_1 + \vec{\ell}_2)\cdot(\vec{s}_1 + \vec{s}_2) + \tfrac{1}{2}\zeta(\vec{\ell}_1 - \vec{\ell}_2)\cdot(\vec{s}_1 - \vec{s}_2). \quad (4.1)$$

The second part of this hamiltonian is antisymmetric in the spins, and so it can have non-vanishing matrix elements between spin states of opposite parity; in other words it can generate a spin symmetric (triplet) state out of a spin antisymmetric (singlet) two-electron state. The orbital operator in the same expression, $\vec{\ell}_1 - \vec{\ell}_2$, is also antisymmetric, and so when $(\vec{\ell}_1 - \vec{\ell}_2)\cdot(\vec{s}_1 - \vec{s}_2)$ operates on an overall antisymmetric spin-orbital state it generates an overall antisymmetric spin-orbital state, but the parity of both space and spin components has been reversed.

Another way of looking at the role of the spin-orbit coupling is as a redistributor of angular momentum. In order for a singlet to turn into a triplet, a spin angular momentum has to be generated, and that requires the operation of a torque. Such a torque is available from the orbital angular momentum, but some of its angular momentum must be destroyed in the process so that the total angular momentum of the system, which is isolated, remains constant. This can be demonstrated formally by noting that J^2, where $\vec{J} = \vec{L}+\vec{S}$, $\vec{L} = \vec{\ell}_1+\vec{\ell}_2$, and $\vec{s} = \vec{s}_1+\vec{s}_2$, commutes with H_{so} and so is a constant of the notion.

The appreciation that any generation of spin angular momentum must be accompanied by a destruction of orbital angular momentum gives a way of understanding why, in molecules, the rate of inter-system crossing is anisotropic, and different triplet sub-states are populated at different rates. Basically this is because in order to generate spin angular momentum oriented in a particular direction in the molecular frame, it is necessary to generate orbital angular momentum of equal magnitude but orientated in the opposite direction, and some orientations in the molecule may permit an orbital circulation more freely than others. In more precise terms, the effectiveness of the ISC perturbation depends both on its matrix elements and on the accessibility of orbital states, and different components are effective to different extents [2].

The anisotropy can be expressed in group theoretical terms as follows. The perturbation $\zeta\vec{\ell}\cdot\vec{s}$, being a scalar in joint orbital and spin spaces, transforms as the totally symmetric irreducible representation of the molecular point group. Write this $\Gamma_{so}(A_1)$. The orbital state of the excited singlet has symmetry $\Gamma_o(S^*)$, and its spin part spans $\Gamma_s(A_1)$ because there is no spin orbital momentum. It follows that the overall symmetry of the combined spin and orbital states of the excited singlet is $\Gamma_o(S^*)\times\Gamma_s(A_1) = \Gamma_{so}(S^*)$. The orbital state of some nearby triplet is $\Gamma_o(T)$, and to construct its overall symmetry we have to know the symmetries of the three spin states: call these $\Gamma_s(T_{xx})$, $\Gamma_s(T_{yy})$, and $\Gamma_s(T_{zz})$. (These spin states transform as rotations; we give examples below.) Then the combined spin-orbital symmetries of the three states of the triplet are $\Gamma_o(T) \times \Gamma_s(T_{qq})$, $q = x,y,z$. The ISC depends upon nonvanishing matrix elements of the form $\langle \text{triplet},T_{qq}|H_{so}|\text{singlet}\rangle$, and by familiar arguments of group theory, this is vanishing unless the direct product of the irreducible representations of the bra, the operator, and the ket contain the totally symmetric representation. Therefore we form $\{\Gamma_o(T) \times \Gamma_s(T_{qq})\} \times \{\Gamma_{so}(A_1)\} \times \{\Gamma_{so}(S)\}$ and inspect whether its decomposition contains $\Gamma_{so}(A_1)$. Since $\Gamma_{so}(A_1) \times \Gamma_{so}(S) = \Gamma_{so}(S)$, this essentially involves deciding whether $\Gamma_o(T) \times \Gamma_s(T_{qq})$ contains $\Gamma_{so}(S)$: clearly this depends on the substrate T_{qq}, and hence we arrive at the group theoretical basis of the ISC anisotropy.

As an example, consider a carbonyl molecule R_2CO, where the local symmetry of the chromophore is C_{2v}. If the initial transition is $\pi^* \leftarrow n$, $\Gamma_o(S^*) = A_2$. In the triplet T_{xx}, T_{yy}, T_{zz} transform as $\Gamma_s = B_2$, B_1, and A_2 respectively (rotations around the x, y, z axes transform as B_2, B_1, A_2 respectively). Suppose the triplet is an $\pi^* - \pi$ orbital state; this has orbital symmetry $\Gamma_o(T) = A_1$. The possible spin-orbital states of the triplet are $\Gamma_o(T) \times \Gamma_s(T_{qq}) = A_1 \times \{B_2,B_1,A_2\} = B_2,B_1,A_2$. Of these, the T_{zz} state of the triplet has a spin-orbital symmetry (A_2) the same as the spin-orbital symmetry of the excited singlet, and so spin-orbit coupling can populate T_{zz} but not T_{xx} and T_{yy}. The underlying physical picture should be clear: T_{zz} has a spin momentum along z, the C-O axis; the switch from $\pi^*- n$ to $\pi^*-\pi$ is a twisting motion around that axis, and is the source of the torque that generates T_{zz}.

The crucial point to emphasize is that the selection rules are devised, and are operative, in the molecular frame, and so anisotropy of population appears in the molecular frame, which might be rotating. This remark will play a significant role when we consider generation of spin polarization by the triplet mechanism, Chapter .

REFERENCES

1. J.G. Calvert and J.N. Pitts, Photochemistry, Wiley, New York (1966).
2. S.P. McGlynn, T. Azumi, and M. Kinoshita, The triplet state, Prentice-Hall, Englewood Cliffs (1969).
3. M. Bixon and J. Jortner, J. Chem. Phys., 50, 3284 (1969).
4. P.J. Robinson and K.A. Holbrook, Unimolecular reactions, Wiley, New York (1972).
5. P.C. Haarhoff, Molec. Phys., 7, 101 (1963).

CHAPTER XI

THE TRIPLET MECHANISM

P.W. Atkins

Physical Chemistry Laboratory, South Parks Road,
Oxford, England.

ABSTRACT. The triplet mechanism of electron spin polarization is described first qualitatively, and then with full dynamical freedom allowing for rotation chemical quenching, and initiation with polarized light. The contents are (1) the underlying physical model, including an estimate of the magnitudes and an argument based on a simple kinetic scheme, (2) the full dynamical problem, treating relaxation, rotation, and reaction consistently.

1. THE UNDERLYING PHYSICAL MODEL

This article is based on the work of Atkins and McLauchlan [1], who treated the generation of electron spin polarization in a static array of triplets, of Wong, Hutchinson, and Wan [2], who introduced the essential step of competition with chemical quenching, and of Atkins and Evans [3], who gave the first account of the full dynamical problem. For reviews see Atkins and Evans [4] and Freed and Pedersen [5].

In Chapter X the point was established that intersystem crossing (ISC) from an excited singlet into a triplet took place in accord with selection rules operating in the molecular frame. If these rules are sufficiently discriminating, the resultant triplet will be formed predominantly in one of its substates, with the spin lying along one of the principal axes of the molecule, Fig. 1. The molecule, however, is tumbling in the solution, and so polarization in the molecular frame need not give rise to polarization in the laboratory frame and appear,

L. T. Muus et al. (eds.), Chemically Induced Magnetic Polarization, 191-203.

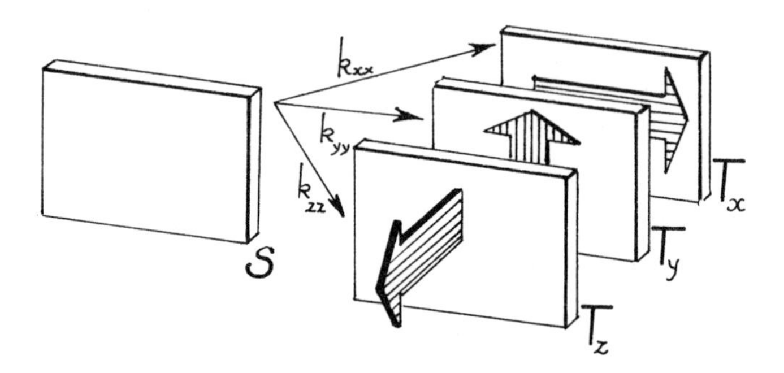

Fig. 1. The generation of magnetization in the molecular frame.

when doublets are formed from the polarized triplets, as a spectrum in emission or enhanced absorption. The first point to establish, therefore, is why the polarization can appear in the static frame even though the selection rules govern the behaviour in the rotating molecular frame.

The answer can be found in the effect of the magnetic field on the three levels of the triplet. This can be established by thinking about the case when the applied field is so strong that it strips the spin vector away from the molecular frame so that the best description of the state of the molecule is in terms of the quantum number M_S, the projection of the spin on to the direction of the applied field. The first point to note is that the three states corresponding to $M_S = 0, \pm 1$ can still be expressed as a mixture of the three states corresponding to quantization in the molecular frame, but the mixtures depend on the orientation of the molecule. For example, when the molecular x-axis is aligned parallel to the applied field, the $M_S = +1$ state is identical to the T_x state, but as rotation turns the molecule so that its y-axis is parallel to the field, M_S +1 becomes pure T_y. This remixing of the states in order to express a single M_S state goes on continuously as the molecule rotates over all its possible orientations.

Now we come to the crucial point. If in the absence of the field T_z lies higher in energy than T_x and T_y, the $M_S = +1$ state (which is the highest energy state in the presence of the field) will be composed of a mixture which, on average, is predominantly T_z. This is because T_z lies closest in energy to $M_S = +1$, and so contributes most character. If the selection rules for ISC favour crossing into T_z, we see that the $M_S = +1$ state will be dominantly populated by virtue of its average dominant T_z character. Hence polarization in the molecular frame may appear in the laboratory frame.

Even though one of the M_s states may be populated preferentially, the spin-lattice relaxation time of the triplet is so short (of the order of nanoseconds) that it might disappear. This is where the chemical quenching agent comes into play. If the triplet can be chemically quenched (by electron or H atom transfer, or some other effective one-electron transfer) before the triplet has time to relax, the polarization will be transferred to the resulting doublet. That doublet has a much longer relaxation time (of the order of microseconds) and the polarization may persist long enough for it to be detected in an e.s.r. experiment. Since diffusion-controlled reactions have rate-coefficients of the order of 10^9 dm^3 mol^{-1} s^{-1}, quencher concentrations of the order of 0.1-1.0 mol/dm^3 give quenching rates comparable to the triplet spin relaxation rates, and so conditions for the preservation of triplet spin polarization may be attained quite readily.

The question that now arises is the order of magnitude of the effect. If the ISC is equally rapid into all molecular frame triplet substates there will be no polarization. Hence we expect the polarization to be proportional to the relative anisotropy of the ISC rate (the absolute crossing rate is irrelevant if it is sufficiently fast); call this dimensionless quantity $\hat{K}$. If the triplet sublevels are degenerate in the absence of the applied field, the M_s levels will be composed of equal contributions from each zero field level, and the earlier discussion suggests that the anisotropy of the population transfer would then not appear in the molecular frame. Therefore it can be anticipated that the polarization in the laboratory frame is proportional to the zero-field splitting, D. If the applied field, B, were infinite, the M_s = +1 level would be so far from the zero-field levels that, even if D $\neq$ 0, M_s = +1 is virtually equally far from each in energy, and hence on average equally composed of each. Hence the laboratory-frame polarization is expected to be inversely proportional to B. Therefore we can write

$$S_z^* - S_z^o \propto D\hat{K}/B \tag{1.1}$$

where S_z^o is the thermal equilibrium magnetization and S_z^* the induced magnetization. We shall see later that the constant of proportionality is of the order of 8/15 for D and B expressed in the same units, and so the triplet polarization, $\gamma_T = (S_z^* - S_z^o)/S_z^o$, is of the order of

$$\gamma_T \sim 8D\hat{K}/15S_z^oB \sim (8/15)(1/3)(1)/S_z^o \sim 1/6S_z^o \; . \tag{1.2}$$

when ISC occurs to only one state ($\hat{K} = 1$). Since $S_Z^o \sim -10^{-3}$, a triplet polarization of the order of −150 can be expected.

The amount of the triplet polarization that will emerge into the laboratory frame depends on the relative values of the triplet relaxation time T_1^* and the quenching rate $k_q[Q]$, where [Q] is the quencher concentration. This can be examined by a simple argument based on the rate laws for the processes going on. If the population of the M_s state of the triplet is denoted N_M^T its kinetics are expressed by

$$\dot{N}_M^T = -W\{N_M^T - N_{M,e}^T\} - k_T N_M^T \tag{1.3}$$

where the first term represents the decay towards thermal equilibrium (when the population is $N_{M,e}^T$) and the second represents the disappearance of the state as a result of ISC (other decay modes being neglected). Note that the 'equilibrium' population is in fact time-dependent because of the ISC process, but since

$$N^T = \sum_M N_M^T = \sum_M N_{M,e}^T = N_e^T \tag{1.4}$$

the overall rate law is simply

$$\dot{N}^T = -k_T N^T \tag{1.5}$$

with a simple exponential solution. The state population at equilibrium is related to the total population by

$$N_{M,e}^T = p_M N_e^T = p_M N^T(0)\exp(-k_T t) \tag{1.6}$$

where p_M is a time-independent Boltzmann factor. Substitution of this into eqn (1.3) leads to a differential equation which has the solution

$$N_M^T(t) = N^T(0)\{q_M + p_M(e^{Wt} - 1)\}\exp\{-(k_T + W)t\} \tag{1.7}$$

where $q_M = N_M^T(0)/N^T(0)$.

The rate of formation of the α-state of the doublet is

$$\dot{N}_\alpha^D = k_T N_{+1}^T + \tfrac{1}{2} k_T N_0^T \tag{1.8}$$

because it emerges from the $M_s = +1$ state with probability 1, but with probability ½ from the $M_s = 0$ state. A similar equation applies to N_β^D. The total magnetization of the doublets when sufficient time has elapsed for all the triplets to become doublets, but at times so short that no doublet spin relaxation has taken place, is

$$S_z^D = \tfrac{1}{2}\int\{dN_\alpha^D - dN_\beta^D\} = \tfrac{1}{2}\int_0^\infty dt\{\dot{N}_\alpha^D - \dot{N}_\beta^D\}$$
$$= \tfrac{1}{2}k_T\int_0^\infty dt\{N_{+1}^T(t) - N_{-1}^T(t)\}. \tag{1.9}$$

Substitution of the solution for the triplet populations leads at once to

$$S_z^D = \left(\frac{N^T(0)}{2(k_T+W)}\right)\{(p_{+1} - p_{-1})W + (q_{+1} - q_{-1})k_T\} \tag{1.10}$$

The last expression may be simplified as follows. When the triplet polarization is strong $|p_{+1} - p_{-1}| << |q_{+1} - q_{-1}|$; but since the quenching rate k_T is often comparable to the triplet spin relaxation rate W, it follows that the term in pW is negligible. Furthermore, the equilibrium doublet magnetization is

$$S_{z,e}^D = \tfrac{1}{2}(p_{\frac{1}{2}} - p_{-\frac{1}{2}})N^D \sim -(\hbar\omega/4kT)N^T(0) \tag{1.11}$$

because the number of doublets formed is equal to the initial number of triplets (on this simplified model). It follows that the polarization of the doublets is

$$\gamma^D = \left\{\frac{S_z^D - S_{z,e}^D}{S_{z,e}^D}\right\}$$

$$= \frac{2(q_{+1} - q_{-1})kTk_T}{(k_T+W)\hbar\omega} - 1 , \tag{1.12}$$

and so, with $k_T = k_q[Q]$, where [Q] is the quencher concentration, and

$$\gamma^D(0) = 2(q_{+1} - q_{-1})kT/\hbar\omega \tag{1.13}$$

we arrive at a Stern-Volmer type plot for the polarization:

$$1/(\gamma^D + 1) = 1/[\gamma^D(0)+1] + 1/[\gamma^D(0)+1]k_q[Q]T_1 \tag{1.14}$$

where $T_1 = 1/W$, the triplet spin relaxation time.

The last equation has been exploited in order to measure triplet spin relaxation times ([6], see Chapter VIII), but our concern here is to assess the magnitude of the spin polarization emerging from the triplet mechanism. When $k_q[Q]T_1 >> 1$ we see that $\gamma^D \sim \gamma^D(0)$, and since we have already seen[†] that $\gamma^D(0) \sim 10^2$, polarizations of the order of 10^2 can be expected. Even when $k_q[Q]T_1 \sim 0.1$ we may still have $\gamma^D \sim \gamma^D(0)/10$, and so substantial polarizations can be expected.

The analysis so far has been in terms of a simple kinetic model combined with dimensional arguments on the magnitude of the initial polarization. What we now turn to is a more detailed version of the whole mechanism, taking into account (sometimes in a disguised form) all the modes of motion of the system.

2. THE FULL DYNAMICAL MODEL

The full dynamical model must take into account a variety of features. First, the formation of the excited singlet might be anisotropic if polarized light is used. Since the sample is normally exposed to a uni-directional beam of light, even if it is unpolarized the illumination is not isotropic (there is no component of the electric vector along the propagation direction) and so a precise theory must take as an initial condition an anisotropic distribution of singlets. The singlet may decay by either of two routes: ISC, with a rate k, or other modes of deactivation with a collective rate k_S. As explained in the introduction, the ISC is anisotropic in the molecular frame, and each zero-field state of the triplet is populated at a different rate. Therefore we should speak of the three components of k, and it is convenient to arrange them in a matrix diagonal in the principal axes of the zero-field interaction $\mathbb{D}$ and with eigenvalues k_{xx}, k_{yy}, k_{zz}. This matrix we write $\mathbb{k}$. Since we have an interest in its anisotropy, we separate out the mean ISC rate $k = \frac{1}{3}(k_{xx} + k_{yy} + k_{zz})$, and call the remainder the kinetic anisotropy matrix, $-\mathbb{K}$:

$$-\mathbb{K} = \mathbb{k} - \mathbb{1}k, \quad k = \frac{1}{3}\,\mathrm{tr}\,\mathbb{k}, \quad \mathrm{tr}\,\mathbb{K} = 0 . \tag{2.1}$$

A simplification in the ensuing calculation is that $\mathbb{K}$ may be expressed as a spin operator. Thus, on writing $K \rightarrow \vec{S}\cdot K\cdot\vec{S}$ it can

---

[†]What we have in fact shown is that $\gamma^T(0) \sim 10^2$. The Stern-Volmer type of expression takes a slightly different form when $\gamma^T(0)$ is used [4], and it is easy to show that if the condition after eqn (1.10) is relaxed

$$4/(3\gamma^D-1) = 1/\gamma^T(0) + 1/\gamma^T(0)k_q[Q]T_1 .$$

be verified that the matrix of the new operator between the zero-field spin states is exactly the matrix $\mathbb{K}$ described above. Now it will be noticed that the zero-field interaction is of the form $-\vec{S}\cdot\mathbb{D}\cdot\vec{S}$, with $\mathbb{D}$ a traceless matrix, and so this interaction and the anisotropy of populating the three levels have a similar nature. This means that when we come to evaluating rotational averages, there will be cross-terms between $\mathbb{K}$ and $\mathbb{D}$ which will not disappear.

The rotation of the triplet modulates the dipolar interaction between the spins, and we shall see that a term in D^2 occurs as a result. This term corresponds to the spin relaxation of the triplet, a rapid process because the dipolar interaction is strong. Finally, there is another type of motion, the chemical evolution of the triplet into the doublet product; this is the process ascribed as rate k_T in the earlier discussion.

All the relevant information about the triplet is contained in its spin density matrix ρ. Each of the processes in the preceding discussion can be traced to a term in the equation of motion of the density matrix. The tactics, therefore, are to set up the equation of motion, and then to solve it for the magnetization of the triplet at some time t. As in the last section, especially as seen by reference to eqn (1.9), the doublet polarization is given by a time integral of the triplet polarization. This greatly simplifies the calculation, because solving for the integral instead of the polarization itself, has the effect of turning the evolution equation, which is a complicated differential equation, into a simple algebraic equation which can be solved almost exactly.

The magnetic evolution of the triplet arise from the presence of its spin-spin interactions, $H_D = -\vec{S}\cdot\mathbb{D}\cdot\vec{S}$, and the torque exerted by the external field, $H_Z = \omega_o S_Z$. Under these influences the density matrix evolves according to

$$(\dot{\rho})_{mag} = i[\rho,H_Z] + i[\rho,H_D(\Omega)] \ . \tag{2.2}$$

(We have indicated the orientation dependence of H_D when $\mathbb{D}$ is referred to the laboratory frame.) The chemical evolution of the triplet can be summarized by a source term, representing its creation by ISC, and a sink term, representing its annihilation by virtue of the attack by the quenching agent. The source is itself time dependent because the singlet is depleted both by ISC and by alternative processes, and is orientation dependent if the excitation was anisotropic. It may be expressed as $kP_S(\Omega,t)$. The decay component of the time dependence of this source is simply $\exp\{-(k+k_s)t\}$, and its evolution by virtue of the molecular rotation can be retained implicitly by allowing P_S to depend on

Ω to be time-dependent. Since k can be expressed as an operator on the spin states we arrive at

$$(\dot{\rho})_{source} = kP_s(\Omega,t)$$
$$= \tfrac{1}{6} kS^2 P_s(\Omega,t) + H_K(\Omega)P_s(\Omega)\exp\{-(k+k_s)t\} \qquad (2.3)$$

where $H_K(\Omega) = -\vec{S}\cdot\mathbb{K}\cdot\vec{S}$. The sink term is simply $-k_T\rho$, because each M_s state is attacked at the same rate by the quenching agent (at least, there is no evidence to the contrary). Collecting these pieces, the overall equation of motion for the density matrix is

$$\dot{\rho} = i[\rho,H_Z] + i[\rho,H_D(\Omega)] - k_T\rho$$
$$+ \{\tfrac{1}{6} kS^2 + H_K(\Omega)\}P_s(\Omega)\exp\{-(k+k_s)t\}. \qquad (2.4)$$

The evolution equation may be developed by introducing an interaction picture based on the transformations

$$\sigma(t) = \exp(k_T t)\exp(iH_Z t)\rho(t)\exp(-iH_Z t) \qquad (2.5a)$$

$$H_X(\Omega_t) = \exp(iH_Z t)H_X(\Omega)\exp(-iH_Z t), \quad X = D,K \qquad (2.5b)$$

When these transformations are applied to eqn (2.4) they yield

$$\dot{\sigma} = i[\sigma,H_D(\Omega_t)] + \{\tfrac{1}{6} kS^2 + H_K(\Omega_t)\}P_s(\Omega_t)\exp\{-(k+k_s-k_T)t\} \qquad (2.6)$$

with $P_s(\Omega_t)$ the interaction transform of $P_s(\Omega)$,

$$P_s(\Omega_t) = \exp(iH_Z t)P_s(\Omega)\exp(-iH_Z t) . \qquad (2.7)$$

The <u>formal</u> solution of eqn (2.6) is

$$\sigma(t) = \sigma(0) + i\int_0^t dt'[\sigma(t'),H_D(\Omega_{t'})]$$
$$+ i\int_0^t dt'\{\tfrac{1}{6} kS^2 + H_K(\Omega_{t'})\}P_s(\Omega_{t'})\exp\{-(k+k_s-k_T)t\} \qquad (2.8)$$

and substitution back into eqn (2.6) gives

$$\dot{\sigma}(t) = i[\sigma(0), H_D(\Omega_t)]$$
$$+i^2\int_0^t dt'[[\sigma(t'), H_D(\Omega_{t'})], H_D(\Omega_t)]$$
$$+i\int_0^t dt'[\{\tfrac{1}{6}kS^2+H_K(\Omega_{t'})\}P_s(\Omega_{t'}), H_D(\Omega_t)]\exp\{-(k+k_s-k_T)t'\}$$
$$+ \{\tfrac{1}{6}kS^2+H_K(\Omega_t)\}P_s(\Omega_t)\exp\{-(k+k_s-k_T)t\} . \qquad (2.9)$$

Now average over the isotropic ensemble. The average of H_D is zero; furthermore, S^2 commutes with H_D; therefore

$$\langle\dot{\sigma}(t)\rangle = -\int_0^t dt'\langle[[\sigma(t'), H_D(\Omega_{t'})], H_D(\Omega_t)]\rangle$$
$$+ i\int_0^t dt'\langle P_s(\Omega_{t'})[H_K(\Omega_{t'}), H_D(\Omega_t)]\rangle\exp\{-(k+k_s-k_T)t'\}$$
$$+ \langle\{\tfrac{1}{6}kS^2+H_K(\Omega_t)\}P_s(\Omega_t)\rangle\exp\{-(k+k_s-k_T)t\}. \qquad (2.10)$$

Now we introduce the first approximation (everything so far has been exact). The first term on the r.h.s. is exactly the expression arrived at in a density matrix treatment of spin relaxation where the relaxing perturbation is the spin-spin interaction. In recognition of this we make the normal Redfield approximation and write is as $R\{\langle\sigma\rangle - \langle\sigma\rangle_o\}$, where $\langle\sigma\rangle_o$ is the thermal equilibrium spin density matrix, and R the relaxation tetradic [7,8]. Note that at this stage the spin relaxation of the triplet enters quite naturally into the theory and does not have to be added in an _ad hoc_ fashion.

Next, we realize that the doublet polarization is given by eqn (1.9), which in density matrix terms can be written

$$S_Z^D = \tfrac{1}{2}k_T N^T(0)\int_0^\infty dt\{\rho_{11}(t) - \rho_{-1-1}(t)\}$$
$$= \tfrac{1}{2}k_T N^T(0)\int_0^\infty dt\{\langle\sigma_{11}(t)\rangle - \langle\sigma_{-1-1}(t)\rangle\}\exp\{-k_T t\}. \qquad (2.11)$$

From now on we write the difference of the two elements of $\langle\sigma\rangle$ as f(t), and from eqn (2.10) find that f satisfies

$$\dot{f}(t) = -W\{f(t) - f_o(t)\}$$
$$+ i\int_0^t dt' \langle P_s(\Omega_{t'})\{[H_K(\Omega_{t'}), H_D(\Omega_t)]_{11} - [H_K(\Omega_{t'}), H_D(\Omega_t)]_{-1-1}\}\rangle$$
$$\times \exp\{-(k+k_s-k_T)t'\} \qquad (2.12)$$

where the first term comes from

$$\dot{\sigma}_{11} - \dot{\sigma}_{-1-1} = \sum_{MM'}\{R_{11MM'}\sigma_{MM'} - R_{-1-1MM'}\sigma_{MM'}\}$$
$$= -(R_{11-1-1} - R_{1111})(\sigma_{11} - \sigma_{-1-1})$$
$$= Wf(t), \qquad (2.13)$$

and where f_o is the corresponding part of the thermal equilibrium density matrix. Note that $W = R_{11-1-1} - R_{1111} = 1/T_1$, the triplet spin relaxation time, as anticipated. It is now possible to see very clearly from eqn (2.12) how the polarization arises from cross-terms in H_K and H_D, as was anticipated in the opening remarks of this section. Furthermore, we are interested, according to eqn (1.9) in the Laplace transform of f(t), not f(t) itself, and so eqn (2.12) can be dealt with very simply.

In order to proceed, we need the time-dependence of the commutators in the second term. Finding this is a straightforward exercise in evaluating matrix elements of angular momentum operators, and is fully described elsewhere [3]. If for the present we assume an isotropic excitation process, $P_s(\Omega_{t'})$ can be set equal to unity, and the final term in eqn (2.11) becomes

$$i\int_0^t dt'\{[\ ]_{11} - [\ ]_{-1-1}\}\exp\{-(k+k_s-k_T)t'\}$$
$$= \frac{4}{15}\int_0^t dt' F(t-t')\exp\{-\omega_2(t-t')\}\{2\sin 2\omega_o(t-t')$$
$$+ \sin\omega_o(t-t')\}\exp\{-(k+k_s-k_T)t'\} \qquad (2.14)$$

where $F(t-t') = DK + 3EI\exp(-at)$,
and the symbols have the following meaning. D and E are the conventional components of the zero-field interaction; K and I are the analogous quantities for the kinetic anisotropy matrix:

$$K = \tfrac{1}{2}(k_{xx} + k_{yy}) - k_{zz}, \qquad I = \tfrac{1}{2}(k_{yy} - k_{xx}). \qquad (2.15)$$

ω_o is the Larmor frequency at the magnetic field of the experiment; ω_2 is an inverse rotational correlation time for a second-rank tensor in a molecule with rotational diffusion coefficients $D_{||}$ and $D_{\perp}$ (the assumption has also been made that the diffusion tensor is axially symmetric in the frame that diagonalizes the z.f. tensor): $\omega_2 = 6D_{\perp}$ and $a = 4(D_{||} - D_{\perp})$. This calculation is based on an anisotropic rotational diffusion equation.

The integral over t' in eqn (2.14) has the form of a convolution, and so the Laplace transform of eqn (2.11) can be taken without any trouble. Evaluating the transform at k_T then gives the polarization S_Z^D immediately:

$$S_Z^D = \left(\frac{k}{2(k+k_s)}\right) \{ S_{Z,e}^T + (S_Z^T - S_{Z,e}^T)\{k_T/(k_T+W)\}\} \tag{2.16}$$

where

$$S_Z^T = \frac{4}{15}\,\omega_o\{D\hat{K}J(0) + 3E\hat{I}J(a)\} \tag{2.17}$$

with

$$J(x) = \frac{4}{4\omega_o^2 + (k_T+\omega_2+x)^2} + \frac{1}{\omega_o^2 + (k_T+\omega_2+x)^2} \tag{2.18}$$

$$\hat{K} = K/k, \qquad \hat{I} = I/k\ . \tag{2.19}$$

Note how the expression for S_Z^T is reproducing the form anticipated in the introduction, $S_Z^T \sim D(K/k)/\omega_o$, when $\omega_o^2 >> (k_T+\omega_2+x)^2$.

Now we proceed to a sequence of simplifications of the general expression. First take $\omega_o >> k_T$. This is certainly valid for X-band experiments.

(1) In the limit of <u>very slow spin relaxation</u> ($T_1 \rightarrow \infty$)

$$S_Z^D = \tfrac{1}{2}kS_{Z,e}^T/(k + k_s) \tag{2.20}$$

and $k/(k+k_s)$ is the proportion of excited singlets that convert to triplets, the latter generating polarized doublets long before spin relaxation destroys any triplet polarization.

(2) When k_s is negligible (so that <u>almost all singlets cross into triplets</u>), the <u>thermal equilibrium magnetization</u> of the triplet is negligible relative to any induced magnetization, and the <u>rotation is isotropic</u>

$$S_Z^D = \frac{k_T(DK + 3EI)\omega_o\tau_2\, j(\omega_o\tau_2)}{\omega_o k_T + \Delta^2 j(\omega_o\tau_2)} \tag{2.21}$$

where
$$j = \frac{2}{15}\left\{\frac{4\tau_2\omega_o}{1 + 4\omega_o^2\tau_o^2} + \frac{\omega_o\tau_2}{1 + \omega_o^2\tau_2^2}\right\} \tag{2.22}$$

$$\Delta^2 = D^2 + 3E^2 .$$

Note that T_1 is still embedded in this expression because $W = \Delta^2 j(\omega_o\tau_2)/\omega_o$. The slow motion limit ($\omega_o\tau_2 >> 1$) of this expression is

$$S_Z^D = \frac{k_T\omega_o\tau_2(DK + 3EI)/\Delta^2}{1 + (15k_T\tau_2\omega_o^2/4\Delta^2)} \tag{2.23a}$$

$$= \{k_T/(k_T+W)\}\,\{\tfrac{4}{15}(DK + 3EI)/\omega_o\} , \tag{2.23b}$$

and in the limit of <u>very fast scavenging</u> and <u>slow spin relaxation</u> ($k_T T_1 >> 1$) this becomes

$$S_Z^D = \left(\frac{4}{15}\right)(D\hat{K} + 3E\hat{I})/\omega_o \tag{2.24}$$

which is also the expression for a static array of triplets [1,2], and is of the form expected from the argument in the opening section.

If $D \sim 0.1\ \text{cm}^{-1}$ and $\omega_o \sim 0.3\ \text{cm}^{-1}$ (at X-band), $D/\omega_o \sim 1/3$, and so eqn (2.23a) gives

$$\gamma^D + 1 = \frac{S_Z^D}{S_{Z,e}^D} \sim -\frac{237\, k_T\tau_2\hat{K}}{0.029 + k_T\tau_2}$$

This suggests that the maximum value of γ_D is about 200, and arises when $k_T\tau_2 >> 1$; in any case, polarizations exceeding unity should be expected when $k_T\tau_2 > 10^{-3}$.

The model described here can be developed to take anisotropic generation into account [4,9], and is clearly well suited for the discussion of anisotropic and motional effects on the extent of polarization.

REFERENCES

1. P.W. Atkins and K.A. McLauchlan, in Chemically Induced Magnetic Polarization, ed. by A.R. Lepley and G.L. Closs, Wiley, New York, 1973.
2. S.K. Wong, D.A. Hutchinson, and J.K.S. Wan, J. Chem. Phys., 58, 985 (1973).
3. P.W. Atkins and G.T. Evans, Mol. Phys., 27, 1633 (1974).
4. P.W. Atkins and G.T. Evans, Adv. Chem. Phys., 35, 1 (1976).
5. J.H. Freed and J.B. Pedersen, Adv. Mag. Reson.,8, 1 (1976).
6. P.W. Atkins, A.J. Dobbs, and K.A. McLauchlan, Chem. Phys. Letts. 29, 616 (1974).
7. C.P. Slichter, Principles of Magnetic Resonance, Harper and Row, New York, 1963.
8. L.T. Muus and P.W. Atkins, eds, Electron Spin Relaxation in Liquids, Plenum, New York, 1972.
9. F.J. Adrian, J. Chem. Phys., 61, 4875 (1974).

CHAPTER XII

LIGHT MODULATED CIDEP EXPERIMENTS*

L.T. Muus

Department of Chemistry, Aarhus University,
Aarhus, Denmark

ABSTRACT. The theoretical background for these experiments is developed in terms of the general theory proposed by Pedersen and Freed. The experiments are performed either in the time domain or in the frequency domain with high intensity light sources. Of crucial importance is the generation of intense light pulses of a shape conducive to quantitative interpretation of the experimental data. The ESR spectrometer is operated at constant magnetic field and with a response time much larger than the radical spin-lattice relaxation time. Time-averaging is necessary for experiments in the time domain. The experiments have demonstrated the existence of both a triplet mechanism (TM) and a radical pair mechanism (RPM) for generation of polarization. Data from recent experiments in the time domain and in the frequency domain are in good agreement with the microscopic theory.

1. THEORY

1.1 General approach

In the notation used by Pedersen and Freed [1-3] the ESR signal intensity produced by radical A in the hyperfine state a is given by the following solution of the Bloch equations [3]

---

* This work was supported in part by a NATO Research Grant.

L. T. Muus et al. (eds.), Chemically Induced Magnetic Polarization, 205-224.

$$M_y^a(t) = (m_o \sin\omega_1 t + y_o \cos\omega_1 t)\exp(-t/T_1)$$
$$+ f_a(t)\exp(-t/T_1)(\omega_1 \exp(t/T_1) - \omega_1 \cos\omega_1 t$$
$$- T_1^{-1}\sin\omega_1 t)(T_1^{-2} + \omega_1^2)^{-1} \quad (1)$$

where

$$f_a(t) = P_a^{\infty}(I)k_o(t) + P_{eq}T_1^{-1}n_A(t) + Fk_2P_a^*n_A(t)n_B(t) \quad (2)$$

We have introduced the assumptions that both a first order chemical decay and a second order decay are much slower than the T_1 relaxation process and that $T_1 \simeq T_2$. Symbols $P_a^{\infty}(I)$, P_{eq} and P_a^* refer to initial polarization, the polarization in thermal equilibrium, and the polarization caused by a spin selective reaction of radical A in hyperfine state a with some other radical.

$P_a^{\infty}(I)$ may be due to either a triplet mechanism (TM) or a radical pair mechanism (RPM). P_a^* can be caused by RPM only. k_2F indicates the second order rate constant in the spin selective reaction assuming that radical pairs can react in the singlet state only. m_o and y_o are the initial values of M_z and M_y, respectively. The unexplained symbols have their usual meaning [1-3].

1.2 Time-resolved experiments

This type of experiment is carried out in the time domain. Let us assume that it is carried out with low observing power, i.e. $\omega_1 T_1 << 1$. The initial values m_o and y_o are related by $m_o = y_o/(\omega_1 T_1)$ [4] and neglecting the constant factor $\omega_1 T_1^2$ we may rewrite Eq. (1) as

$$M_y^a(t) = M_y^{oa}(1 + t/T_1)\exp(-t/T_1)$$
$$+ f_a(t)(1 - (1 + t/T_1)\exp(-t/T_1)) \quad (3)$$

where t = 0 represents the beginning of the particular experimental period, i.e. the instant when light goes on or off. The signal at t = 0 of a particular period M_y^{oa} equals the signal at the end of the preceding period. Eq. (3) shows that the signal M_y^{oa} dies out during a time of the order T_1 in any particular period, and that the signal thereafter follows $f_a(0)$ and finally $f_a(t\to\infty)$.

Let us now assume that the radical decay is caused by a second order reaction and that $n_A(t) = n_B(t) = n(t)$. We have then

$$dn(t)/dt = k_o(t) - k_2F\, n(t)^2 \qquad (4)$$

In time-resolved experiments the light intensity may be given by the rectangular pulse [4]

$$k_o(t) = \begin{cases} k_o & \text{in the light period} \\ \varepsilon^2 k_o & \text{in the "dark" period} \end{cases} \qquad (5)$$

where ε^2 is the fraction of maximum light intensity in the "dark" period.

The solutions for Eq. (4) under the conditions in Eq. (5) are

$$n(t)/n_L^{ss} = (1 - \alpha\exp(-2\beta t))/(1 + \alpha\exp(-2\beta t)) = \ell(t) \qquad (6)$$

for the light period, and

$$n(t)/n_D^{ss} = \varepsilon^2(1 + \alpha\exp(-2\beta\varepsilon t))(1 - \alpha\exp(-2\beta\varepsilon t)) = \varepsilon d(t) \quad (7)$$

for the "dark" period. n_L^{ss} and n_D^{ss} are the steady state concentrations in the light and "dark" periods, respectively ($n_D^{ss} = \varepsilon n_L^{ss}$). α and β are given by

$$\alpha = (1 - \varepsilon)/(1 + \varepsilon)$$

$$\beta = (k_o k_2 F)^{\frac{1}{2}} = k_o/n_L^{ss} \qquad (8)$$

For the steady state in the light period we have from Eq. (2)

$$f_a(t\to\infty) = P_{eq}T_1^{-1}n_L^{ss} + P_a^{\infty}(I)k_o + k_2FP_a^*(n_L^{ss})^2 \qquad (9)$$

Combination of Eq. (9) with Eqs. (8) gives

$$f_a(t\to\infty) = T_1^{-1}n_L^{ss}(P_{eq} + T_1(k_ok_2F)^{\frac{1}{2}}(P_a^{\infty}(I) + P_a^*))$$

$$= T_1^{-1}n_L^{ss}\, P_{eq}(1 + V(I) + V(R)) \qquad (10)$$

where the enhancements V(I) and V(R) are given by

$$V(I^a) = P_a^{\infty}(I)\beta T_1/P_{eq} = P_a^{\infty}(I)T_1(k_o k_2 F)^{\frac{1}{2}}/P_{eq}$$

$$V(R^a) = P_a^{*}\beta T_1/P_{eq} = P_a^{*}T_1(k_o k_2 F)^{\frac{1}{2}}/P_{eq} \qquad (11)$$

Note that V(I), V(R) and β (Eqs. (8) and (11)) depend on the light intensity k_o.

For the "dark" period we obtain in an analogous fashion

$$f_a(t\to\infty) = P_{eq}T_1^{-1}n_D^{ss} + P_a^{\infty}(I)\varepsilon^2 k_o + k_2 F P_a^{*}(n_D^{ss})^2 \qquad (12)$$

and with combination of Eq. (12) and Eq. (8)

$$f_a(t\to\infty) = T_1^{-1}n_D^{ss}(P_{eq} + T_1\varepsilon(k_o k_2 F)^{\frac{1}{2}}(P_a^{\infty}(I) + P_a^{*})$$

$$= T_1^{-1}n_L^{ss}P_{eq}(\varepsilon + \varepsilon^2(V(I^a) + V(R^a)) \qquad (13)$$

where $V(I^a)$ and $V(R^a)$ are the enhancements for the light period given by Eqs. (11).

The general expression for $M_y^a(t)$ in the light period is now obtained by combination of Eq. (3) with the following expressions derived from Eqs. (2), (6), and (13)

$$M_y^{oa} = n_L^{ss}\, P_{eq}T_1^{-1}(\varepsilon + \varepsilon^2(V(I^a) + V(R^a)) \qquad (14)$$

$$f_a(t) = n_L^{ss}P_{eq}T_1^{-1}(V(I^a) + \ell(t) + V(R^a)(\ell(t))^2)$$

For the "dark" period we get in a similar fashion

$$M_y^{oa} = n_L^{ss}P_{eq}T_1^{-1}(1 + V(I^a) + V(R^a)) \qquad (15)$$

$$f_a(t) = n_L^{ss}P_{eq}T_1^{-1}(\varepsilon^2 V(I^a) + d(t) + V(R^a)(d(t))^2)$$

Finally, we have to convolute $M_y^a(t)$ with the spectrometer response function r(t). In many experiments the response function may be taken as $r(t) = t_s^{-1}\exp(-t/t_s)$, where t_s is the spectrometer response time, i.e. the response time to a step function signal. For the observed signal intensity we obtain

$$I^a(t) = \int_{-\infty}^{t} M_y^a(\tau)\; r(t - \tau)d\tau \qquad (16)$$

In experiments with steady states both in the light and in the "dark" period we have $M_y^a(t) = M_y^{oa}$ in an extended period, say $-10\ t_s \leq t \leq 0$, and Eq. (16) may then be reduced to

$$I^a(t) = M_y^{oa}\exp(-t/t_s) + t_s^{-1}\exp(-t/t_s)\int_0^t M_y^a(\tau)\exp(\tau/t_s)d\tau \quad (17)$$

Eq. (17) may be solved by numerical integration using Simpson's rule.

Fig. 1 is a typical time-resolved curve calculated from Eq. (17) with t_s = 100 μs. The light period starts at time 0 and the "dark" period (ε^2 = 0.10) at time 15 msec. The steady state intensities $(1 + V(I^a) + V(R^a))$ and $(\varepsilon + \varepsilon^2(V(I^a) + V(R^a)))$ are accurate. The two extremas are exact only when $\beta^{-1} >> t_s$. The steep slopes preceding and immediately following the extremas are determined by t_s.

With continuous light of the <u>same</u> intensity as that in the light period the relative intensity (h_a) of the hyperfine line a in the ordinary ESR spectrum (i.e. corrected for linewidth and degeneracy is given by

$$h_a \propto f_a(t\to\infty) = n_L^{ss}P_{eq}T_1^{-1}(1 + V(I^a) + V(R^a)) \quad (18)$$

according to Eq. (10).

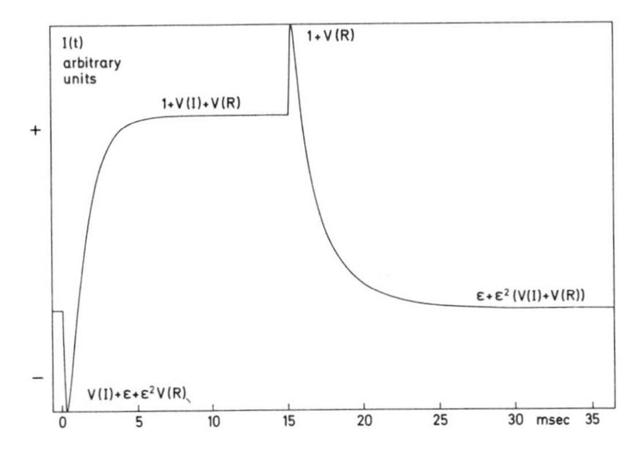

Fig. 1. Time-resolved curve calculated from Eqs. (3) - (17) with ε^2 = 0.10 and t_s = 100 μs. Light goes on at t = 0 and the "dark" period starts at t = 15 msec. The sign convention for the signal intensity is + for absorption, - for emission. The two steady state values are accurate, i.e. independent on t_s, whereas the two peak values are reduced depending on t_s.

1.3 Attenuation experiments

We want to reduce the light by a factor f both in the light and in the "dark" period. Following the same procedure as previously we obtain for the steady state in the light period

$$f_a(t\to\infty)f^{-\frac{1}{2}} = P_{eq}T_1^{-1}n_L^{ss} + n_L^{ss}(k_o k_2 F)^{\frac{1}{2}} f^{\frac{1}{2}}(P_a^{\infty}(I) + P_a^*) \quad (19)$$

or

$$f_a(t\to\infty)f^{-\frac{1}{2}} = P_{eq}T_1^{-1}n_L^{ss}(1 + f^{\frac{1}{2}}(V(I^a) + V(R^a)) \quad (20)$$

where n_L^{ss}, $V(I^a)$ and $V(R^a)$ refer to the experiment with unattenuated light. The radical decay rate β' in the attenuated experiment is given by $\beta' = f^{\frac{1}{2}}\beta$, where β is for the experiment with unattenuated light.

Similarly, we get for the steady state in the "dark" period

$$f_a(t\to\infty)f^{-\frac{1}{2}} = P_{eq}T_1^{-1}n_L^{ss}(\varepsilon + \varepsilon^2 f^{\frac{1}{2}}(V(I^a) + V(R^a))). \quad (21)$$

The complete time-resolved curve for an attenuated experiment is similar to Fig. 1 with the obvious substitutions $V(I^a) \to f^{\frac{1}{2}}V(I^a)$, $V(R^a) \to f^{\frac{1}{2}}V(R^a)$ and $\beta \to f^{\frac{1}{2}}\beta = \beta'$.

Eq. (19) shows that the relative line intensity $f_a(t\to\infty)$ divided by $f^{\frac{1}{2}}$ and plotted against $f^{\frac{1}{2}}$ is linear, and that the slope divided by the intercept $P_{eq}T_1^{-1}n_L^{ss}$ is a measure of $(P_a^{\infty}(I) + P_a^*)$ [5,6]. These plots have a common intercept for hyperfine lines with the same T_1.

1.4 Calibrated kinetic experiments

In these experiments we have $k_o(t) = k_o$ in the light period, $k_o = 0$ in the dark period and introduce three additional assumptions for the low-field hyperfine line a and the corresponding high-field line a' with identical $|\sum_i a_i m_i|$, viz.

T_1 is the same

$$P_a^{\infty}(I) = P_{a'}^{\infty}(I)$$

$$P_a^* = -P_{a'}^*$$

From Eq. (10) we obtain for the steady state in the light period

$$\tfrac{1}{2}(f_a(t\to\infty) + f_{a'}(t\to\infty)) = P_a^\infty(I)k_o + P_{eq}T_1^{-1}n_L^{ss} \quad (22)$$

and from Eq. (2) for any time τ in the dark period

$$\tfrac{1}{2}(f_a(\tau) + f_{a'}(\tau)) = P_{eq}T_1^{-1}n(\tau) \quad (23)$$

For both light and dark period we have at time t

$$\tfrac{1}{2}(f_a(t) - f_{a'}(t)) = Fk_2P_a^*(n(t))^2 \quad (24)$$

For the dark period we have accordingly

$$f_a(\tau) - f_{a'}(\tau) = (Fk_2P_a^*T_1^2/2P_{eq}^2)(f_a(\tau) + f_{a'}(\tau))^2 \quad (25)$$

A graph showing $f_a(\tau) - f_{a'}(\tau)$ vs. $(f_a(\tau) + f_{a'}(\tau))^2$ for the dark period may therefore by extrapolation to $\tau = 0$, i.e. the start of the dark period, give the true radical concentration at this time. This radical concentration is identical to the steady state concentration n_L^{ss} in the light period. The presence of initial polarization $P_a^\infty(I)$ may thus be revealed by comparison with Eq. (22), i.e. the sum of the relative intensities of lines a and a' in the steady state of the light period [12,13]. These procedures require careful calibration, e.g. in terms of absolute radical concentrations.

1.5 Harmonic analysis experiments

In principle, the same information may be derived from experiments carried out in the frequency domain as from those in the time domain. As a simple, however, realistic example of an harmonic analysis experiment we assume that the modulation frequency of the light is ω_L and the function $k_o(t)$ (Eq. 4) given by [5]

$$k_o(t) = k_o(1 - \cos\omega_L t)/2 \quad (26)$$

and accordingly

$$dn(t)/dt = k_o(1 - \cos\omega_L t)/2 - k_2F(n(t))^2$$

for second order radical decay.

For slow modulation in the limiting case $\omega_L\tau \ll 1$, where τ is the radical "mean lifetime", we have that $dn(t)/dt = 0$ during the entire experiment and accordingly

$$n(t) = (k_o/2k_2F)^{\frac{1}{2}}(1 - \cos\omega_L t)^{\frac{1}{2}} \tag{27}$$

Fourier expansion to the second harmonic of Eq. (27) gives

$$n(t) = (k_o/k_2F)^{\frac{1}{2}}\left(\frac{2}{\pi} - \frac{4}{3\pi}\cos\omega_L t - \frac{4}{15\pi}\cos^2\omega_L t\right) \tag{28}$$

Introducing Eq. (28) into Eq. (2) we obtain for the signal intensity $f_a(t)$

$$\begin{aligned} f_a(t) = {} & \frac{2}{\pi} A_o (P_{eq} + \frac{\pi}{4} T_1 (k_o k_2 F)^{\frac{1}{2}} (P_a^{\infty}(I) + P_a^*)) \\ & - \frac{4}{3\pi} A_o (P_{eq} + \frac{3\pi}{8} T_1 (k_o k_2 F)^{\frac{1}{2}} (P_a^{\infty}(I) + P_a^*)) \cos\omega_L t \\ & - \frac{4}{15\pi} A_o P_{eq} \cos 2\omega_L t \end{aligned} \tag{29}$$

where $A_o = (k_o/k_2F)^{\frac{1}{2}}T_1^{-1}$. The quantity $(P_a^{\infty}(I) + P_a^*)$ may therefore be found by comparing amplitudes at frequencies ω_L and $2\omega_L$.

The general theory for harmonic analysis experiments is extended to fast modulation and mixed first and second order radical decay by Paul [5]. Of particular diagnostic value is the finding that the signal intensity amplitude of the fundamental (Z_1) for fast light modulation is given by $Z_1^{-2} \propto (\omega_L^2 + 2k_2Fk_o)k_o^{-2}$ for $P_a^{\omega}(I) = 0$ and pure second order decay.

2. EXPERIMENTAL CONFIGURATIONS

2.1 ESR spectrometers

The ESR spectrometer is magnetic field modulated (100 kc/s - 2Mc/s) and the ESR signal is observed in the first derivative mode. The magnetic field is adjusted to one of the extremas of a particular hyperfine line during the entire observation, preferably by means of a field-frequency lock.

The most critical parameter of the ESR spectrometer is its response time. With 100 kc/s magnetic field modulation the response time is in the range 30 - 200 μs, usually 100 - 150 μs. The response time may be reduced somewhat by replacing the spectrometer 100 kc/s lock-in amplifier with one having a smaller time constant [7,14]. Application of a 2Mc/s field modulation unit may reduce the response time to about 1 μs, albeit at cost of a reduced signal-to-noise ratio [14-17].

Without exception, light modulated CIDEP experiments are made with ESR spectrometers operating at the X-band. Supplementary Q-band experiments might give information on the magnetic field dependence of the polarization but are difficult to carry out due to the small dimensions of the cavity.

2.2 Time resolved experiments

Quantitative interpretation of the experimental data is facilitated by using light pulses that are close approximations to rectangles (square-wave pulses). Light pulses are produced either by mechanical chopping of the light from a variety of D.C. continuous U.V. light sources (150 - 1000 W) with a rotating sector or by modulating the light source electronically. Suitable optical filters are usually placed in front of the cavity to avoid unnecessary heating of the sample solution and extraneous chemical reactions. For the same reasons, it is advisable to flow the sample solution. In general, the solution is purged with helium or argon to remove oxygen.

A major problem in sector design is reduction of the time required for the sector to turn the light on or off. A small rise or fall time of the light pulse may be achieved by using an optical train of quartz lenses to focus the light on a narrow stationary slit and placing the rotating sector close to this slit. The dark-to-light ratio of most sectors is in the range 3-20.

Shorter rise times may be obtained by using large diameter (about 60 cm) high speed (1800 r.p.m.) rotating sectors with only few cut out areas and placing a mask with a slot between the U.V. lamp and the high speed rotating sector [7,8]. Rise times in the range 120 - 240 μs may be obtained with this arrangement. In addition, the loss of light intensity in a quartz optical system is avoided.

In the alternative procedure the light pulses are produced by electronic modulation of the U.V. light source. A suitable lamp is the Eimac VIX 150 Xenon Illuminator (Eimac Division, Varian Associates, Palo Alto, Calif.) [9]. Modulation is accomplished by a feed-back system comparing the light intensity as measured by a photo-diode to a reference square-wave. The rise time of the light is about 25 μs. To secure dependable operation of the Xenon illuminator the light intensity in the "dark" period can be reduced only to about 10% of that in the light period [4]. To record the signal in the true dark period, a two-bladed propeller may be placed between the Xenon illuminator and the ESR cavity. The two-bladed propeller is driven by a step motor locked to the timing signals of the light modulation unit and cuts the light completely off during part of the "dark" period [10]. The true dark signal may be found also by making the field-frequency lock switch the magnetic field rapidly between the first derivative peak and a point of the base line and accumulate the same number of signals at both positions [11]. This electronic modulation system allows easy change both of the time scale of the experiment and the relative lengths of the light and "dark" periods. Typical lengths of the light period are 1-30 ms and of the "dark" period 3-70 ms. The fact that the Xenon illuminator operates between 100% and about 10% light intensities is no disadvantage. On the contrary, this feature allows calculation of (V(I) + V(R)) from stationary states only [10].

The SPP 1000 High Pressure Mercury Capillary Arc (Philips, Eindhoven, Holland) is designed for use only in the pulsed mode. The prominent part of the D.C. power supply is a bank of 14 separate 50 μF 2,5 kV capacitors in parallel. Modulation is accomplished by a high-frequency thyristor-chopper circuit. For stable operation the capillary arc must remain ignited between pulses and current for this purpose is derived from a separate power circuit. The light output during the "dark" period is negligible [12,13]. The rise time of the light intensity of the SPP 1000 is quite long, about 500 μs. The maximum and minimum lengths of the light period are 0.25 - 2.5 ms and the duration of the "dark" period is maximum 15 - 20 times that of the light period. The Philips SPP 1000 has the special advantage of a high output at wave lengths below 300 nm [12,13].

It is an important feature of most U.V. lamps for both pulsed and continuous operation that power inputs considerably greater than the power rating can be applied in the pulsed mode.

The signal from the spectrometer lock-in amplifier is sampled at successive intervals in the range 100 ns - 50 μs and usually converted from analogue to digital form. The converted data are stored at successive addresses in a magnetic core store. Each address is accessed once and a single number stored in it during one sequence of light and dark periods. The experiment is now repeated and the numbers thus obtained added to those already stored at the particular address. The timing pulses are synchronized with the light pulses and the numbers are added coherently in the data store. The time-averaged data may be observed on an oscilloscope screen or on a graphic display terminal and later plotted. The literature describes a variety of data acquisition systems at different levels of speed, sophistication and investment [e.g. 4, 7, 8, 12, 13, 16]. Application of a medium size computer for time-averaging offers the advantage of simultaneous display of an experimental curve with its simulated curve [4].

2.3 Calibrated kinetic experiments

This type of experiment is a modification of the procedure for obtaining complete time-resolved curves. The duration of the light pulse is chosen so that a steady state signal is reached within the pulse, and the signal intensity is observed only during the last part of the light period and the entire subsequent radical decay in the "dark" period [12,13]. Absolute radical concentrations may be determined conveniently by comparison with suitable calibration standards in a double cavity.

2.4 Harmonic analysis experiments

The harmonically modulated light (Eq. 26) may be produced by chopping a parallel beam from a suitable light source (e.g. a 1 kW Xe-Hg-lamp) with a rotating sector. The rotating sector is placed close to a mask with an aperture with lower and upper edges both having the contours of a half period sine curve [5]. A quartz lens behind the aperture focusses the light on the flat sample cell in the cavity. Both the open and covered parts of the rotating sector cover exactly the aperture resulting in a continuous train of identical harmonically modulated light pulses. The frequency ω_L (Eq. 26) may be varied from 200 sec^{-1} to 13000 sec^{-1}.

The output voltage from the 100 kc/s field modulation unit is fed into a heterodyne phase-sensitive detector locked to a photo pick-up in the rotating sector. The phase-sensitive detector is adjusted to $n\omega_L$ (n = 1,2 ...). After phase optimization, the ESR spectrum may be displayed in the usual fashion. The amplitude sensitivity depends on ω_L and must be calibrated with an amplitude modulated 100 kc/s signal from an audio oscillator. The experiment may be designed to determine rate constants if radical concentrations are measured by comparison with a standard in a double cavity [5].

3. TYPICAL RESULTS

3.1 Time-resolved experiments

The first time-resolved curve from a light modulated experiment was produced by Wong and Wan in 1972 on the 1,4-naphthosemiquinone radical in 2-propanol [18]. These experiments were extended to other quinones and other solvents [19,20]. The time-resolved curves have the general shape of Fig. 1, although the light period of the rotating sector was too short to obtain a steady state in this period. The experiments strongly indicated that the emissive polarization observed in the initial light period could not be due to a RPM [19]. The strongest indication was the observation that both the 1,4-naphthosemiquinone radical and the radical from the hydrogen donor both were observed in emission during and just after the light period [20].

Experiments with steady states both in the light and the dark period on quinones gave time-resolved curves with the appearance of Fig. 1, both for hyperfine lines of the 2,6-di-t-butyl-p-benzosemiquinone radical and for the radical derived from 2,6-di-t-butylphenol used as hydrogen donor, thus again demonstrating initial emission for both radicals [17]. The polarization ratio [21] was found to depend on the hyperfine line for experiments on the 1,4-naphthosemiquinone radical in 2-propanol at $-50^{\circ}C$ and to be strongly hyperfine dependent at room temperature for the same radical in t-butylalcohol with 4-hydroxybenzophenone added as hydrogen donor [22]. Measurements of T_1 showed that the hyperfine line dependency of the polarization ratio could not be explained in terms of the variation of T_1 among hyperfine lines [17]. The experiments prove the coexistence of two polarization mechanisms, i.e. both a TM and a RPM mechanism are operating, the TM playing the dominant part under the experimental conditions.

Rotating sector experiments on aqueous solutions of tartaric acid gave time-resolved curves similar to Fig. 1 for hyperfine lines from the radical $HOOCCHOH\dot{C}HOH$. This radical was rearranged to $HOOC\dot{C}HCHO$, probably in an acid-catalyzed first order reaction [7]. The initial emissive character of a low-field line was more pronounced than that for the corresponding high-field line. For all hyperfine lines, the initial emissive character decreased with decreasing tartaric acid concentration giving characteristic phase changes of the ESR spectrum [7]. Sector experiments on the secondary radical $HOOC\dot{C}HCHO$ showed also strong initial emissive character thus demonstrating the transfer of polarization from its precursor [7].

A room temperature time-resolved curve for one of the hyperfine lines from the p-benzosemiquinone radical ($PBQH^{\bullet}$) in ethylene glycol is shown in Fig. 2 together with its computer simulated curve for fitted values of V(I), V(R), and β (Eqs. (3), (14), (15), and (17) and the experimentally determined "dark"-to-light ratio ε^2 [4]. Similar good agreement between experimental and computer simulated curves was obtained also for a different hyperfine line. The simulated curves for both lines were fitted to $V(I) = -0.66$ and $\beta = 1.10 \cdot 10^3\ sec^{-1}$. However, V(R) was different for the two lines. Assuming V(I) and β identical for all hyperfine lines, V(R) was calculated for each centerline of the six

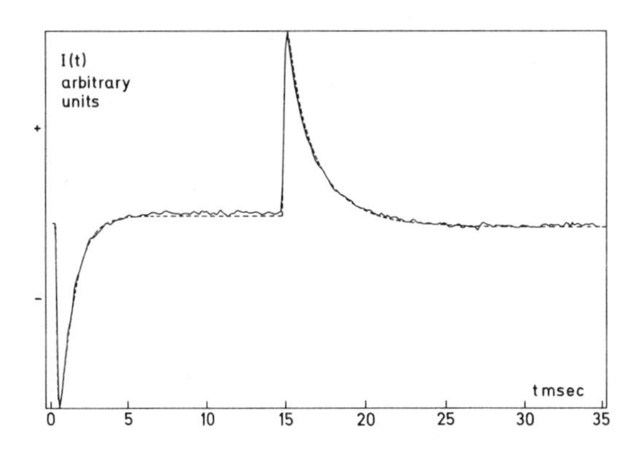

Fig. 2. Comparison between a computer simulated time-resolved curve (broken line) and an experimental curve (full line) for a PBQH$^{\bullet}$ line with ethylene glycol as solvent [4].

separated triplets using the proportionality between (1+V(I)+V(R)) and the relative line intensities (corrected for line width and degeneracy in the continuous light ESR spectrum Eq. (18)). The values for V(R) and P_a^* are shown in Table I and are seen to vary in the way predicted for a second order spin sorting reaction between identical PBQH$^{\bullet}$ radicals, most probably the disproportionation 2PBQH$^{\bullet}$ $\rightarrow$ PBQ + $PBQH_2$ [4].

Table I

RPM ENHANCEMENT AND POLARIZATION FOR PBQH$^{\bullet}$ IN ETHYLENE GLYCOL [4]

Line	2	5	8	11	14	17
V(R)	-0.17	-0.16	-0.05[a)]	-0.06	0.20	0.20[a)]
P_a^{*}[b)]	-0.027	-0.026	-0.008[a)]	-0.010	0.032	0.032[a)]

a) Determined by simulation of time-resolved curves.

b) Calculated from Eq. (11) with $\beta = 1.1 \cdot 10^{-3}$ sec^{-1} and $T_1 = 7.8$ μs. V(I) = -0.66 corresponds to $P^{\infty}(I)$ = -.107 [4].

Time-resolved curves for PBQH$^\bullet$ in 2-propanol showed at room temperature a small initial polarization V(I) = -0.12 and no detectable hyperfine line dependency of the polarization [4,16]. Experiments on PBQH$^\bullet$ in ethanol and methanol did not reveal any polarization at room temperature [4].

A time-resolved curve from recent room temperature experiments on PBQH$^\bullet$ in ethylene glycol is shown in Fig. 3 together with the simulated curve [10]. The true dark signal was obtained by having a two-bladed propeller cut the light in the "dark" period of the electronically modulated Xenon illuminator. As seen from Fig. 3 (V(I) + V(R)) may be obtained from steady states only leaving V(R) and β to be fitted by computer simulation. The experiments showed that V(I) and β were essentially constant for the six selected hyperfine lines and that (1 + V(I) + V(R)) was proportional to the relative line intensities in the continuous light ESR spectrum extrapolated to the same light intensity as used in the modulated mode. Attenuation experiments were performed both with modulated and continuous light using wire gauzes. Time-resolved curves for one hyperfine line with light ratios (f) 1, 0.5 and 0.25 showed that the sum of enhancements and the second order decay constant both were proportional to $f^{\frac{1}{2}}$ (p. 210). The continuous light experiments gave for all six selected hyperfine lines relative line intensities that followed Eq. (20) to a good approximation with slopes essentially proportional to (V(I)+ V(R)) [10]. The attenuation experiments thus confirm that radical termination is by a second order reaction.

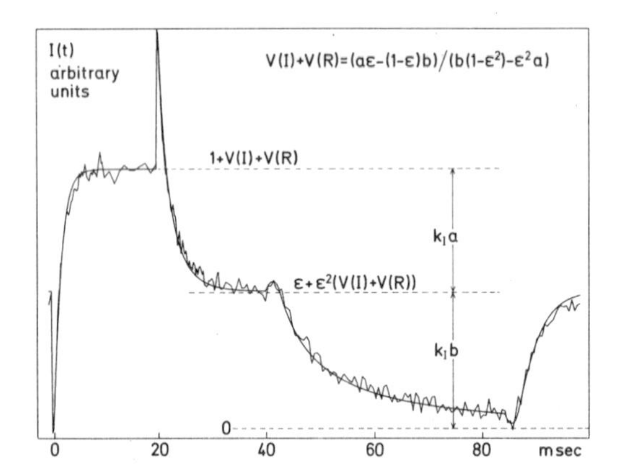

Fig. 3. Computer simulated (smooth curve) and experimental time-resolved curve (with wiggles) for a PBQH$^\bullet$ hyperfine line with ethylene glycol as solvent. The true dark signal (0) is obtained by extrapolation of the experimental curve for the period during which a two-bladed propeller has cut the light in the "dark" period. Note that (V(I) + V(R)) as indicated can be calculated from steady state values only.

3.2 Calibrated kinetic experiments

Experiments of this type were performed on the biacetyl semidione radical $CH_3\dot{C}OHCOCH_3$ and the semidione radical from pyruvic acid $CH_3\dot{C}OHCOOH$ with cyclohexanol and 2-propanol as solvents and hydrogen donors [13]. The biacetyl radical experiments gave evidence both of initial polarization $P_a^{\infty}(I)$ and polarization P_a^* from a RPM [13]. The initial emissive polarization was several times larger in cyclohexanol with the higher viscosity. Experiments on the pyruvic acid semidione radical as well as on the radicals derived from ethanol, ethylene glycol and cyclohexanol by hydrogen abstraction showed P_a^* polarization only [13,29].

3.3 Harmonic analysis experiments

The α-tetrahydrofuryl radical $\underbrace{CH_2CH_2CH_2\dot{C}HO}$ was produced by illuminating tetrahydrofuran and di-t-butyl-peroxide at -93°C. The inverse square of the line intensities for the fundamental frequency ω_L showed a linear dependency of ω_L^2 in an extended frequency range for the centerlines of the 6 groups each containing 9 lines. This observation indicated second order radical decay and the absence of initial polarization. Measurement of T_1 for each line as well as absolute determination of $(k_2k_oF)^{\frac{1}{2}}$ allowed calculation of the ratio P_a^*/P_{eq} for each of the 6 lines selected for the harmonic analysis experiments. These values were compared to the values of P_a^*/P_{eq} obtained from attenuation experiments with continuous light. The agreement between the results from the two types of experiments was excellent, and P_a^*/P_{eq} exhibited the hyperfine line dependency expected for a spin selective reaction between two identical radicals. Similar experiments on the $\dot{C}H_2OH$ radical formed by illuminating t-butyl peroxide in methanol gave no evidence of either initial of RPM polarization [5].

4. MICROSCOPIC THEORY

We address now ourselves to the problem of quantitative interpretation of the experimental data from light modulated experiments in terms of the microscopic theory. Only two series of experiments lend themselves readily to this exercise. These are the time-resolved experiments on the p-benzosemiquinone radical ($PBQH^•$) [4,10] and the harmonic analysis experiment on the α-tetrahydrofuryl radical ($THF^•$) [5].

The generally accepted steps in the photolysis of quinones may be applied to PBQ as follows (e.g. [16])

$$PBQ + h\nu \rightarrow {}^1PBQ^* \quad (30a)$$

$${}^1PBQ^* \rightarrow {}^3PBQ \quad (30b)$$

$${}^3PBQ + RH \rightarrow PBQH^\bullet + R^\bullet \quad (30c)$$

$$2PBQH^\bullet \rightarrow PBQ + PBQH_2 \quad (30d)$$

Superscripts 1 and 3 indicate excited singlet and triplet states. RH is the hydrogen donor, which may be identical to the solvent. The step (30b) is the intersystem crossing.

The room temperature experiments in ethylene glycol showed an initial emissive polarization $P_a^\infty(I) = -0.107$ essentially independent of the hyperfine line. This emissive polarization might conceivably be caused by a RPM, since the value of V(I) is within the theoretical limits for a RPM [1]. However, the g factor for radical $R^\bullet$ must be much lower (< 1.98) than that for $PBQH^\bullet$ (2.00463) to result in emissive initial polarization for all $PBQH^\bullet$ hyperfine lines. This requirement is not fulfilled for the likely hydrogen donor radicals in the experiments. Actually, the g factor for $\dot{C}HOHCH_2OH$ is 2.00302 [23]. Therefore, we can discard the possibility that a RPM generates the initial polarization. This leaves us as alternative the triplet mechanism (TM) [24,25].

The polarization $P_a^\infty(I)$ by TM emerges from nonuniform rates w_x, w_y, and w_z of populating the triplet zero-field states T_x, T_y, and T_z (x,y,z in the molecular frame) in the intersystem crossing step (30b). The nonuniform populations persist in the laboratory frame for the rotating triplet depending upon the rotational correlation time (τ_R), the zero-field splitting parameters (D and E), the magnetic field and the triplet "lifetime" k^{-1} [24,25].

The triplet "lifetime" is determined both by the rate of hydrogen abstraction to give a pair of radicals in the pseudo-first order reaction (30c) with rate constant k_H and by several other processes, e.g. decay to the ground singlet state. These other processes are summarily treated as a first order reaction with rate constant k_S. The triplet "lifetime" k^{-1} is accordingly $k^{-1} = (k_H + k_S)^{-1} = p/k_H$. According to theory [24,25] $P_a^\infty(I)$ by TM is proportional both to p and to $W = (w_z - \frac{1}{2}(w_x + w_y))$.

Fig. 4 shows $P^\infty(I)/Wp$ for different values of D and E = 0 as a function of the triplet "lifetime" k^{-1} for X-band measurements in ethylene glycol and 2-propanol at room temperature [26]. The rotational correlation times for the two solvents were derived from viscosity measurements using the relation $\tau_R = 4\pi\eta a^3/3kT$, where the "hydrodynamic radius" a of $PBQH^\bullet$ was taken as 0.32 nm,

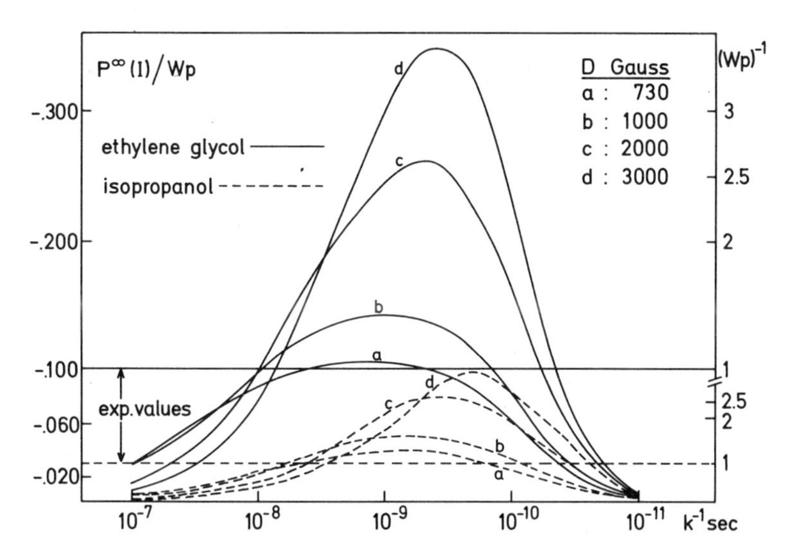

Fig. 4. $P^{\infty}(I)$ by TM as a function of the triplet "lifetime" k^{-1} for ethylene glycol (full lines) and 2-propanol (broken lines). Letters a - d refer to the zero-field splitting parameter D. The magnetic induction is 3300 Oe. Horizontal lines show $P^{\infty}(I)$ observed for PBQH• [4]. The curves were calculated from Eq. (34) in Ref. 25 for Wp = 1. For Wp < 1 displace the experimental lines to the proper value for $(Wp)^{-1}$ on the right-hand scale and multiply the value for $P^{\infty}(I)$ on the left-hand scale by Wp.

and η is the solvent viscosity in poise. Straight lines in Fig. 5 present the observed values for $P^{\infty}(I)$ in the two solvents.

Fig. 4 shows that the observed values for $P^{\infty}(I)$ are compatible with the calculated curves for $D \geq 700$ Gauss and k^{-1} in the range 10^{-8} - 10^{-10} sec. A comparison for Wp < 1 may be facilitated by displacing the lines to the proper value of $(Wp)^{-1}$ on the right-hand scale. The range of the triplet "lifetime" 10^{-8} - 10^{-10} sec is in agreement with the inability to observe the triplet in flash photolysis experiments with 10 nsec time resolution [27]. The zero-field splitting parameter D of PBQ is not known with certainty. Liquid helium temperature measurements with PBQ incorporated in host crystals have given values as different as 730 and 1890 Gauss [30,31]. The low value would require $Wp \simeq 1$, which appears unlikely, and the "lifetime" 1 nsec in both solvents. A high value of about 2000 Gauss suggests $(Wp)^{-1} \simeq 2.5$ which is more likely and a "lifetime" of about 0.3 nsec. The values of $P^{\infty}(I)$ for the p-benzosemiquinone radical in ethylene glycol is essentially the same for all hyperfine lines. This lack of hyperfine line dependency may be due to a short lifetime (< 2μs) of the $\dot{C}HOHCH_2OH$ radicals indicating that they react before completing a RPM with PBQH$^\bullet$. The inability to observe $\dot{C}HOHCH_2OH$ radicals in the ESR spectrum is in agreement with this argument. It is noteworthy that the experiments on the 1,4-naphthosemiquinone radical [17] appear to give data in qualitative agreement with those just discussed for PBQH$^\bullet$, i.e. V(I) with essentially no hyperfine line dependency and V(R) characteristic of a radical disproportionation [22].

The polarization P_a^* is due to a spin selective reaction of a radical pair. At least one of the participating radicals must be PBQH$^\bullet$ since the polarization is generated in a reaction depleting them. Assuming that this reaction is the disproportionation (30d) P_a^* is given by the equation

$$P_a^* = X^{-1} \sum_{a'} \frac{P^\infty}{F}(Q_{aa'})X_{a'} \tag{31}$$

where a is the PBQH$^\bullet$ hyperfine line under consideration, $X_{a'}$ is the degeneracy of line a', and $X = \Sigma X_{a'}$ [1-4]. The quantity $P^\infty/F(Q_{aa'})$ is a function of $Q_{aa'}$, where $Q_{aa'}$ is half the difference of the ESR frequencies between hyperfine line a and a'. $P^\infty/F(Q_{aa'})$ depends strongly also upon the relative diffusion constant 2D, the distance of closest approach d and the details of the exchange interaction J(r). We assume that $J(r) = J_o\exp(-\lambda(r-d))$, where r is the radical separation and that $\lambda = 5r_{ex}^{-1}\ell n10$. r_{ex} is the separation that reduces J(R) with the factor 10^5. As in Eq. (2.17) in Ref. 25 we define

$$\tau_1(\lambda) \simeq (d/D\lambda)(1 + (\lambda d^{-1})) \tag{32}$$

For PBQH$^\bullet$ we take d as twice the "hydrodynamic radius" a, i.e. d = 0.64 nm and $r_{ex} \simeq d$. The relative diffusion constant 2D is derived from the measured solvent viscosity η using Stoke's law ($D = kT/6\pi a\eta$). For ethylene glycol we obtain $\tau_1(\lambda) = 4\cdot10^{-10}$ sec at room temperature. It is most probable, therefore, that $2J_o\tau_1(\lambda) >> 1$ corresponding to $J_o >> 10^9$ sec^{-1}. This argument suggests a polarization by RPM that is independent of J_o (Eq. (2.18) in Ref. 25). Finally, P_a^* was derived using Fig. 8 in Ref. 1 for $r_{ex}/d = 1$. For the disproportionation (30d) we obtained for PBQH$^\bullet$ lines 8 and 17 $P_a^* = -0.009$ and $P_a^\infty = 0.035$, respectively.

The close agreement between these values and those derived from experiment (Table 1) may to some extent be fortuitous, since later experiments on PBQH$^\bullet$ in ethylene glycol do not appear to give such close agreement [10].

Following essentially the same procedure as outlined above P_a^* was calculated for the α-tetrahydrofuryl radical at -93°C in tetrahydrofuran [5]. The calculated values were smaller than those derived from experiment by a factor 2-3. This discrepancy is by no means surprising considering the many assumptions involved. It is noteworthy that the same procedure gave essentially correct results for the values of P_a^* observed in a radiolysis experiment [2,28].

There is an obvious need, however, for additional accurate experiments to test the details of the microscopic theory for both the TM and the RPM.

REFERENCES

1. J.B. Pedersen and J.H. Freed, J. Chem. Phys. 58, 2746 (1973).
2. J.B. Pedersen and J.H. Freed, J. Chem. Phys. 59, 2869 (1973).
3. J.B. Pedersen, J. Chem. Phys. 59, 2656 (1973).
4. J.B. Pedersen, C.E.M. Hansen, H. Parbo, and L.T. Muus, J. Chem. Phys. 63, 2398 (1975).
5. H. Paul, Chem. Phys. 15, 115 (1976).
6. H. Schuh, E.J. Hamilton, H. Paul, and H. Fischer, Helv. Chim. Acta, 57, 2011 (1974).
7. R. Livingston and H. Zeldes, J. Mag. Res. 9, 331 (1973); J. Chem. Phys. 54, 4891 (1973).
8. K.Y. Choo and J.K.S. Wan, J. Am. Chem. Soc. 97, 7127 (1975).
9. H. Levanon and S.I. Weissman, J. Am. Chem. Soc. 93, 4309 (1971).
10. L.T. Muus, S. Frydkjær, and K. Bondrup Nielsen, (to be published).
11. E.J. Hamilton and H. Fischer, J. Phys. Chem. 77, 722 (1973).
12. P.B. Ayscough, T.H. English, and D.A. Tong, J. Phys. E. 9, 31 (1976).
13. P.B. Ayscough, G. Lambert, and A.J. Elliot, J.C.S. Faraday I, 72, 1770 (1976).
14. P.W. Atkins, K.A. McLauchlan, and A.F. Simpson, J. Phys. E. 3, 547 (1970).
15. P.W. Atkins, K.A. McLauchlan, and P.W. Percival, Mol. Phys. 25, 281 (1973).
16. H.M. Vyas, S.K. Wong, B.B. Adeleke, and J.K.S. Wan, J. Am. Chem. Soc. 97, 1385 (1975).
17. B.B. Adeleke and J.K.S. Wan, J.C.S. Faraday I, 72, 1799 (1976).
18. S.K. Wong and J.K.S. Wan, J. Am. Chem. Soc. 94, 7197 (1972).
19. S.K. Wong, W. Sytnyk, and J.K.S. Wan, Can. J. Chem. 50, 3052 (1972).
20. S.K. Wong, D.A. Hutchison, and J.K.S. Wan, Can. J. Chem. 52, 251 (1974).
21. The polarization ratio P in Ref. 17 is equal to $-V(I)(1+V(R))^{-1}$ in our notation. This identification is approximate only since no allowance is made for spectrometer response.

22. If we make the identification in Ref. 22 and further assume that $-V_\ell(R) = V_{\ell'}(R)$ where $-\ell$ and $+\ell$ are corresponding low and high-field lines we can calculate the following values for the 1,4-naphthosemiquinone radical in Ref. 17, Table I:

line:	1	2	3	4	4'	3'	2'	1'
V(I):	-.33	-.32	-.35	-.27	-.27	-.35	-.32	-.33
V(R):	-.65	-.48	-.17	.15	-.15	.17	.48	.65

23. R. Livingston and H. Zeldes, J. Chem. Phys. 44, 1245 (1966).
24. P.W. Atkins and G.T. Evans, Mol. Phys. 27, 1633 (1974).
25. J. Boiden Pedersen and J.H. Freed, J. Chem. Phys. 62, 1706 (1975).
26. Calculated from Eq. (34) in Ref. 25.
27. D.R. Kemp and G. Porter, Proc. Roy. Soc. (London) A 326, 117 (1971).
28. N.C. Verma and R.W. Fessenden, J. Chem. Phys. 58, 2501 (1973).
29. P.B. Ayscough, T.H. English, G. Lambert, and A.J. Elliot, Chem. Phys. Lett. 34, 557 (1975).
30. H. Veenvliet and D.A. Wiersma, J. Chem. Phys. 60, 704 (1974).
31. A.I. Attia, B.H. Loo, and A.H. Francis, Chem. Phys. Lett. 22, 537 (1973).

CHAPTER XIII

LOW-FIELD EFFECTS AND CIDNP OF BIRADICAL REACTIONS*

G. L. Closs

Department of Chemistry, The University of Chicago,
Chicago, Illinois 60637, USA

ABSTRACT. The field dependence of the CIDNP effect based on the radical pair theory is discussed in terms of simple theory. Much use is made of perturbation arguments. Several experimental examples are taken from the literature and compared with theory. Next, the theory of CIDNP derived from biradical reactions is developed based on a simple model for the dynamic processes. The field dependence of the signals is discussed with the aid of experiments involving Norrish Type I cleavage of cycloalkanones. Some deficiencies of the model are pointed out. Two other reactions are also included in the discussion.

1. LOW-FIELD EFFECTS IN RADICAL PAIR BASED CIDNP

Most CIDNP experiments are carried out at magnetic field strengths corresponding to the operating fields of modern high-resolution nmr spectrometers. Since there is quite a difference between the operating fields of say, a Varian A-60 on one end of the scale and a superconducting magnet on the other, it is important to understand how the CIDNP effect depends on magnet field strength. In addition to these practical considerations additional information can be gained in many cases if the effect is studied over a range of field strength including very low fields. There are two aspects which distinguish low field experiments from those run in regular high fields, one experimental and one theoretical.

The experimental problem arises from the fact that nmr spectra recorded at low fields become less and less resolvable and

* This work has been supported by the National Science Foundation.

L. T. Muus et al. (eds.), Chemically Induced Magnetic Polarization, 225-256.

second-order effects begin to predominate in many cases. At fields of less than 1 Kgauss it may be impossible to distinguish between signals arising from different nuclei if there is substantial spin spin coupling. This problem can be overcome by running the reaction in a reaction field of low strength and then sampling the magnetization in a conventional nmr spectrometer operating at high field. The problems associated with this procedure will be discussed in more detail further below.

The theoretical problem associated with the low-field CIDNP arises from the breakdown of the so-called S-T_0 approximation. Most CIDNP spectra arising from radical pair reactions in high fields can be easily interpreted by considering mixing of only two electronic states of the radical pair, the singlet state, S, and the triplet level with $m_z = 0$, T_0. This is, of course, due to the fact that in a radical pair where the components are separated by several Å the T_0 and S levels are degenerate while the other two triplet levels, T_+ and T_-, are separated by the electron Zeeman splitting. As that splitting becomes smaller with decreasing field, mixing among all four states becomes more important and has to be incorporated into any realistic model. While it is possible to give explicit expressions for the polarization within the S-T_0 approximation, this is no longer the case for the complete mixing scheme involving all four states. Instead it becomes necessary to rely on numerical solutions which can be obtained by computer.

1.1 Theoretical considerations of low-field CIDNP[1]

In this section we want to explore the field dependence of CIDNP spectra with special emphasis on low magnetic fields (< 1 Kgauss). First, we will develop a qualitative picture of the physical reason for the field dependence. A more quantitative treatment, using elementary theory, will be given for the field dependence of biradical polarizations. A quick summary of the salient features of the radical pair theory is in order to facilitate discussion on the field dependence.[2]

We may distinguish among a number of events which are required to give nuclear polarization in the products of radical pair reactions. First, a radical pair is generated from a precursor by breaking a bond or by an electron transfer step. The time scale of this event is in the order of a molecular vibration of 10^{-12} seconds. By definition, a radical pair can be looked upon as two doublet states in which the two unpaired electrons interact between themselves only to a minor extent or not at all. In terms of a stationary state description of the radical pair states, this requires that the electronic wavefunction is a mixture of singlet and triplet components. However, when the pair is born it is not in a stationary state because the time scale of its formation is much shorter than the inverse of the interactions which mix singlet and triplet. As far as the spin function is concerned, at the time of the birth of the pair they are describable by the spin function of the precursor molecule, that is, either as singlet or as triplet (depending on the spin-multiplicity of the precursor). To put it in different terms, at its birth the radical pair is not in an eigenstate of the spin Hamiltonian describing it. If we use the electron spin singlet and triplet functions as a basis set to describe the system, the result of this sudden formation of the pair will be that the wavefunction will oscillate between the zero-order basis set functions. The frequency of the oscillations are determined by the interaction elements which mix the zero-order states. In radical pairs whose components are separated by several molecular diameters the main interactions mixing singlet and triplet functions are the hyperfine interaction, and in the presence of a magnetic field, the differences in the Zeeman energy of the individual components of the pair. The magnitude of these interactions in typical organic radical pairs is of the order of 10^8 rad/sec, giving us oscillation frequencies of 10^8 sec^{-1} between singlet and triplet functions of the pair. The next assumption of the radical pair theory is that the probability of product formation from the pair by either combination or disproportionation is proportional to the singlet character of the pair. This is a good assumption for all cases where the pair annihilation step involves the formation of a sigma bond (the assumption may break down in electron transfer reactions). Of course, the second requirement for pair annihilation is the necessity of an encounter of the two components. The probability of a geminate encounter is given by a pair distribution function which is approximated by the diffusion theory of Noyes. Since this function diminishes rapidly with time, it is clear that those pairs which acquire (or retain) singlet character most efficiently will predominate in the product derived from geminate pair collapse. It is this competition between the development of the wavefunction and diffusion which is the basis of the effect. The hyperfine interaction provides the mechanism to make the process dependent on the nuclear spin states.

To examine the effect of variation of the magnetic field we write the spin Hamiltonian describing the radical pair

$$\mathcal{H} = \beta H_o \cdot (g_1 S_1 + g_2 S_2) - J(\frac{1}{2} + 2\, S_1 \cdot S_2) + \sum_i a_i S_1 \cdot I_i + \sum_k a_k S_2 \cdot I_k \qquad (1.1)$$

where the first term allows for different Zeeman energies on the two components of the radical pair by defining two independent g-factors and S_1 and S_2 are the electron spin operators for the two components. The next term describes the exchange interaction which is the energy separation between the singlet and triplet state of the radical pair, more precisely the separation between T_o and S. In radical pairs J is effectively zero because the components have separated far enough on a time-scale short relative to the time scale of the evolution of the wavefunction. However, as we will see, J is an important nonvanishing parameter in the description of biradicals. The last two terms describe the isotropic part of the hyperfine interaction in the two components of the pair.

The spin functions of the radical pair states are generated from a basis set which is the direct product of the electron spin functions ϕ_v (v: S, T_+, T_o, T_-) and the nuclear spin function χ_n which are conveniently chosen as the basic product functions of the nuclear spins. In this basis set the spin Hamiltonian is not diagonal and the off-diagonal elements determine the time evolution of the wavefunction.

For example, the off-diagonal elements between T_o and S states take the form

$$\langle S\ \chi_n | \mathcal{H} | T_o \chi_n \rangle = \frac{1}{2}\ \beta\ H_o(g_1 - g_2) + \frac{1}{2} \sum_i a_i m_i - \frac{1}{2} \sum_k a_k m_k \qquad (1.2)$$

where m_i and m_k are the spin quantum numbers of the ith and k^{th} nuclear spin. If we apply the S-T_o approximation these are the only off-diagonal elements and the problem factors into a set of 2^n matrices of dimension 2 where n is the number of nuclei of spin 1/2. We see however that even in this case the system has a field dependence because of the first term in (1.2) which measures the difference in Zeeman energy between the two components of the radical pair. To elaborate on this, consider a radical pair with only one nuclear spin. Then the off-diagonal elements, H_{STo}, are

$$H_{STo,1} = \frac{1}{2}\ \beta\ H_o(g_1 - g_2) + \frac{1}{4}\ a \qquad (1.3)$$

$$H_{STo,2} = \frac{1}{2}\ \beta\ H_o(g_1 - g_2) - \frac{1}{4}\ a\ . \qquad (1.4)$$

Since in the diffusion model of Adrian, the probability of forming the geminate product in the nth nuclear spin state is related to $|H_{STo,n}|^{\frac{1}{2}}$, we see that the greatest difference between the populations in the product states is obtained when the first term in 1.3 and 1.4 equals the second. Within the S-T_o approximation then we expect for a given hyperfine interaction and $\Delta g = g_1 - g_2$, a field strength where the polarization goes through a maximum. This has the slightly amusing experimental consequence that a proton CIDNP experiment run at 60 MHz may yield good signal-to-noise ratios whereas the signals may be barely noticeable in a much more expensive 300 MHz system.

If the S-T_o approximation were valid down to very low field, we see that the polarization of a radical pair with one proton (or a set of protons with identical chemical shifts) would become less and less and finally vanish at zero field. In reality the situation is more complicated. The off-diagonal elements connecting the S states with T_+ and T_- must also be taken into account. The selection rules are such that there will be off-diagonal elements in the

Hamiltonian matrix between zero-order states which obey the relationship

$$\Delta m_S + \Delta m_I = 0, \qquad \Delta m_S = 0 \text{ or } 1 \tag{1.5}$$

which means that a change in the z-component of the electron spin angular momentum is accompanied by a change of opposite sign in the z-component of the nuclear spin-angular momentum. The magnitude of these elements is given by

$$|H_{ST\pm,i}| = \frac{1}{\sqrt{8}} a_i \tag{1.6}$$

where a_i is the hyperfine constant for the nucleus which undergoes the z-component change.

To examine when this problem of the breakdown of the S-T_0 approximation arises, we remember from perturbation theory

$$\phi_n = \phi_n^\circ + \sum_{m \neq n} \frac{H_{nm}}{H_{nn}^\circ - H_{mm}^\circ} \phi_m^\circ , \tag{1.7}$$

stating that the zero-order functions receive admixtures of the other zero-order functions weighted by the off-diagonal elements divided by the zero-order energy differences. In our case the zero-order energy difference between S and $T_\pm$ is essentially the Zeeman energy and the off-diagonal elements are of the magnitude of the hyperfine interaction. In a 60 MHz spectrometer the Zeeman energy is $\sim 4 \times 10^{10}$Hz while $H_{ST\pm} \sim 10^7$Hz for a typical organic radical. So we see why there is little mixing at high field. But if we drop the field significantly below 1 Kgauss the Zeeman energy becomes small enough that we have to consider all four electronic states.

Now let us develop a qualitative picture of what will happen if the Zeeman splitting becomes comparable to the hyperfine interaction. This will happen around a few hundred gauss. Again we use a one-proton radical pair, giving rise to 8 states. Assuming positive hyperfine coupling we can construct the energy diagram as shown in Figure 1a. Keeping in mind the selection rules for mixing we find that state $|S,\alpha\rangle$ mixes with $|T_+,\beta\rangle$, and $|S,\beta\rangle$ mixes with $|T_-,\alpha\rangle$, in addition to the mixing between T_0 with S, and T_0 with $T_\pm$. The latter does not concern us here. States $|T_+,\alpha\rangle$ and $|T_-,\beta\rangle$ are not mixed and remain pure. Inspecting the figure we see that for the case of J = 0 (T_0 and S degenerate) the energy separation between the two sets of mixed states is not equal. Keeping in mind equation 7 we therefore know that the degree of mixing between the S states and the T_+ and T_- states are not equal. Since the transition probability is related to the degree of mixing, the rate of $|T_+,\beta\rangle \rightleftharpoons |S,\alpha\rangle$ is greater than $|T_{-1},\alpha\rangle \rightleftharpoons |S,\beta\rangle$. If the precursor of the radical pair was in a triplet state, the population of $|S,\alpha\rangle$ will be greater than $|S,\beta\rangle$. This leads to enhanced absorption in the product ($|\beta\rangle$ is the upper state in the product). If the hyperfine interaction is negative the opposite situation is the case (Figure 1b).

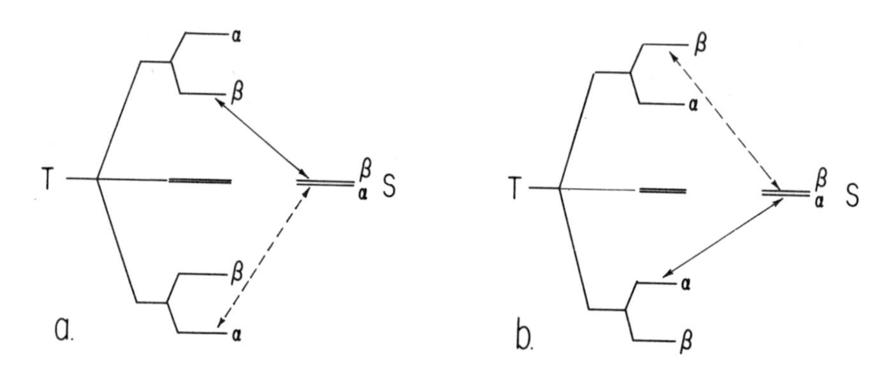

Fig. 1. Energy levels of a 1-proton radical pair; a) $a_i>0$; b) $a_i<0$. Predominant transitions are indicated with solid arrows.

At this point it is interesting to note that geminate product and escaping radicals carry the same polarization. This is easily verified considering that in the case discussed above the $|T_+,\beta>$ state is depleted more than $|T_-,\alpha>$ and the remaining states, $|T_+,\alpha>$ and $|T_-,\beta>$ carry equal population. This leads to excess radicals with $|\alpha>$ spin states, just as in the coupling product. This is in direct contrast to what happens at high field where predominant S-T_0 mixing leads to opposite polarization for geminate product and escape product.

To obtain a quantitative expression for the field dependence the problem has to be solved by taking all mixings into account. Since this is similar to what has to be done for biradicals, we will postpone the discussion of it until biradicals are discussed. However, we can summarize at this point what we expect to happen in a one proton radical pair simply on the basis of predictions from perturbation theory.

i. Singlet precursors give opposite polarization from triplet precursors and uncorrelated pairs at <u>all</u> fields.

ii. Change in the sign of the hyperfine interaction changes the sign of the polarization at <u>all</u> fields.

iii. The sign of Δg influences the sign of the polarization only at high fields.

iv. Escape products have opposite polarization from geminate combination product at high field and the same polarization at low fields.

v. For the case of a one-proton radical pair, and more general, for radical pairs in which the nuclear spins are not coupled among themselves the polarization goes to 0 at zero-field.

Some idealized field dependence curves are shown in Figure 2. The figure caption lists the parameters.

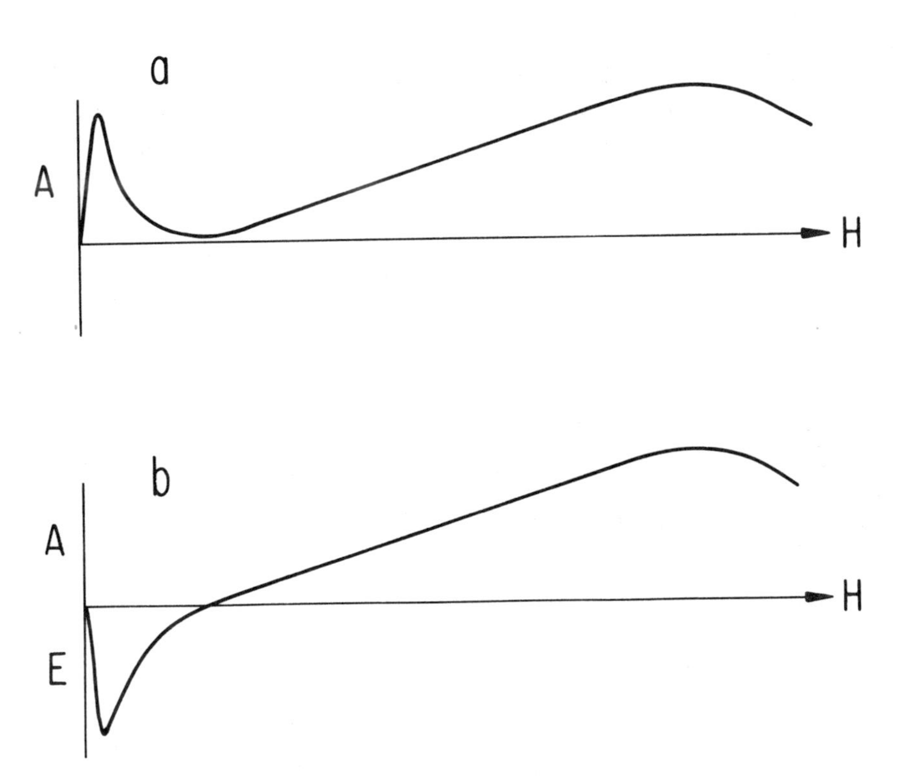

Fig. 2. Schematic representation of the polarization predicted for the geminate product derived from a one-proton radical pair. Assuming a triplet precursor the field dependence shown in a) requires $\Delta g>0$ and $a_i>0$; the curve b) is for $\Delta g<0$ and $a_i<0$.

As noted under (v) the polarization vanishes at zero-field if there is only one nuclear spin. This is no longer true for a pair containing two or more nuclear spins which carry hyperfine interaction and are coupled in the product by a nuclear spin-spin coupling constant J_{ij}. It is worthwhile to remember that at zero-field the nuclear spin Hamiltonian consists only of the spin-spin interaction which in the absence of dipolar interactions reduces to

$$\mathcal{H} = \sum_{i>j} J_{ij}\, I_i \cdot I_j \tag{1.8}$$

Similarly the radical pair Hamiltonian contains only the hyperfine interaction; assuming a vanishing electron exchange coupling this leaves us with

$$\mathcal{H}_{RP} = S_1 \cdot \sum_i a_i I_i + S_2 \cdot \sum_k a_k I_k \quad . \tag{1.9}$$

If we define a total nuclear spin operator $K = \sum_i I_i$ then the product states are eigenstates of K^2 and K_z and are characterized by the quantum numbers K and M_k. For example a two-proton system gives three levels, $|1,1\rangle$; $|1,0\rangle$ and $|1,\bar{1}\rangle$ with K = 1 (triplet) and one, $|0,0\rangle$ with K = 0 (singlet). The eight levels of a three proton system factor into 4 quartet and 4 doublet states and so on. It can be shown that at zero field the radical pair model predicts that levels which belong to the same K branch are equally populated independent of M_k, but levels which have different K may differ in their population. While this can be proven rigorously, it suffices to say here that is it intuitively reasonable. The radical pair model considers only isotropic interactions and within this assumption the levels within each K branch are degenerate and indistinguishable from one another at zero-field. There is no preferred axis for quantization!

If a spin system prepared this way at zero-field is adiabatically transferred into a nmr spectrometer operating at high field the resulting spectrum consists of the so-called "n-1 multiplets."

This is best illustrated by considering a two-proton radical pair at zero field. The degenerate triplet levels are assumed to receive all the excess population. Adiabatic transfer to high field gives a state diagram as shown below in Figure 3. Only two of the four lines will carry measurable intensities because there is no population difference associated with the other transitions.

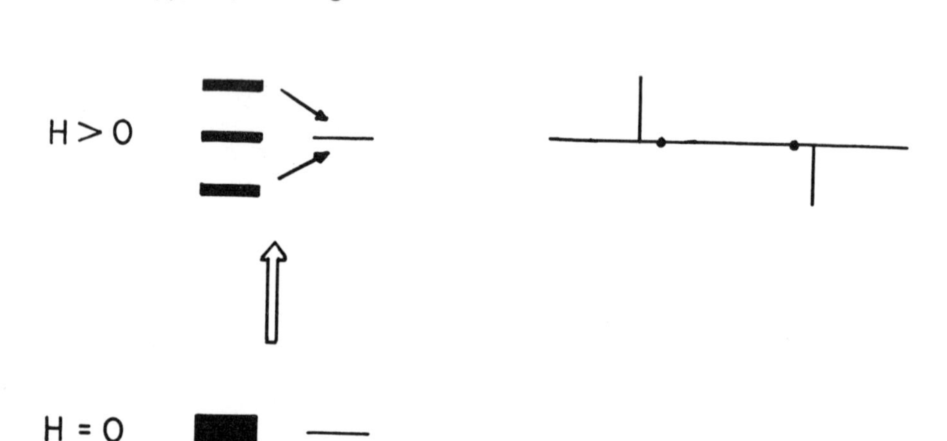

Fig. 3. Energy level scheme showing the origin of the "n-1 multiplet" effect for a two-proton radical pair.

It is generally true for simple multiplets that one set of lines arising from spin set i has the opposite polarization as the other associated with spin set j. Accordingly Kaptein was able to extend his simple rule to zero-field multiplets.

$$\mu \; \varepsilon \; A_i A_j J_{ij} \sigma_{ij} \} \begin{array}{l} + \text{EA} \\ - \text{AE} \end{array} \qquad (1.10)$$

Here EA stands for emission of the low field multiplet and absorption for the high field multiplet and AE is defined accordingly. All other parameters have the same meaning as defined for the high-field multiplet effect.

1.2. Experimental considerations

As has been pointed out in the Introduction there are experimental problems associated with low-field experiments. About the only practical way to get information on polarization at low field is to run the reaction in a separate magnet and then transfer the sample to a conventional high-resolution spectrometer. However, there are two problems associated with this technique. The first has to do with the fact that the polarization decays with the spin lattice relaxation time which is typically of the order of seconds. This means the sample has to be transferred and the nmr spectrum has to be recorded within that relaxation time. Mechanical transfer by hand is possible within a period of three to five seconds but it must be remembered that this time has to be repeatable to the highest possible accuracy because one is usually interested in comparing peak intensities at various field strengths.

The second problem is associated with the necessity that the reaction has to be stopped after it has been run at the low reaction field and the sample is being analyzed at high field. This creates serious problems in thermochemical reactions. Since the cooling of a nmr sample tube by tens of degrees will take at least several seconds, this time requirement competes seriously with the requirement to make the measurement no later than a few T_1's. On the other hand in photochemical reactions this problem does not exist since by definition the reaction stops when the light is turned off.

The problem of making the transfer time as repeatable as possible can be solved in two ways. The first involves activating a computer timing sequence at the moment the sample is taken out of the low reaction field. That timing sequence also controls the spectrometer, either starting a sweep in CW mode, or controlling the rf-pulse sequence in Fourier transform spectroscopy.

Another, perhaps more accurate way of transferring the sample from the reaction field to the measuring field is via a flow system.[3] If the diameter of the flow tube is kept small enough transfer times of less than 2 or 3 seconds can be achieved. Laboratories that routinely want to get low-field information are well advised to invest in the relatively small expenditure of building a flow system.

There is a potential problem with transfer experiment which so far seems to be mostly theoretical. If a spin system is transferred from one magnet to another, it is a question whether the transfer takes place adiabatically or not. To put it in other words, the question is whether the populations of the eigenstates at low field will be transferred without change to the high-field eigenstates with which the low field-states correlate. For example, if we have a product with two protons coupled with a coupling constant J_{ij} of a few Hz, there will be little mixing at high field between the states $|\alpha,\beta>$ and $|\beta,\alpha>$ if $\delta_i - \delta_j > 10\ J_{ij}$. At low field however $\delta_i - \delta_j$ will no longer be much larger than J_{ij} and substantial mixing of these states occurs. To transfer such a population adiabatically, the field change at the low field end has to occur slower than a time which is the inverse of $2\pi J_{ij}$. In principal, at least with small coupling constants, this condition may not be met and transitions between the two states could be induced. In practice no such experiment has yet been reported although recognition of this effect might easily escape attention.

1.3 Illustrative examples

A set of nice low field experiments were carried out by Kaptein and den Hollander.[4] Among the many low field experiments reported by them two are selected for discussion because they demonstrate particular well the points made in the discussion. The n-1 multiplet effect is beautifully demonstrated by the reaction of propionylperoxide in the presence of CCl_3Br. The reaction sequence is the photochemical decomposition of the peroxide yielding a geminate pair of ethyl radicals in the singlet state and the escaping radicals are trapped by CCl_3Br to give ethyl bromide. It is the A_3B_2 spectrum of ethyl bromide which shows the n-1 multiplets at 0.5 gauss shown in Figure 4.

$$(C_2H_5COO)_2 \xrightarrow{h\nu} \overline{2C_2H_5} \xrightarrow{CCl_3Br} C_2H_5\text{-}Br + \cdot CCl_3$$
$$\overline{2C_2H_5} \longrightarrow C_4H_{10} \qquad (1.11)$$

Application of rule 1.10 predicts an AE multiplet as is found. As the field is gradually increased, the inner lines begin to appear. While the methylene spectrum does not change very much up to sixty gauss, the CH_3 lines undergo a complicated transformation. Using the simple perturbation argument advanced above, one might have

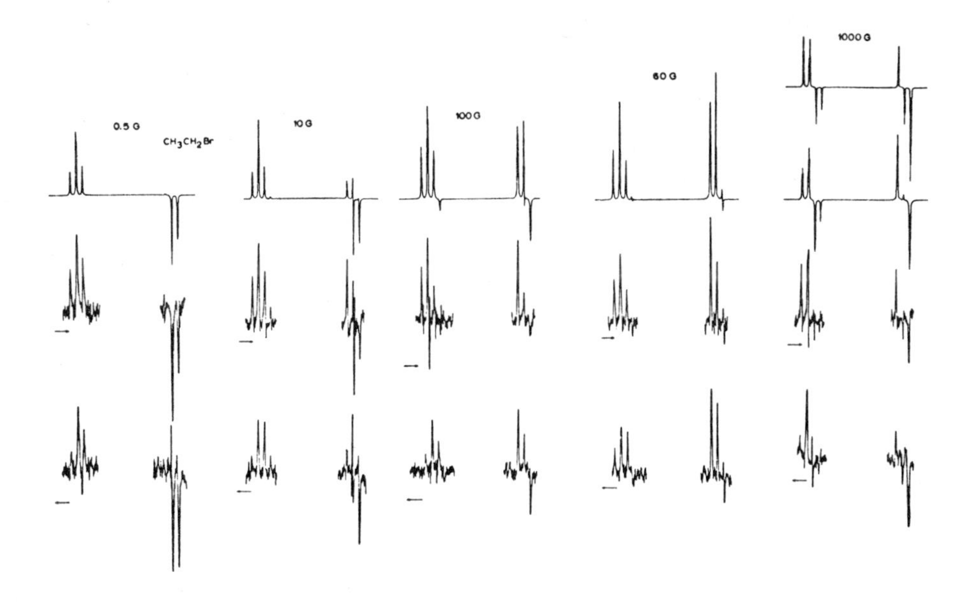

Fig. 4. The spectra of ethyl bromide obtained from the photolysis of propionyl peroxide in CCl_3Br. The reaction field strength is indicated above the calculated spectra. The experimental spectra are shown in both sweep directions. The calculations included a finite exchange term of -5×10^8 rad/sec. Later work showed this to be unnecessary when all the nuclear spins were taken into consideration (see reference 4, footnote 24).

expected that the CH_2 lines are in enhanced absorption because negative a_i should give predominant T_- mixing. This is in agreement with experiment. However the complicated behavior of the CH_3 lines is not easily unraveled from simple two-state perturbation arguments. Numerical calculations are needed to simulate this behavior. It is simply not possible to make perturbation arguments for a system of 4×2^{10} states! At high field the spectrum goes over into the normal AE multiplet effect predicted from the simple rules.

The complexity of the behavior of the polarization at low field is demonstrated with the photolysis of di-isopropyl ketone in CCl_4. The reaction proceeds apparently from a singlet state exciplex containing the CCl_4 giving rise to the isopropyl-trichloromethyl geminate pair. The disproportionation products are chloroform and propene (R(-H)). Although cloroform has only one proton the problem strains available computer programs because the pair contains seven spin 1/2 and 3 spin 3/2 nuclei, the latter distributed among two isotopes with different coupling constants.

$$R{-}CO{-}R + CCl_4 \xrightarrow{h\nu} \overline{R \cdot CCl_3} \longrightarrow \text{diff.} \xrightarrow{CCl_4} R{-}Cl$$

$$\downarrow \quad \downarrow$$

$$R{-}CCl_3 + R(-H) + CHCl_3 \qquad (1.12)$$

R = isopropyl

The low field dependence of the chloroform resonance is shown in Figure 5 and shows two sign changes. The change at 500 gauss is

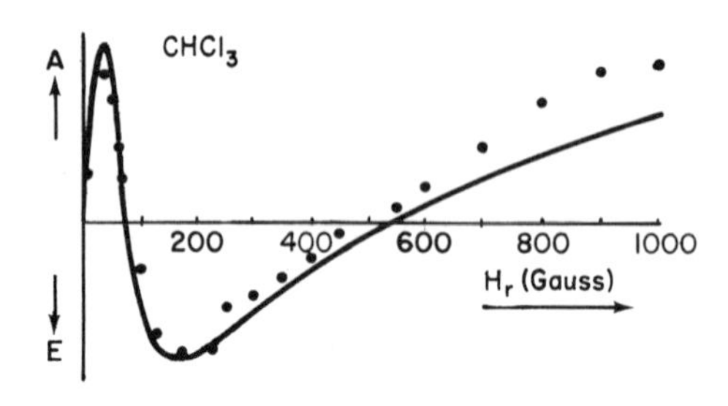

Fig. 5. Field dependence of the CCl_3H proton generated in reaction (1.12).

caused by the onset of the Δg effect while the low-field change comes from the complicated mixing of the thousands of states. Interference effects among these transition make an analysis by all but numerical methods impossible.

Many other low-field experiments on radical pairs have been reported[5] but for illustrative purposes the two examples should suffice.

One additional point is worthwhile mentioning. Although we have added an exchange coupling term into the Hamiltonian it appears that none of the experimental spectra require this term for their quantitative explanation. In the early papers simulations were often carried out with assuming a finite J. But it developed that when the calculations were repeated including all the nuclei (including spins of halogen atoms) the exchange term was no longer needed. As pointed out above this is expected for the diffusion model. However, there are CIDNP spectra which require the inclusion of an exchange term for an explanation and those are spectra generated from biradical reactions to which we shall turn our attention now.

2. CIDNP GENERATED FROM BIRADICALS

Biradicals are similar to radical pairs by having two weakly interacting electrons, giving rise to near degeneracy of singlet and triplet states. However, there are two important differences which must be taken into consideration in any successful theoretical treatment. The first one is the fact that the electron exchange interaction does not vanish during the lifetime of the biradical as it does in diffusing radical pairs. The second, closely related, difference arises from the fact that electron spin correlation is not lost by a diffusion process. This has the consequence that the oscillating spin functions, which are damped by a radical pair distribution function in the radical pair model, must undergo damping by some other process in biradicals. At first sight it may appear that a simple experimentally decaying function characterized by the lifetime of the biradical may be a suitable function for this purpose, but we will see that this view is too simplistic since the lifetime of the radical pair is determined by more than one process.

2.1 Theoretical considerations[6]

To develop a model for biradical based CIDNP we assume the simple chemical sequence

$$\text{precursor} \longrightarrow \text{BR} \longrightarrow \text{product} \qquad (2.1)$$

in which BR stands for a biradical. Just as for radical pairs we assume that the bond breaking step leading to BR is too fast to proceed adiabatically with reference to the electron spin coordinates. At the time of its birth BR will be in a spin state that was an eigenstate of the spin Hamiltonian of the precursor but is not an eigenstate of the BR Hamiltonian. Another assumption taken over from the radical pair theory is that of nonadiabatic product formation. Since the product is almost always a singlet state the probability of the product formation will be proportional to the singlet character of the wavefunction. The oscillations between the singlet and triplet function will again be determined by the parameters of the spin Hamiltonian which is taken over from the radical pair theory (equation 1.1), but we specifically retain the exchange interaction. Here we encounter our first new problem. Consider the biradical to be a flexible hydrocarbon chain with the two electrons localized on the terminal carbon atoms. During the lifetime of BR the chain will undergo random conformational changes which will influence the magnitude and perhaps even the sign of the exchange coupling constants. This means J is a function of time. Since it is very difficult to evaluate the time dependence of that function in a proper way, we replace it by a time averaged constant function $\langle J \rangle$.

The next difficulty arises from the fact that intersystem crossing between the zero-order states of BR can be expected to occur via other non nuclear spin dependent processes as well. Spin orbit coupling is probably one of the more important mechanism. The efficiency of that process depends on the proximity of the two radical centers and thus is also time dependent. Instead of evaluating this function we treat intersystem crossing by a rate constant, k_{so}.

One more assumption is needed, this is the rate of depletion of the singlet state. It is reasonable to assume that this process is characterized by a rate constant, k_V, characterizing the lifetime of the biradical if it were completely in the singlet state. The processes are summarized in scheme (2.2) where the biradical states are connected via the spin Hamiltonian and the BR decay

Precursor ⟶ ≡ $\overset{\mathcal{H}}{\longleftrightarrow}$ —

k_{so} ↘ $\quad$ ↓ k_v

product $\quad$ (2.2)

processes are indicated by the rate constants k_{so} and k_V. The probability of formation of the individual nuclear spin states in the product is then given by the vector $\underset{\sim}{P}$ which is related to the populations in the RP state by

$$\frac{d}{dt}\underset{\sim}{P} = k_v\, |\underset{\sim}{C}_S(t)|^2 + k_{so} \sum_j |\underset{\sim}{C}_{T_j}(t)|^2 \qquad j = T_+;\ T_o;\ T_- \tag{2.3}$$

where $|C_S|^2$ and $|C_t|^2$ are the vectors describing the singlet and triplet character as a function of time. To obtain the time dependence of these vectors we have to define an operator containing the spin Hamiltonian and a provision for the diminishing populations in the RP states. This is conveniently done by defining an effective Hamiltonian of the type

$$\mathcal{H}_{eff} = \mathcal{H}_{sp} + i\Gamma \tag{2.4}$$

where Γ is a diagonal damping matrix containing as element for the singlet states $k_V \cdot 2\pi$, and for the triplet states $k_{so} \cdot 2\pi$.

The remainder of the treatment is straightforward and equivalent to that of the radical pair theory except that the operator is no longer hermitian and will give complex eigenvalues of the type

$$\omega'_r = \omega_r - i\gamma_r \, , \tag{2.5}$$

where the γ's are close to the rate constants for well separated zero-order states. Using the same basis set as in the radical pair theory and using the time dependent Schrödinger equation

$$i \frac{\partial}{\partial t} \Psi = \mathcal{H}\Psi \tag{2.6}$$

we can write as a solution for the coefficients

$$|C(t)|^2 = \tilde{Q} \, Z^* \, e^{i\omega' t} \, Z^{-1*} \, |C(0)|^2 \, Z \, e^{-i\omega' t} \, Z^{-1} \, Q \tag{2.7}$$

Here Z and Z* are the eigenvector matrices and their complex conjugates, $|C(0)|^2$ is the initial population vector of RP (singlet or triplet) and Q is an operator whose product with its transpose will project the radical pair states onto the nuclear eigenstates of the product.

The evaluation of the elements of (2.7) is given by

$$|C_n(t)|^2 = \sum_{ij} \sum_m \sum_r |Q_{mn}|^2 |C_{rr}(0)|^2 \, Z^*_{mi} \, Z^*_{mj} \, Z^{-1}_{ri} \, Z^{-1}_{rj} \, e^{-i\omega_{ij} t} \, e^{-\gamma_{ij} t} \tag{2.8}$$

where the real parts of the eigenvalues have been separated from the imaginary, and

$$\begin{aligned} \omega_{ij} &= \omega_i - \omega_j \\ \gamma_{ij} &= \gamma_i + \gamma_j \, . \end{aligned} \tag{2.9}$$

Inspection shows that there is not only an oscillatory term but also an exponential decay term as is required for proper dissipatory behavior. Equation (2.8) has to be evaluated for all states, singlet and triplet and substitution into (2.3) and integration gives the population rates of the nuclear eigenstates in the product as

$$P_n = \sum_{ij} \sum_m \sum_r |C_r(0)|^2 \, \{k_v |Q_{s,mn}|^2 \, Z^*_{mi} Z^*_{mj} Z^{-1}_{ri} Z^{-1}_{rj} \gamma_{ij} / (\gamma^2_{ij} + \omega^2_{ij}) + k_{so} |Q_{T,mn}|^2 \, Z^*_{ml} Z^*_{mj} Z^{-1}_{ri} Z^{-1}_{rj} \, \gamma_{ij} / \gamma^2_{ij} + \omega^2_{ij})\} . \tag{2.10}$$

This equation is not very instructive from an experimentalist's point of view. While it is an exact solution of the problem as defined and is valid for all field strength its predictive value is limited by the assumptions in the model. Let us examine the deficiencies of the model in some more detail before we go to some predictions based on some more qualitative arguments.

First, the derivation is based on scheme (2.2) which treats spin orbit coupling induced intersystem crossing as irreversible. This restricts the model to cases where the biradical is generated from a triplet precursor. However, this deficiency is not as serious as one might think at first sight because it follows from the principal of microscopic reversibility that in a purely intramolecular reaction no polarization can result from a singlet precursor. This arises from the fact that the triplet state is only a branch on the reaction coordinate which within our model does not lead to product. Only if there is another exit channel from the triplet, as for example rearrangement or intermolecular reaction, is it possible to obtain polarization with a singlet precursor.

Another deficiency of the model is the failure to allow for relaxation among the three triplet levels. The consequences of this deficiency can qualitatively be explained in the following way. If the exchange interaction is of the order of the Zeeman splitting one of the triplet levels, T_+ or T_- depending on the sign of J, is mixed with the S level much more strongly than the other two. Consequently, in a steady state there will be a substantial population difference between the levels in the absence of relaxation. If there is an efficient mechanism for relaxation so that $1/T_1$ become comparable or greater than the intersystem crossing rate, the population difference gets diminished and the whole system may drain through the strongly mixed level. This can be a real problem because relaxation in biradicals may be much more efficient than in radicals because of the modulation of the dipolar electron-electron interaction by molecular tumbling. The magnitude of the dipolar interaction depends on the distance of the two electrons with an $1/r^3$ dependence. Simple calculations show that if the electrons are in the average 5 Å or less apart, this dipolar mechanism may provide for relaxation rates which become competitive with the intersystem crossing rates. To complicate matters even further, the relaxation rates are field dependent and become larger at low fields. This field dependence is given by

$$\frac{1}{T_1} = \frac{2}{15} \; {}^{*}D^2 \; \tau_c \left[\frac{1}{1 + \omega^2\tau_c^2} + \frac{4}{1 + 4\omega^2\tau_c^2}\right] \qquad (2.11)$$

where $^{*}D$ is the zero-field splitting parameter, measuring the average distance of the two electrons, τ_c is the rotational correlation time and ω is the electron Larmor frequency ($\omega = g\beta H_o$). At typical nmr field strength of $\sim$20 Kgauss and correlation times of 10^{-11} seconds, $\omega^2\tau_c^2 >> 1$. This provides a field dependence down to low field where $\omega^2\tau_c^2 \ll 1$.

It is possible to incorporate equation (2.11) into a model to give a revised equation for P_n, making that expression even more complicated. Instead of doing that we are more interested to make some simple predictions based on perturbation arguments as to the qualitative features of spectra obtained from biradical reactions.

With the aid of a hypothetical one proton biradical let us examine how the polarization depends on the parameters of the spin Hamiltonian and precursor multiplicity. Assuming the singlet state to be more stable than the triplet state we get the state diagram (2.12) in which positive hyperfine interaction has been assumed.

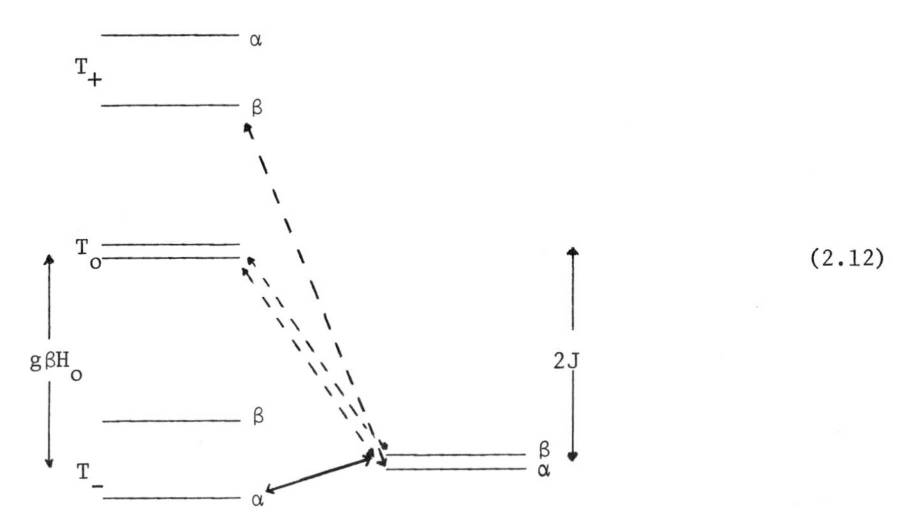

Since mixing between states is inversely proportional to the energy separation, $|T_-,\alpha\rangle \leftrightarrow |S,\beta\rangle$ mixing will predominate. In the dynamic model this corresponds to the largest transition probability. So if we enter the system from a triplet precursor, we expect the product nuclear states to predominate in β spins corresponding to emission signals. If we change the hyperfine interaction to negative, this will interchange $|T_-,\alpha\rangle$ with $|T_-,\beta\rangle$ and an equivalent interchange in the T_+ levels, but will have no consequence on the predominance of T_-,S mixing. Consequently the spectrum will still be emissive.

Next, let the sign of J become positive corresponding to a triplet ground state of the biradical. In this case the smallest energy difference will be between S and T_+ giving the greatest transition probability for $|T_+,\beta\rangle \leftrightarrow |S,\alpha\rangle$. The result will be enhanced absorption for a triplet precursor.

If we enter via the singlet level we need a different exit channel from the triplet levels to get any polarization at all and their signs will be reversed from those observed for a triplet precursor. We can summarize the predictions:

i. For negative J and triplet precursor: emission

ii. For positive J and triplet precursor: enhanced absorption

iii. Singlet precursor give opposite signs of polarization than triplet precursors.

iv. The sign of the hyperfine interaction has no effect on the spectrum.

An interesting question is what happens to the molecules in $|T_-,\beta>$ and $|T_+,\alpha>$ which are not mixed with S? One way for them to leave the triplet state is to undergo relaxation transitions, $|T_-,\beta> \rightarrow |T_+,\beta>$ and $|T_+,\alpha> \rightarrow |T_-,\alpha>$, and then cross to the S states via the hyperfine coupling allowed transition. If that were the only mechanism there would be no polarization left in S (except that what remains from the Boltzmann distribution). The molecules in the T_o states will not contribute to the polarization either, since the $T_o \leftrightarrow S$ transitions do not involve a nuclear "spin flip." So we need another competing mechanism and this is the spin orbit induced intersystem crossing which is nuclear spin independent and has been included in our model by the rate constant k_{so}. This process is very closely related in function to the diffusion process in the radical pair mechanism.

Next we examine the field dependence in radical pair spectra. From the state diagram (2.12) we see that the energy separation between S and T_-, E_{ST-}, is given by

$$g\beta H_o + 2J = E_{ST_-}, \tag{2.13}$$

and

$$g\beta H_o - 2J = E_{ST_+} \tag{2.14}$$

gives the separation between S and T_+. Since H_o is an experimental variable it should be possible to make S degenerate with either T_- or T_+ depending on the sign of J. In the case of degeneracy the mixing is the strongest and we predict a maximum of the polarization if

$$g\beta H_o - 2J = 0. \tag{2.15}$$

This is an important finding because not only do CIDNP spectra of biradicals allow us to determine the sign of J but from the field dependence we can get the magnitude as well. Of course there is an experimental limitation on the available field strength. Assuming a superconducting magnet with a field strength corresponding to a proton frequency of 360 MHz (the present upper limit of nmr spectrometers) the maximum value for the singlet-triplet splitting to be determined this way is 7.9 cm^{-1} or 22.6 cal/mole.

There is a complication factor in this simple consideration arising from the uncertainty principle. The lifetime of the singlet state in biradicals can be expected to be very short. Of course it will be dependent on the structure but in small chains the closure rate may exceed $10^{10}sec^{-1}$. This will lead to a substantial lifetime broadening of the S level. At a 10^{-10} seconds lifetime this corresponds to a broadening of 600 gauss and the polarization maxima will be smeared out correspondingly. On the other hand, this opens up the possibility of obtaining estimates for the singlet closure rate, k_v, in biradical reactions. The broadening of the triplet state is much less because of its much longer lifetime. Nevertheless, there is some effect on the shape of the field dependence curves if k_{so} is made longer. The effect of k_v and k_{so} on the polarization-field curves is shown in Figures 6 and 7 where each of the rate constant in equation (2.10) has been varied while the other has been kept constant.

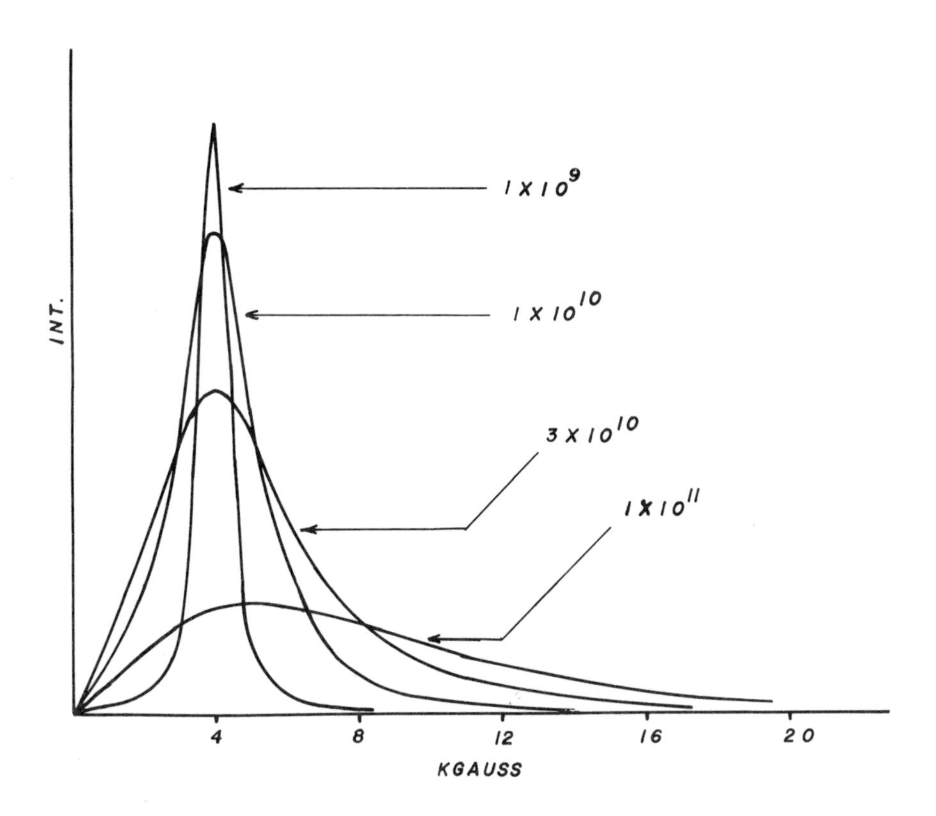

Fig. 6. The effect of varying k_v on the polarization-field curves. 2J and k_{so} are kept constant at 4000 gauss and 1 x 10^6 rad/sec (k_v is in units of rad/sec).

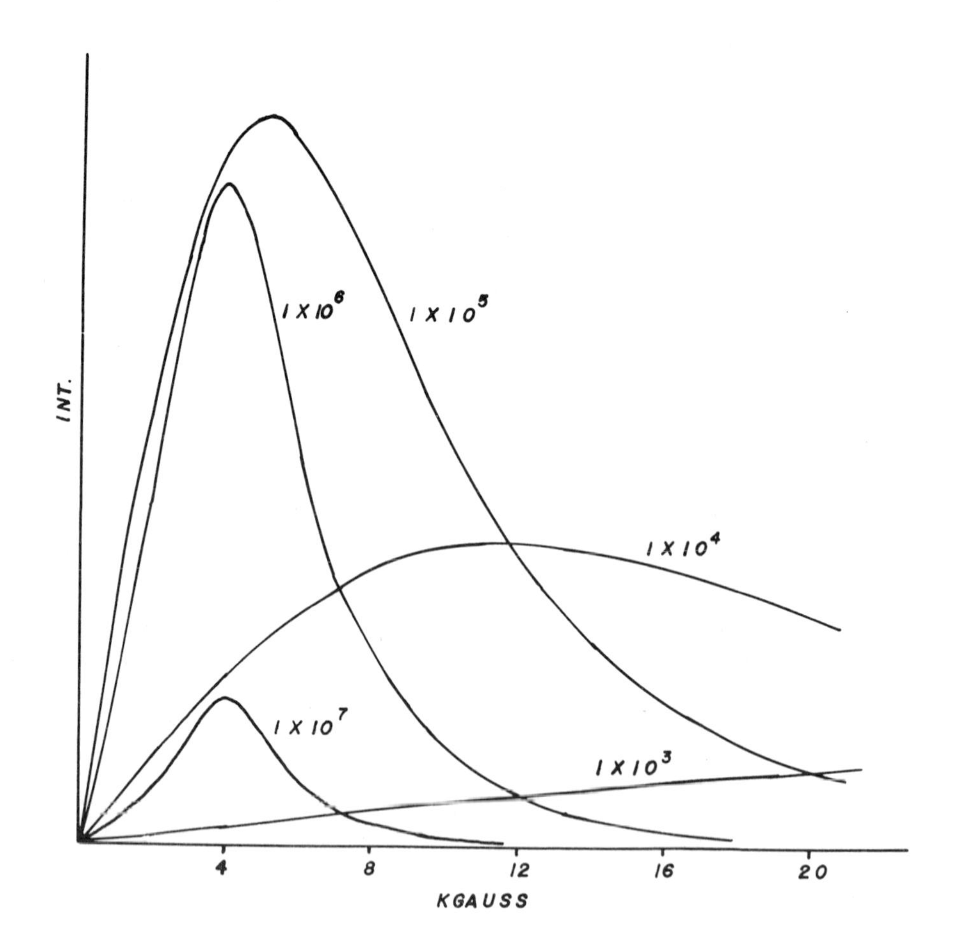

Fig. 7. The effect of varying k_{so} on the polarization-field curves. 2J and k_v are held constant at 4000 gauss and 1×10^{10} rad/sec (k_{so} is in units of rad/sec).

The model for CIDNP in biradical reactions is a crude one in its present form, and many improvements can be made. However, it does account for the major features of CIDNP spectra and their field dependence. We shall now discuss some examples.

2.2 Experimental data and discussion

A particularly well suited system to study biradicals by CIDNP is the Norrish Type I cleavage reaction of alicyclic ketones.[7,8] As summarized in scheme 2.16, the ketone is photoexcited into its first excited singlet state and crosses over efficiently into the triplet state, which undergoes α-cleavage to give the biradical. The biradical can partition between regenerating ground state ketone and disproportionation to the unsaturated aldehyde.

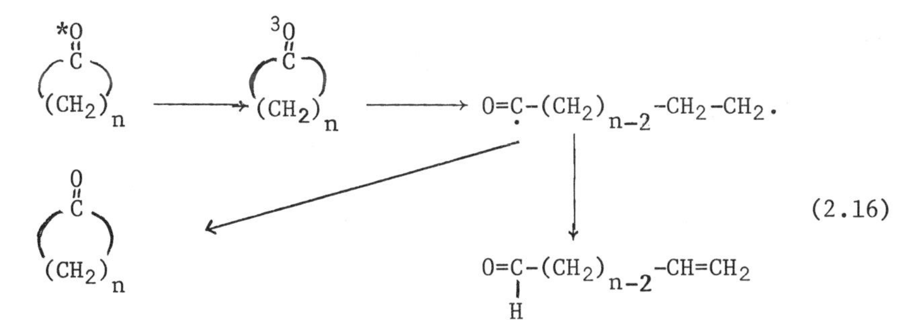

When this reaction is run inside an nmr spectrometer, strong polarizations are observable for ketones from n = 4 - 10. A good example is shown in Figure 8 which shows the spectra obtained from cycloheptanone in the dark and light. A number of conclusions can be drawn just by inspection.

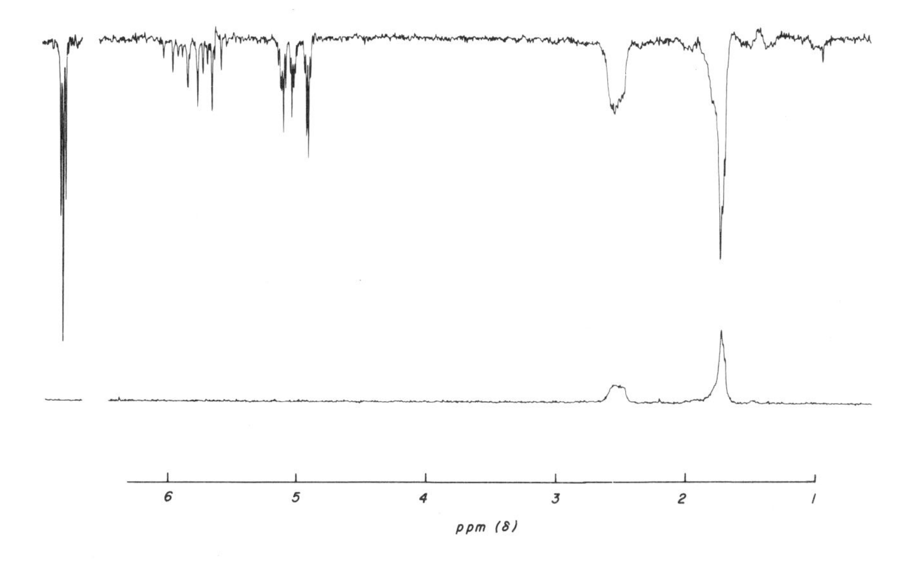

Fig. 8. Spectra obtained from cycloheptanone in $CDCl_3$. Lower trace is the dark spectrum, upper trace is the spectrum obtained under uv-irradiation.

First, both cyclic ketone and unsaturated aldehyde are formed from the biradical. Integration of the respective signal areas (corrected for the dark signal) divided by the corresponding nuclear relaxation times will give the ratio of cyclization to disproportionation. This ratio is rather difficult to determine by other means because the cyclization product is identical with the reactant.

Next, we note that all signals are in emission regardless of the signs of the hyperfine coupling constants (the methine and the methylene protons experience hyperfine coupling of opposite sign in the biradical). Since the reaction is known to proceed through a triplet state, this is indicative of predominant T_--S mixing. J must be negative and the singlet state is the ground state of the biradical.

The series of cyclic ketones from cycloheptanone to cycloundecanone give essentially similar spectra. In all cases T_--S mixing predominates indicating singlet ground states for the biradicals. The spectra show decreased polarization with increasing ring size partly due to side reactions. A 1,6-diradical generated from the bicyclic ketone (2.17) also behaves this way, although

(2.17)

very little ring closure product is formed in this case. Figure 9 shows that the tertiary α-proton (at δ 2.8) does not carry any polarization in the ketone spectrum. Presumably ring closure is slowed down somewhat because of strain in the bicyclic compound. Cyclohexanone and cyclopentanone behave differently and will be discussed below.

The field dependence of the aldehyde proton signal is easy to measure by the manual transfer method because of the long T_1. The results are shown in Figure 10, where all intensities have been normalized to given identical heights at their maxima. Two characteristics of these curves are apparent on inspection. First the maxima occur at lower fields with increasing ring size of the ketone. From the qualitative considerations we expect decreased singlet-triplet splittings to move the polarization maxima to lower fields. We find, in agreement with chemical intuition, that the larger the biradical chains become, the more the exchange interaction is reduced. The second prominent feature characterizing the curves is their width. As has been discussed above, the width of the polarization-field curves is related to the lifetime of the singlet state. It is expected that the longer chains have

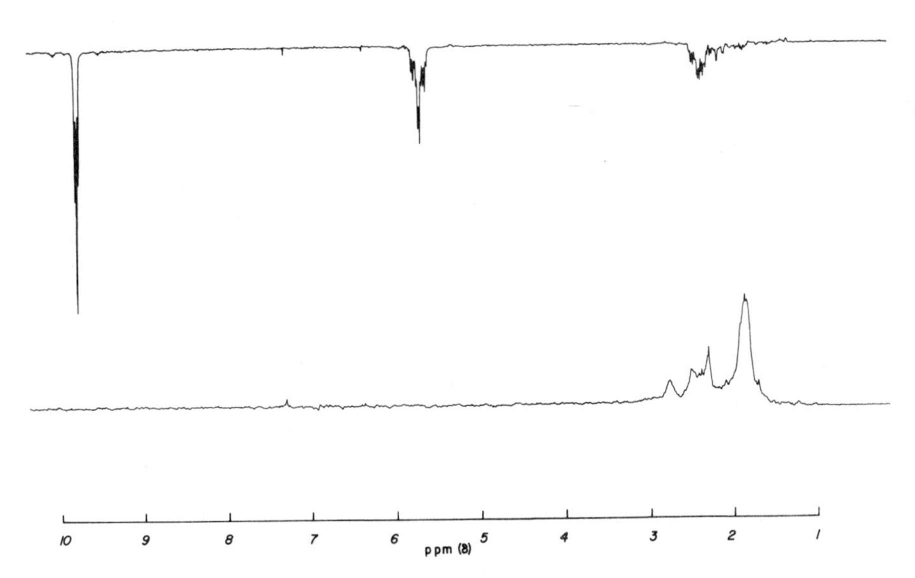

Fig. 9. Spectra of the bicyclic ketone (2.17) in $CDCl_3$. Lower trace: dark spectrum; upper trace: light spectrum.

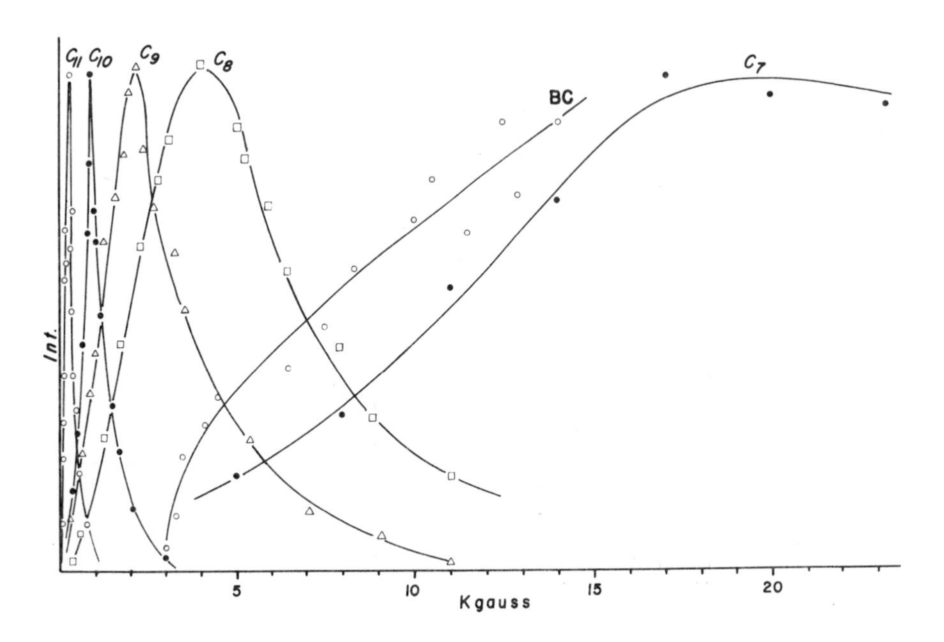

Fig. 10. Normalized intensity vs. field curves of aldehyde proton signals obtained on photolysis of cycloalkanones and bicyclic ketone (2.17) labelled BC. All signals are in emission.

a smaller probability per unit time of having their ends within reacting distance than the shorter chains. The relative width of the curves reflect the anticipated behavior.

To get more quantitative information the experimental data were fitted to equation 2.10 by adjusting the three variables, J, k_V and k_{SO}. The results for three characteristic examples are shown in Figure 11, which shows a satisfactory fit of the data points with the calculated polarization. The parameter used for these fits are summarized in the Table.

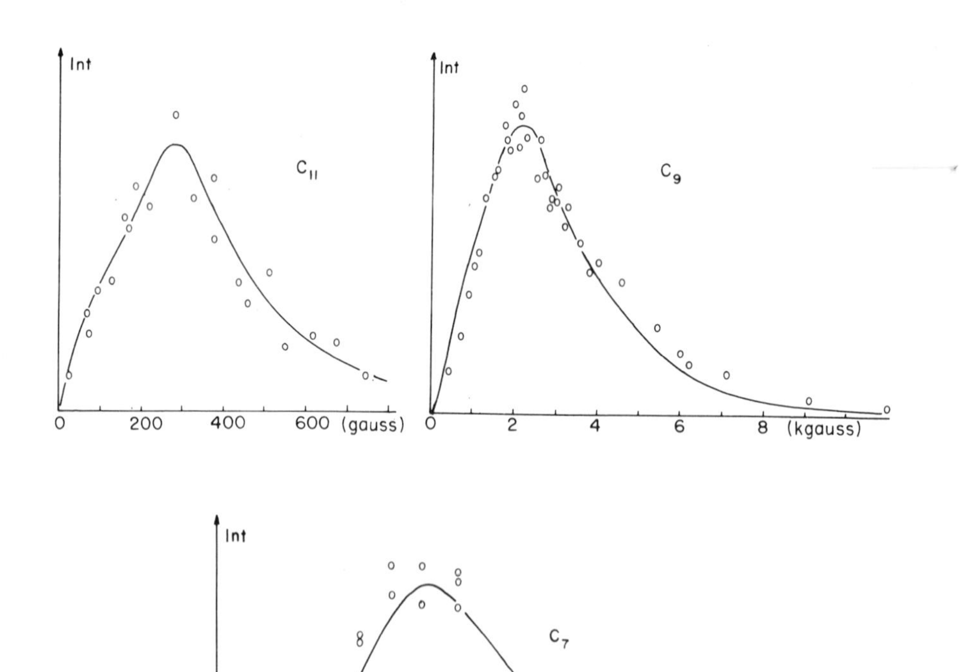

Fig. 11. Calculated intensity vs. field curves for the aldehyde protons obtained on photolysis of three characteristic cyclic ketones. The circles are the experimental points.

FITTING PARAMETERS FOR EFFECTIVE HAMILTONIANS OF BIRADICALS DERIVED FROM CYCLOALKANONES[a]

Chain length of biradical	$2J(cm^{-1})$	$k_v(sec^{-1})$	$k_{so}(sec^{-1})$
7	-1.87 ± 0.28	$4.8 \pm 1.6 \times 10^{10}$	$8.0 \pm 4.8 \times 10^5$
8	-0.374 ± 0.037	$1.2 \pm 0.2 \times 10^{10}$	$6.4 \pm 0.7 \times 10^5$
9	-0.196 ± 0.009	$4.8 \pm 0.8 \times 10^9$	$6.4 \pm 0.7 \times 10^5$
10	-0.085 ± 0.003	$3.2 \pm 0.5 \times 10^8$	$6.4 \pm 0.7 \times 10^5$
11	-0.026 ± 0.005	$1.0 \pm 0.4 \times 10^8$	$6.4 \pm 0.7 \times 10^5$

[a]The hyperfine coupling constants for the two α protons and two β protons were set at -22 and 33 G, respectively, in analogy to the experimentally measured values of the n-propyl radical.

Considering the simplifications inherent in the model it is necessary to comment on the physical significance of the obtained parameters. We believe the values for 2 J are probably realistic estimates of the average exchange coupling. However, it must be remembered that this coupling is not only a function of the chain length but also of the geometry and the average represents the statistically weighted average over all conformations. The two rate constants are to be interpreted with caution and should be viewed as representing trends rather than giving exact numbers. It should be noted that k_{so} is of the order of electron relaxation in doublet states, indicating so at these chain lengths the dipolar interaction may be too small to contribute much to relaxation. This may make the simplification of neglecting relaxation in the triplet manifold a less severe assumption.

Cyclohexanone and cyclopentanone photolysis do not fit the picture developed for the higher cycloalkanones. Figure 12 shows the spectrum of the unsaturated aldehyde obtained from photolysis of cyclohexanone. A similar spectrum can be obtained from cyclopentanone although the polarization is somewhat weaker. Clearly this spectrum shows opposite polarization for the methine and methylene protons and has the general appearance of a radical pair spectrum rather than a biradical spectrum. Application of the net effect rules to this biradical predicts exactly the observed polarization. One suspects therefore that T_0-S mixing somehow is the predominant process here, and the question is how this can arise. As was pointed out above, T_0-S processes cannot contribute to biradical based polarization as long as scheme 2.2 is the proper description of all processes going on in the pair.

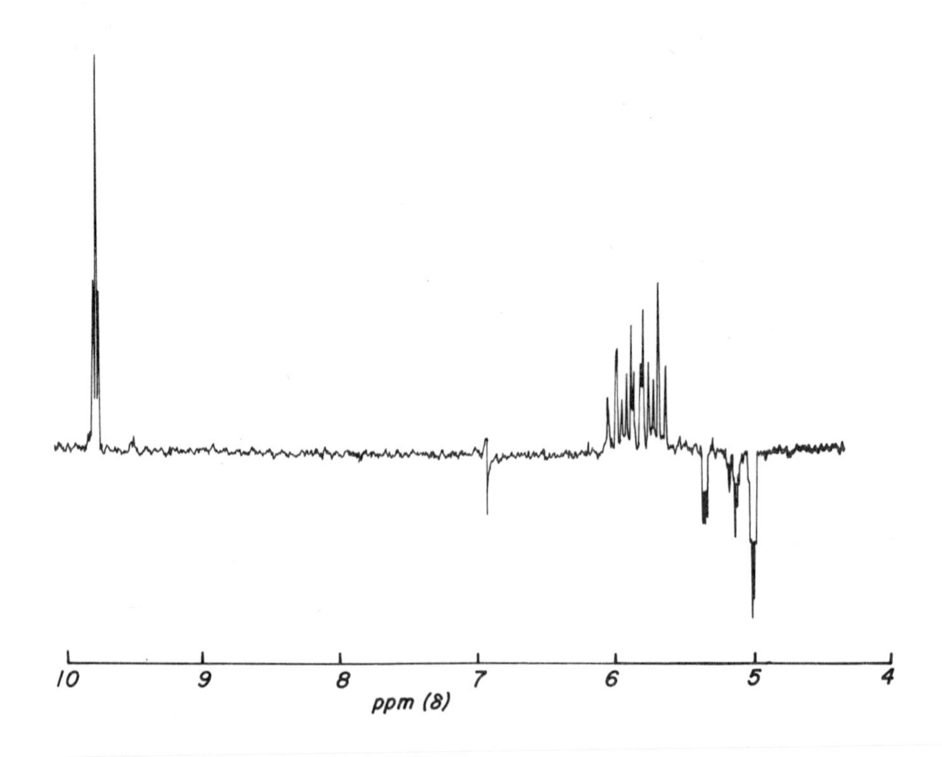

Fig. 12. Spectrum of the unsaturated aldehyde formed on photolysis of cyclohexanone.

The solution to the problem is found in the polarization of the cyclohexanone in which the α-protons are in enhanced absorption and the β-protons in emission. This is the exact opposite pattern as is observed for the aldehyde! This situation is reminiscent of the relationship of geminate product and escape product in the radical pair theory. To make T_0-S processes observable in biradical reactions we need two different products and we have to postulate that one product forms preferentially from the triplet state and the other from the singlet state of the biradical. This allows the "spin-sorting" mechanism which is required for T_0-S contributions to come to play. In the case at hand the aldehyde is formed in the normal manner from the singlet state and the ketone is formed from the triplet state via a spin orbit coupling mechanism. Reasonable arguments can be advanced why this should be so, but space does not permit us here to go into these speculations.

The question remains why the T_0-S mechanism predominates over the T_--S mechanism. One possible answer is that the trend for J listed in the Table continues and J becomes quite large in magnitude relative to the Zeeman splitting. It can be shown that under these circumstances contributions from T_+-S and T_--S processes become more equal and lead to cancellations. Also the magnitude of the T_0-S matrix elements is larger.

In 1,4 and 1,3 biradicals we expect the singlet-triplet splitting to become still larger and the hyperfine induced mixing to become less important relative to spin-orbit coupling processes. Therefore, CIDNP may not be observable in these cases. However there is one exception to this: when the triplet state of the biradical is the ground state and the product has a singlet multiplicity there must be a region in the surface where there is a degeneracy between zero order triplet and singlet state. If spin orbit coupling is not too important CIDNP can be observed. In line with this prediction the 1,3-cyclopentadiyl which had been shown by esr to have a triplet ground state, generates CIDNP in the ring closure product bicyclopentane.[9] On photosensitized photolysis of the azo compound (2.18), bicyclopentane is formed with all lines in enhanced absorption confirming the esr determination of a triplet ground state for the biradical.

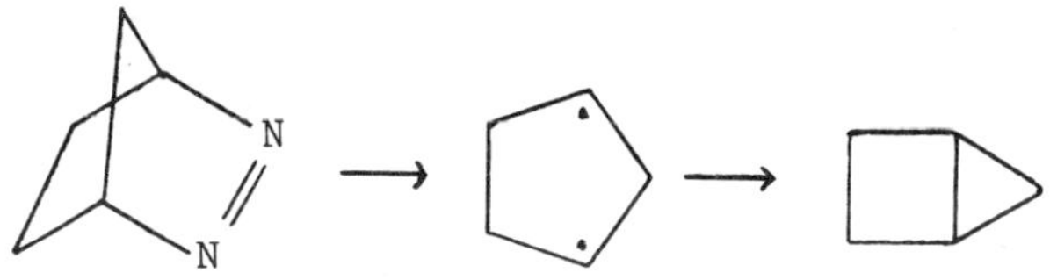

N N (2.18)

Finally, a thermochemical reaction should be briefly mentioned in which biradicals generate CIDNP although the biradical is generated in the singlet state. The pyrolysis of the bis-spiro compound shown in sequence (2.19) produced CIDNP in both the starting material and the product.[10] Enhanced absorption is observed for the CH_2 and

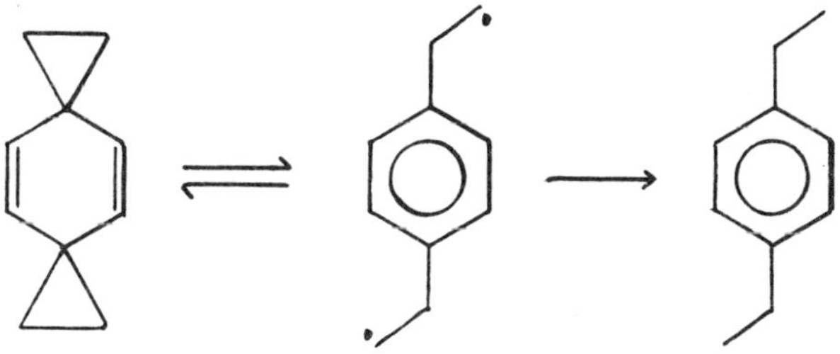

(2.19)

CH_3 protons in p-diethyl benzene and the cyclopropyl protons in the reactant. This can easily be explained in terms of sequence (2.19) if it is assumed that the starting material is regenerated from the singlet state and the trapping of the biradical with hydrogen donors (solvent) occurs preferentially from the triplet state. The latter assumption is entirely reasonable in view of the longer lifetime of the triplet state. Predominant $S \rightarrow T_-$ transitions explain the observations, indicating a singlet ground state for the biradical. An interesting complication occurs by the simultaneous operation of a nuclear Overhauser effect which is pumped by the CIDNP polarization and gives rise to emission signals for the olefinic protons in the reactant.[11]

Lastly, mention should be made of some C-13 CIDNP generated from biradicals. A sequence of cyclic ketones, similar to the one discussed gives strong emission signals analogous to the proton spectra.[8]

2.3 Appendix

Shortly before this Study Institute was scheduled to begin a group of Dutch workers completed calculations based on a much more realistic model for flexible biradicals.[12] The method is based on the solution of the stochastic Liouville equation which has been amended to include the dynamic behavior of the biradical as well as relaxation and reaction processes. We shall briefly discuss the salient features of this model and give a short summary of the results of the numerical calculations.

The equation of motion describing the spin dynamics of biradicals in this model is written in terms of the density matrix $\rho(t)$

$$\frac{\partial \rho(t)}{\partial t} = - i\mathcal{H}^x \rho(t) + \mathcal{R}\rho(t) + \mathcal{W}\rho(t) + \mathcal{K}\rho(t) \tag{2.20}$$

In the model $\rho(t)$ is a function of the spin state variables of the biradical and the reaction product and of the distance between the ends of the biradical chains r. $\mathcal{H}^x$ is the Liouville operator associated with the spin Hamiltonian

$$\mathcal{H}^x\rho(t) = \mathcal{H}\rho(t) - \rho(t)\mathcal{H} \tag{2.21}$$

$\mathcal{R}$ is the Redfield relaxation matrix inducing transitions between elements of $\rho(t)$ with the same value of r but different spin variables as does $\mathcal{H}^x$. $\mathcal{W}$ describes the intramolecular motion, that is the change of r, and $\mathcal{K}$ the chemical reactions. Both induce transitions among elements of $\rho(t)$ with the same spin functions. The operator $\mathcal{K}$ couples the elements of the products with the elements of the biradical in which the ends are within reacting distance r_d, and which have singlet spin functions. At the time t=0 the biradical is formed with an end to end distance close to or equal to r_d and a given precusor multiplicity. It will undergo dynamic motion changing the value of J, the exchange coupling, and $\mathcal{H}^x$ and $\mathcal{R}$ will change the spin variables. It will finally react to diamagnetic products and we are interested in the product elements of $\rho(t)$, which gives us the nuclear polarization.

We will briefly discuss the form each of the four operators take in equation 2.20. The Liouville operator contains the spin Hamiltonian (eq. 1.1) with its exchange interaction J which is a function of r. In the present model J is assumed to vary exponentially although any function can in principal be used.

$$J(r) = J_o e^{-ar} \tag{2.22}$$

If we restrict ourself to a one-proton radical pair, a basis set of 8 biradical states can be constructed, two of which, $|T_+\alpha\rangle$ and $|T_-\beta\rangle$ are not mixed with any other states by the spin Hamiltonian. The remaining six states gives rise to two 9 x 9 matrices associated with $\mathcal{H}^x$. Of course these matrices will be different for each conformation of the biradical because of the dependence of J on r. If there are m conformations there will be 2xm such matrix blocks.

The relaxation problem contained in $\mathcal{R}$ is treated in two independent parts. First, interactions which are uncorrelated at the two radical sites and which are identical to relaxation in free radicals induce transitions between singlet and triplet levels as well as mong the triplet levels. Second, the modulation of the electron dipole-dipole interaction between the two unpaired electrons induces transition among the triplet levels only.

The matrix elements for the uncorrelated relaxation are evaluated by lumping together all mechanisms, such as spin rotation, g-factor anisotropies and hyperfine interactions into a fluctuating local field H whose x, y and z components are taken to fluctuate independently. With this simplication they are of the general form

$$\gamma^2\hbar^2 \; |H|^2 \; \tau_u/(1+\omega_n^2 \; \tau_u^2) \tag{2.23}$$

where τ_u is the correlation time for the molecular motion and ω_n is the energy difference between connecting states. H is treated as an empirical parameter adjusted to give relaxation times typical for free radicals.

The treatment of the dipolar relaxation follows standard proceedures. Using the dipolar interaction Hamiltonian

$$\mathcal{H}_d = \frac{g^2\beta^2}{\hbar^2} \{S_a \cdot S_b/r^3 - 3(S_a \cdot r)(S_b \cdot r)/r^5\} \tag{2.24}$$

with r the radius vector from electron a to electron b, matrix elements are calculated for each conformation of the biradical. The transition probabilities, all of which show and r^{-6} dependence, are evaluated for a rotational correlation time τ_c in each conformation. Transitions caused by end-to-end distance changes in the biradical are neglected.

The treatment of the dynamical behavior of the biradical chain is the most interesting part of the model. In principal the problem has been solved for polymethylene chains with the rotational isomeric state (RIS) model, in which the energies of all rotational isomers are evaluated and transitions between the rotational states are assumed to obey the Arrhenius equation. Although this model has been used for a C_7 chain it was found to be too cumbersome for longer chains. Instead the authors developed what they call a restricted diffusion (RD) model. In this approach the equilibrium distribution C(r) of the end-to-end distance is calculated using a Monte Carlo computer program for n-alkanes. The potential functions used are Buckingham potentials for short range and Lennard-Jones potentials for long range interaction energies. For a large number of conformations the distribution function of r has been calculated

$$C(r) = \sum_j{}' \exp(-U_j/kT)/\sum_i \exp(-U_i/kT) \qquad (2.25)$$

where $\sum_j'$ is the summation over the conformation for which $r<r_j<r+\Delta r$ and $\sum_i$ is the summation over all conformations. Having obtained an equilibrium end-to-end distribution function by this method the normalized distribution C(r) is divided into m segments (typically 20) with equal areas (equal probability). It is now assumed that the end-to-end motion in the chain can be described by diffusional jumps from one segment to the next. Only jumps between neighboring segments are allowed. Probabilities for backward and forward jumps between two segments are taken equal

$$W_{kl} = W_{r_k \to r_l} = W_{lk}. \qquad (2.26)$$

In analogy to Brownian motion the magnitude of the matrix elements are taken to be

$$W_{kl} = D'/(\bar{r}_\ell - \bar{r}_k)^2 \quad \text{for } k = \ell \pm 1. \qquad (2.27)$$

D' is a diffusion coefficient for the restricted diffusion of the biradical ends.

The chemical reaction operator $K\rho(t)$ can be broken up into intra and intermolecular scavenging reactions. For intramolecular reactions

$$K\rho(t) = \frac{-k_p}{2}\{0^{S,r_d}\,\rho(t) + \rho(t)\,0^{S,r_d}\} \qquad (2.28)$$

where k_p is the reaction rate at end-to-end distance r_d, and $0^{S,r_d}$ is a projection operator selecting elements at distance r_d and singlet electronic states. Evaluation of the matrix elements of K is straightforward and leads to nonzero off-diagonal elements of magnitude k_p only between the singlet biradical and the product states.

Finally the nuclear spin polarization in the product is obtained by looking at $t = \infty$ at the difference in the product density matrix elements (nuclear relaxation in the product has been neglected). This is accomplished via the LaPlace transformation as explained elsewhere.[13]

Numerical calculations were carried out with the aim to reproduce the experimental field dependence measured for the Norris Type I cleavage of cycloalkanones from C_7 to C_{11}. A number of parameters enter the calculations and it was attempted to simulate the data for the whole series with one set of parameters. The treatment was quite successful provided the parameters were chosen judiciously.

A critical parameter is the jump rate from one segment to another. This rate is determined by the effective diffusion coefficient D'. For slow jump rates the field dependence shows more than one maximum corresponding to distinct conformations. Since this is not found in the experimental curves diffusional jumps must be quite fast. This suggests possible experiments in very viscous solvents where one might observe additional maxima. The best parameters for the exchange interaction were found to be $J_o = 4.5 \times 10^{16}$ rad/sec and $a = 1.90$ Å^{-1}.

REFERENCES

1. The low-field theory was developed independently by: F.J. Adrian, Chem. Phys. Lett., 10, 70 (1971); J.I. Morris, R.C. Morrison, D.W. Smith, and J.F. Garst, J. Am. Chem. Soc., 94, 2406 (1972); R. Kaptein and J.A. den Hollander, J. Am. Chem. Soc., 94, 6269 (1972). See also, S. Glarum in Chemically Induced Magnetic Polarization, A. Lepley and G.L. Closs ed., Wiley Interscience, 1973.
2. See Chapter I for a discussion of the general theory of CIDNP.
3. M. Lehnig and H. Fischer, Z. Naturforschung, A, 25, 1963 (1970).
4. R. Kaptein and J.A. den Hollander, J. Am. Chem. Soc., 94, 6269 (1972).
5. M. Lehnig and H. Fischer, Z. Naturforschung, A, 24, 1771 (1969); H.R. Ward, R.G. Lawler, H.Y. Loken and R.A. Cooper, J.Am. Chem. Soc., 91, 4928 (1969); J.I. Morris, R.C. Morrison, D.W. Smith, and J.F. Garst, J. Am. Chem. Soc., 94, 2406 (1972).
6. G.L. Closs, Adv. Mag. Res., J. Waugh ed., Vol. 7 pp. 216-229 (1974); also, R. Kaptein, P.W. N.M. van Leeuwen, and R. Huis, Chem. Phys. Lett., 41, 264 (1976).

7. G.L. Closs and C.E. Doubleday, J. Am. Chem. Soc., 94, 9248 (1972); ibid., 95, 2735 (1973).
8. R. Kaptein, R. Freeman and M.D.W. Hill, Chem. Phys. Lett., 26, 104 (1974).
9. S.L. Buchwalter and G.L. Closs, J. Am. Chem. Soc., 97, 3857 (1975).
10. T. Tsuji and S. Nishida, J. Am. Chem. Soc., 96, 3649 (1974).
11. G.L. Closs and M.S. Czeropski, Chem. Phys. Lett., 45, 115 (1977).
12. F.J.J. de Kanter, J.A. den Hollander, A.H. Huizer and R. Kaptein, Mol. Phys., in press, 1977.
13. J.H. Freed and J.B. Pedersen, Advances in Magnetic Resonance, ed.,J.S. Waugh, Vol. 8, p. 1, 1976.

CHAPTER XIV

PAIR SUBSTITUTION EFFECTS IN CIDNP

R.Kaptein

Physical Chemistry Laboratory, University of Groningen, The Netherlands.

1. INTRODUCTION

CIDNP effects are generated in a time span of about 10^{-9} to 10^{-7} sec determined by the magnitude of the hyperfine interactions and the geminate recombination of radical pairs. Chemical reactions, which change the nature of the radicals, taking place on this time scale will affect the polarization from the original pair and in addition polarization can be expected arising from the newly formed pair. From these so-called pair substitution CIDNP effects rapid radical reactions can be characterized and in favourable cases their rates can be determined.
As an example let us consider the decarboxylation of acyloxy radicals

$$\underset{\text{I}}{\overline{2RCO_2^{\cdot}}} \longrightarrow CO_2 + \underset{\text{II}}{\overline{R^{\cdot} + RCO_2^{\cdot}}} \longrightarrow 2CO_2 + \underset{\text{III}}{\overline{2R^{\cdot}}} \qquad (1)$$

The actual process of the reaction (loss of CO_2) takes place in a very short vibrational time 10^{-13} - 10^{-14} sec). This is so short that the phase of the electron spin state will hardly be disturbed. Thus, if pair I is generated as a singlet pair and no S-T mixing occurs in this pair, pair II also starts in the singlet state.
A similar phase conservation holds for the transition to pair III. As a consequence net polarization due to a Δg effect in pair II is potentially present at the birth of pair III. This net polarization will show up in the recombination products of pair III in spite of the fact that this pair consists of equivalent radicals. This "memory effect" [1] has been observed, for instance,

L. T. Muus et al. (eds.), Chemically Induced Magnetic Polarization, 257-266.

in the decomposition of acetyl peroxide and will be discussed in section 5.

Apart from fragmentation reactions such as reaction (1) pair substitution effects have been observed in radical rearrangements [2] (cf.section 5)

$$\overline{2R\cdot} \longrightarrow \overline{R\cdot + R'\cdot} \longrightarrow \overline{2R'\cdot} \qquad (2)$$

and in fast scavenging reactions* [3 - 5]

$$\overline{2R\cdot} + HX \rightarrow RH + \overline{R\cdot + X\cdot} \qquad (3)$$

Several theoretical descriptions of pair substitution CIDNP have been given [1,6-8]. The first theory [1] could account for the memory effect by treating S - T mixing separately for the pairs I, II and III, eg. in the case of reaction 1 .
This was later refined [6-8] by including the effect of phase continuity at the transitions I → II and II → III. Presently the most detailed analysis is that of den Hollander [8]. He showed that this phase continuity leads to so-called cooperative effects of successive radical pairs, which could explain some hitherto puzzling phenomena. The following treatment is based on the work of den Hollander [8].

2.THEORY OF PAIR SUBSTITUTION

Let us consider the reaction sequence of Scheme 1.

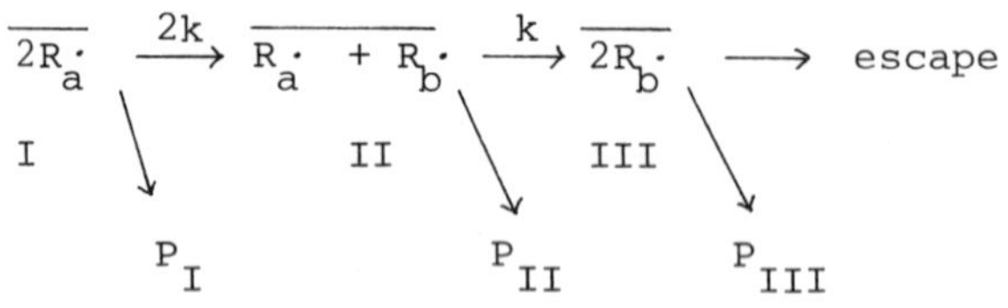

$$\underset{\text{I}}{\overline{2R_a\cdot}} \xrightarrow{2k} \underset{\text{II}}{\overline{R_a\cdot + R_b\cdot}} \xrightarrow{k} \underset{\text{III}}{\overline{2R_b\cdot}} \longrightarrow \text{escape}$$

$$\text{I} \searrow P_I \qquad \text{II} \searrow P_{II} \qquad \text{III} \searrow P_{III}$$

Scheme 1

k is the (pseudo) first order rate constant for the reaction $R_a\cdot \longrightarrow R_b\cdot$ and P_I, P_{II} and P_{III} are the products from the radical pairs I, II and III respectively. The high field approximation is assumed to be valid (S - T_o mixing only). The pairs have their own mixing coefficients Q_{In}, Q_{IIn} and Q_{IIIn} as defined in Chapter 1. Q_{IIn} is switched on at time t' , when the transition from pair I to pair II takes place. Similarly, starting at time t'' the S-T mixing is governed by Q_{IIIn}. It will further be assumed that the radicals $R_a\cdot$ and $R_b\cdot$ have approximately the same size, so that their diffusion rates are the same. We shall consider the

* Scavenging reactions that compete with geminate recombination have sometimes been called "cage wall" reactions. However, no such things as cage walls exist on the CIDNP time scale.

case of a singlet precursor, because sofar experiments have been reported only for this case. The extension to a triplet precursor is straightforward.

2.1 The products of pair I

The probability that pair I born at t=O still exists at time t is exp (-2kt). Therefore the recombination probability of pair I with nuclear state n is now given by

$$P_{In} = \lambda_I \int_0^\infty e^{-2kt} \; f(t) \, |C^I_{Sn} (t)|^2 \, dt \tag{4}$$

where λ_I is the steric factor for pair I and $f(t) = m\, t^{-3/2}$ exp $(-\pi m^2/p^2\, t)$ as discussed in Chapter 1.
With the singlet probability given by

$$|C^I_{sn}(t)|^2 = \cos^2 Q_{In} t \tag{5}$$

eq. (4) yields

$$P_{In} = \lambda_I p - \lambda_I m\pi^{\frac{1}{2}} \left\{ \left[(k^2 + Q_{In})^{\frac{1}{2}} + k \right]^{\frac{1}{2}} + (2k)^{\frac{1}{2}} \right\} \tag{6}$$

This expression is valid for reaction rates k smaller than the frequency of diffusive displacements ($k < m^{-2} \approx \tau_D^{-1}$). When the reaction is faster than $S\text{-}T_o$ mixing ($k >> Q_{In}$) eq. (6) becomes

$$P_{In} = \lambda_I \left[p - 2m (2\pi k)^{\frac{1}{2}} \right] - \tfrac{1}{2}\lambda_I m\pi^{\frac{1}{2}} Q^2_{In} (2k)^{-3/2} \tag{7}$$

The first term describes the decreased product yield, whereas the CIDNP effect results from the second term. It should be noted that the polarization now depends on Q^2 instead of $Q^{\frac{1}{2}}$ as is predicted for free diffusion. This Q^2 dependence shows up, for instance in relative line intensities in multiplets. However, it cannot be used as a diagnostic test for pair substitution, because it can have a different origin such as the presence of other nuclei in the pair [9] or a short electron T_2.

2.2 The products of pair II

The probability that pair II is formed in the time interval (t',t' + dt') is 2k exp (-2kt')dt'. The combined probability that pair II born at time t' still exists at time t in the singlet state is

$$\int_0^t 2k\, e^{-2kt}\; e^{-k(t-t')}\, |C^{II}_{S_n}(t,t')|^2\, dt' \tag{8}$$

Assuming phase continuity at the transformation it can be easily shown[7,8] that the singlet probability of pair II is

$$|C^{II}_{Sn}(t,t')|^2 = \cos^2 [Q_{In}t' + Q_{IIn}(t-t')] \tag{9}$$

The recombination probability for pair II can now be written as

$$P_{IIn} = \lambda_{II} \int_0^\infty f(t)\, e^{-kt}\, dt \int_0^t 2ke^{-kt'}\, |C^{II}_{Sn}(t,t')|^2 dt' \tag{10}$$

This expression can be worked out exactly[8] , but here we shall give an approximation valid in the fast reaction region ($m^{-2} > k > Q_{In}, Q_{IIn}$), where eq.(10) becomes

$$P_{IIn} = \lambda_{II} 2m\pi^{\frac{1}{2}} [(2k)^{\frac{1}{2}} - (k)^{\frac{1}{2}}] -$$

$$-\tfrac{1}{2}\lambda_{II}\, m\pi^{\frac{1}{2}} (k)^{-3/2} [0.13\, Q^2_{In} + 0.31 Q^2_{IIn} + 0.20 Q_{In} Q_{IIn}] \tag{11}$$

Again the first term represents the product yield and the second term the polarization. It can be noted that the polarization of product P_{II} depends not only on Q_{IIn} but also on Q_{In}, the $S\text{-}T_o$ mixing coefficient of pair I (memory effect). Interestingly however, it is not just a sum of contributions from pair I and II because of the cross-term $Q_{In}\, Q_{IIn}$. This latter term gives rise to the co-operative effects, which will be discussed in the next section.

The analysis for product P_{III} can be carried out along the same lines as for P_{II} [8]. However, the formulae tend to become rather unwieldy and will not be given here.

3. CO-OPERATIVE EFFECTS

The effect of the cross-term $Q_{In}\, Q_{IIn}$ is to generate polarization with the combined properties of the pair I and II, although they are separated in time. This is true even in the important case where pairs I and II alone would not give rise to polarization. For instance, consider the following one-proton radical pairs

$$RP_1 \longrightarrow RP_2 \qquad (12)$$

$$A = 0,\ \Delta g > 0 \qquad\qquad A > 0,\ \Delta g = 0$$

On their own these pairs would not give rise to CIDNP, in the first case because $A = 0$, in the second case because $\Delta g = 0$ and a multiplet effect is not possible for one spin. However, when consecutively formed CIDNP can arise from these pairs. This can be understood by considering the precession of the electron spin vectors as is shown in Figure 1, where a singlet precursor has been assumed. For RP_1 pairs with + and - nuclear states precess at equal rates and hence no CIDNP is possible for the products of this pair. Upon the transition to RP_2 at time t_1 the + and - pairs start to precess at different rates. At time t_2 we have arrived at a situation similar to that of a pair with $A > 0$ and $\Delta g > 0$ (cf.Chapter 1, Fig.3). Hence emission is predicted for the recombination product of this pair. The net effect rule can still be applied in this case if the parameters are taken as a combination of those for the pairs RP_1 and RP_2. Thus for the recombination product the net effect rule would become

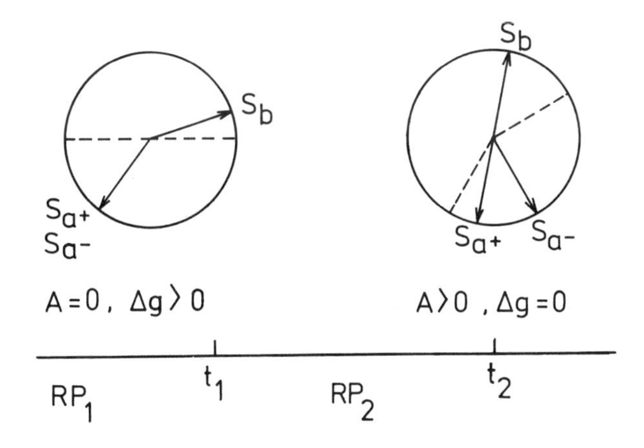

Fig.1 Precession of electron spin vectors of radical pairs RP_1 and RP_2, where radical <u>a</u> contains one proton. RP_2 is formed at time t_1. + and - denote nuclear states.

$\Gamma_n(H) = \mu\ \varepsilon\ \Delta g\ A = -{+}{+}{+} = -\ (E)$

Experimentally the co-operative effect has been observed in the decomposition of acetyl benzoyl peroxide [8,10].

$$\mathrm{PhC(=O)\text{-}OC(=O)CH_3} \longrightarrow \longrightarrow \overline{\mathrm{PhCO_2\cdot + CH_3\cdot}} \longrightarrow \mathrm{CO_2} + \overline{\mathrm{Ph\cdot + CH_3\cdot}} \longrightarrow \qquad (13)$$

$$\longrightarrow \mathrm{PhCH_3}$$

For toluene emission lines have been observed for the phenyl and methyl protons with similar intensities. This must arise from a co-operative effect of both pairs shown in reaction (13), because the hyperfine coupling constants of the phenyl protons in the benzoyloxy radical is negligible. In the phenyl radical A(ortho-H) = + 17.4 G. Applying the net effect rule for the phenyl protons we obtain

$$\Gamma_n(\mathrm{Ph}) = -{+}{+}{+} = -\ (E)$$

and for the methyl group (memory effect)

$$\Gamma_n(\mathrm{CH_3}) = -{+}{-}{-} = -\ (E)$$

The same effect can explain the phenyl proton polarizations observed in the decomposition of phenylazotriphenyl methane [11]

$$\mathrm{Ph\text{-}N{=}N\text{-}C(PH)_3} \longrightarrow \overline{\mathrm{Ph\text{-}N{=}N\cdot + \cdot C(Ph)_3}} \longrightarrow$$

$$\longrightarrow \mathrm{N_2} + \overline{\mathrm{Ph\cdot + \cdot C(Ph)_3}} \longrightarrow \text{products} \qquad (14)$$

Again, the phenyl proton coupling constants in the phenyl diazenyl radical (σ-radical) are probably negligible. The interpretations given in ref. [11] need revision.

4. GEMINATE F-PAIRS

If one of the radicals in a geminate pair has a very short electron spin-spin relaxation time T_2 the spin correlation is rapidly lost. The polarization from such a pair could be calculated from eq. (6) with $k = T_2^{-1}$. When this pair is the first in a pair substitution sequence such as that of Scheme 1, the second pair would start with uncorrelated electron spins. If the original pair is formed from a S precursor the polarization of the second pair could be reversed to F-type polarization. This effect might occur for instance, in the decomposition of metal alkyls

$$MR_n \longrightarrow MR^{\bullet}_{n-1} + R^{\cdot} \xrightarrow{k} MR_{n-2} + 2R^{\cdot} \qquad (15)$$

where the metal containing radical may have a very short T_2. Calculations indicate [12] that the reversal could occur when T_2 is very short ($T_2 < 10^{-10}$sec) and $k < T_2^{-1}$ or when $k < \Delta g \beta \hbar^{-1} H_o$ (the spin correlation would also be lost when Δg of the first pair would be very large).

Kinetically geminate F-pairs are different from random encounter pairs. They could be distinguished by their different behaviour towards scavengers. Thusfar no unambiguous cases of this type of polarization have been encountered.

5. EXAMPLES

5.1 Radical fragmentation: acetyl peroxide

The CIDNP spectrum of the decomposition of acetyl peroxide was shown in Chapter 1 (Fig.7). All polarizations arise from the acetoxy-methyl radical pair

$$\overline{CH_3CO_2^{\cdot} + CH_3^{\cdot}} \xrightarrow{k} CO_2 + \overline{2CH_3^{\cdot}} \xrightarrow{RCl,RH} CH_3Cl, CH_4 \qquad (16)$$

$$\searrow CH_3CO_2CH_3 \qquad \searrow C_2H_6$$

Remarkably, the emission line of ethane (memory effect) has a much larger intensity than that of methyl acetate in spite of the fact that this latter product is directly formed from the polarizing pair. From these relative intensities the rate constant for decarboxylation of acetoxy radicals could be determined to be $k = 2 \times 10^9$ sec^{-1} at 110° [13].

Similarly the rate of decarboxylation of benzoyloxy radicals has been determined from CIDNP [7,8]. The rate constant is about 10^8 sec^{-1} at 110°, several orders of magnitude larger than hitherto thought.

5.2 Radical rearrangement: cyclopropaneacetyl peroxide

The rate of rearrangement of the cyclopropylcarbinyl radical is fast enough to be amenable to CIDNP studies [2]. The 100 Mc spectrum obtained during decomposition of cyclopropaneacetyl peroxide in hexachloroacetone at 80° is shown in Figure 2. The main polarization is observed for 4-chloro-1-butene, which is an escape product of the rearranged radical formed in the following reactions

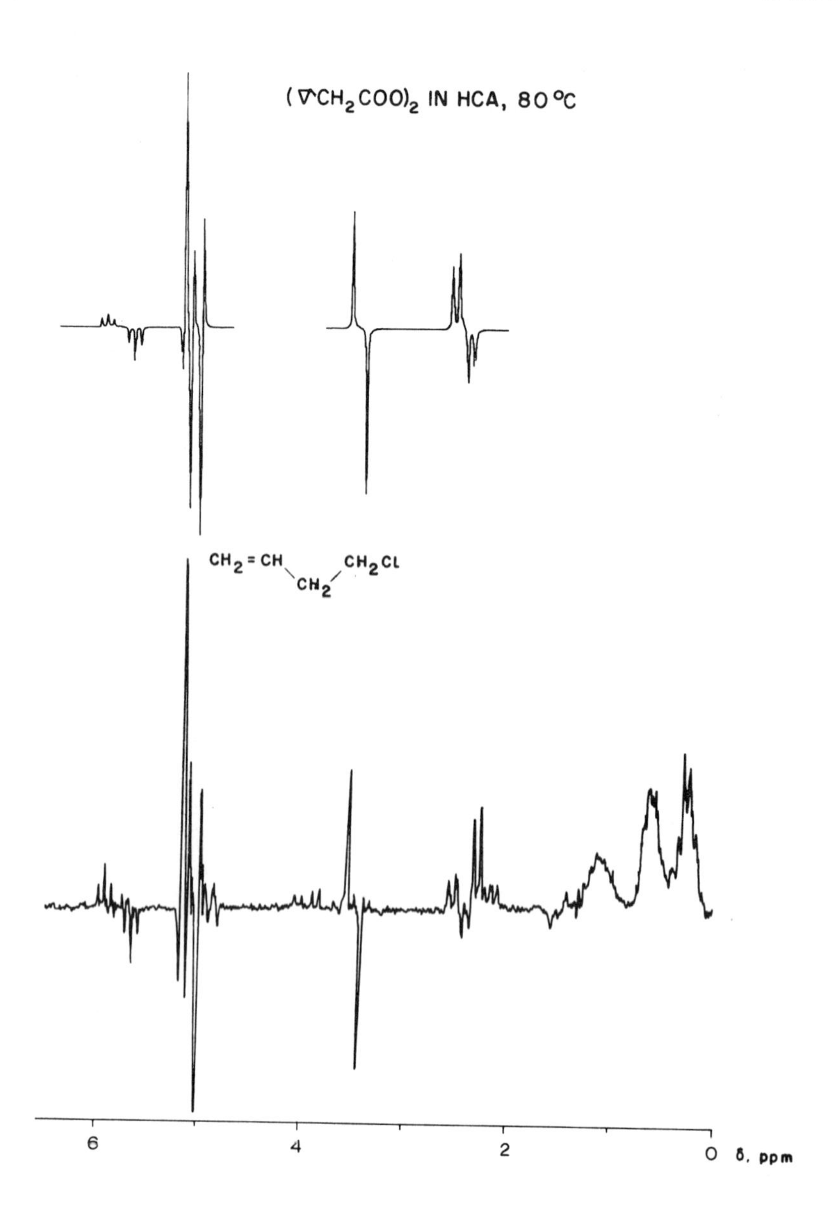

Fig.2 100Mc ^{1}H CIDNP spectrum of the thermal decomposition of cyclopropaneacetyl peroxide in hexachloroacetone at 80°. A simulated spectrum of 4-chloro-1-butene is shown on top.

$$(\triangledown\!-CH_2COO)_2 \longrightarrow \overline{2\,\triangledown\!\cdot} \xrightarrow{2k} \overline{\triangledown\!\cdot + \cdot\!\diagdown\!\diagup\!=} \longrightarrow$$

radical coupling products

$$\xrightarrow{k} \overline{2\cdot\!\diagdown\!\diagup\!=} \xrightarrow{RCl} Cl\diagdown\!\diagup\!= \quad (17)$$

This product shows polarization from both radicals. From the relative contributions and the known hyperfine coupling constants the rate constant for the rearrangement could be determined [1]. A simulated spectrum based on a ratio 8:1 for the contributions of the cyclopropylcarbinyl and butenyl radicals is shown in Figure 2. This ratio corresponds with a rate constant $k = 3 \times 10^7$ sec^{-1}.

5.3 Radical scavenging

When scavenging reactions are fast enough to interfere with geminate recombination, pair substitution effects may be expected. This situation occurred in the decomposition of isobutyryl peroxide in the presence of bromotrichloromethane as a scavenger [14].

$$(RCO_2)_2 \longrightarrow \overline{2R\cdot}^S \xrightarrow{CCl_3Br} RBr + \overline{R\cdot + CCl_3\cdot}^S \longrightarrow \text{escape} \quad (18)$$

R-R, RH, R(-H) $\qquad$ $RCCl_3$, R(-H), $CHCl_3$

$$R\cdot + CCl_3\cdot \longrightarrow \overline{R\cdot + CCl_3\cdot}^F \longrightarrow \text{escape} \quad (19)$$

R CCl_3, R(-H), $CHCl_3$

where $R = CH(CH_3)_2$

At high concentrations of scavenger the spin correlation is conserved as indicated in reaction (18). The A effect observed for $CHCl_3$ is then due to a singlet pair R + CCl_3.

At low concentrations of scavenger (< 0.1 M) a sign reversal occurs for the chloroform polarization indicating competition between the S pairs (reaction 18) and F pairs (reaction 19).

Other applications of pair substitution effects may be envisaged. It should be noted that rates of fast radical reactions, which are hard to measure by other means, can often be accurately determined from CIDNP because of the built-in reference clock of the hyperfine interactions known from ESR.

REFERENCES

1. R.Kaptein, J.Amer.Chem.Soc., 94, 6262 (1972).
2. R.Kaptein, in "Chemically Induced Magnetic Polarization, Ed.A.R.Lepley and G.L.Closs, Wiley, New York, 1973, p.137.
3. H.R.Ward, Accounts Chem.Res. 5, 18 (1972).
4. R.Kaptein, F.W.Verheus and L.J.Oosterhoff, Chem.Commun., 877 (1971).
5. J.A.den Hollander, A.J.Hartel and P.H.Schippers, Tetrahedron, 33, 211 (1977).
6. A.V.Kessenikh, A.V.Ignatenko, S.V.Rykov and A.Y.Shteinshneider, Org.Magn.Res., 5, 537 (1973).
7. R.E.Schwerzel, R.G.Lawler and G.T.Evans, Chem.Phys.Letters, 29, 106 (1974).
8. J.A.den Hollander, Chem.Phys., 10, 167 (1975).
9. G.L.Closs, in "Advances in Magnetic Resonance", ed. J.S. Waugh, Academic Press, New York, 1974, Vol.7, p.157.
10. A.L.Buchachenko, S.V.Rykov, A.V.Kessenikh and G.S.Bylina, Dokl.Akad.Nauk SSSR, 190, 839 (1970).
11. K.G.Seifert and F.Gerhart, Tetrahedron Letters, 829 (1974).
12. J.A.den Hollander and R.Kaptein, Chem.Phys.Letters, 41, 257 (1976).
13. R.Kaptein, J.Brokken-Zijp and F.J.J.de Kanter, J.Amer.Chem.Soc., 94, 6280 (1972).
14. R.Kaptein, F.W.Verheus and L.J.Oosterhoff, Chem.Commun., 877 (1971).

CHAPTER XV

CIDNP FROM BIMOLECULAR REACTIONS OF ORGANOMETALLIC COMPOUNDS*

R. G. Lawler

Department of Chemistry, Brown University
Providence, R. I. 02912 USA

Radical-producing reactions between alkyllithium reagents and organohalides played an important role in the historical development of CIDNP. Reactions of this type

$$RLi + R'X \xrightarrow{-LiX} R\cdot \ \cdot R'$$

$$R_2^*,\ R^*(\pm H) \longleftarrow R\cdot \ \cdot R' \xrightarrow{R'X} R^*X$$

were among the first found to produce CIDNP [1].

In the years since the initial observation of CIDNP in the products of these reactions, a number of other examples of such CIDNP producing organometallic reactions have been discovered and studied. The purpose of this lecture is to describe the way in which CIDNP has been applied to the study of this particular type of reaction. It is intended to illustrate the ability of the phenomenon as a tool in mechanistic organic chemistry. In doing so, we shall not discuss either recent reports of CIDNP during the thermal and photochemical decomposition of organometallic compounds in the absence of organohalides [2a] or the work of Garst et. al. on reactions of radical anions with organohalides [2b].

---

* This work was supported in part by grants from the National Science Foundation.

L. T. Muus et al. (eds.), Chemically Induced Magnetic Polarization, 267-274.

1. DETECTION OF RADICALS VIA CIDNP

Although indirect evidence has existed for many years [3] that free radicals could be formed from the reaction of an alkyllithium compound with an organohalide, it was not until the advent of CIDNP that proof of their intermediacy existed. Shortly thereafter, however, radicals were also detected directly by esr during these reactions [4]. We present below some examples of CIDNP spectra obtained from reactions of two of the most common organometallic reagents: organolithiums and Grignard reagents.

1.1 Organolithium - organohalide reactions

In most of the examples found to date, CIDNP is generated in geminate radical pairs and observed in combination and scavenging products. Figures 1 and 2 show examples of such spectra. In Figure 1 net effects are observed in the olefinic products formed when two different alkyllithiums react with 1,1-dichloro-2,2-dimethyl cyclopropane

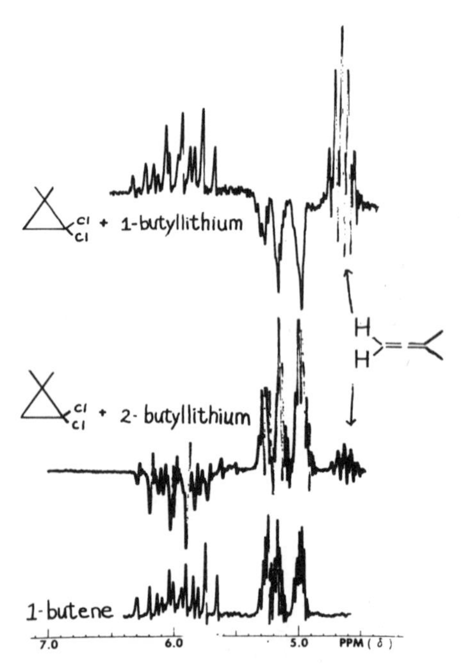

Figure 1. CIDNP observed during the reaction between 1,1-dichloro-2,2-dimethylcyclopropane and two isomeric butyllithiums in hexane. An unpolarized reference spectrum of 1-butene is shown at the bottom.

$$\text{(cyclopropyl)}Cl_2 \xrightarrow{CH_3CH_2CH_2CH_2Li} \text{(cyclopropyl)}\dot{}Cl \;\; \cdot CH_2CH_2CH_2CH_3 \rightarrow \overset{*}{C}H_2{=}\overset{*}{C}HCH_2CH_3 \quad (E\;\;A)$$

$$\text{(cyclopropyl)}Cl_2 \xrightarrow{CH_3CH_2CHLiCH_3} \text{(cyclopropyl)}\dot{}Cl \;\; CH_3\dot{C}HCH_2CH_3 \rightarrow \overset{*}{C}H_2{=}\overset{*}{C}HCH_2CH_3 \quad (A\;\;E)$$

The difference in sign of the net effect for corresponding protons in the two reactions simply reflects the different signs for proton hyperfine splitting in α-and β-alkyl positions since all other aspects of the two radical pairs are approximately the same. Figure 2 also shows CIDNP from 1-butene, but a multiplet effect now occurs because the two radicals formed are identical.

$$n\text{-BuLi} + n\text{BuBr} \longrightarrow 2\,n\text{-Bu}\cdot \longrightarrow \overset{*}{C}H_2{=}\overset{*}{C}HCH_2CH_3 \quad (EA)$$

In both examples the CIDNP is consistent with the formation of a singlet-born radical pair in the initial step of the reaction.

1.2 Grignard reagent-organohalide reactions

In contrast to the behavior of organolithiums, it is fairly rare that Grignard reagents exhibit CIDNP from geminate radical pairs. Only with very reactive halides is such behavior observed. One of the few such examples is the production of enhanced absorption in both isobutylene and chloroform formed during the reaction of t-butylmagnesium bromide and $BrCCl_3$.

$$(CH_3)_3CMgBr + BrCCl_3 \xrightarrow{-MgBr_2} (CH_3)_3C\cdot\,\cdot CCl_3 \longrightarrow (CH_3)_2C{=}\overset{*}{C}H_2 + \overset{*}{H}CCl_3 \quad (A\;\;A)$$

With less reactive halides one usually observes multiplet effects arising from diffusive encounters of identical radicals. An example of such behavior is the report [5] of CIDNP from both isobutylene

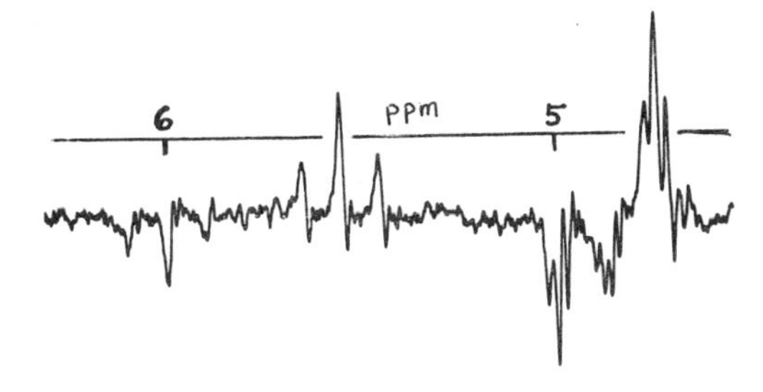

Figure 2. Enhanced nmr spectrum obtained from 1-butene formed during the reaction between n-butyllithium and n-butyl bromide in hexane with a catalytic amount of diethyl ether added.

and isobutane formed in the reaction between t-butylmagnesium bromide and t-butyl bromide.

$$(CH_3)_3CMgBr + (CH_3)_3CBr$$
$$\downarrow -MgBr_2$$
$$\text{DIFFUSIVE ENCOUNTER}$$
$$\downarrow$$
$$2\,(CH_3)_3C\cdot$$
$$\downarrow$$
$$(C\overset{*}{H}_3)_2C{=}C\overset{*}{H}_2 + (C\overset{*}{H}_3)_3CH^* $$
$$\text{AE} \qquad\qquad \text{AE}$$

1.3 Metal-catalyzed Grignard-organohalide reactions

The peculiar tendency of Grignard reagents to form diffusive encounter pairs led us to investigate the role which trace metal impurities might play in the reaction with halides [6]. By using specifically deuterated Grignard and organohalide reactants we were able to determine that in the presence of even very low concentrations of iron salts the radicals producing CIDNP from diffusive encounters arise solely from the organohalide [7]. This is consistent with earlier suggestions of Tamura and Kochi [8] based on kinetic measurements and product analyses. The production of radicals from a reaction between an organohalide and a transition metal compound is also supported by the observation of CIDNP from diffusively formed radical pairs during the reaction between $Pd(PEt_3)_3$ and isopropyl iodide [9].

2. DETECTION OF S_H2 REACTIONS

In an early publication dealing with CIDNP we reported [10] that alkyl iodide reagents may become enhanced during their reaction with organolithiums. This prompted us to speculate that halogen-metal exchange may take place by a radical pair mechanism.

$$\overset{*}{R}X + LiR' \rightleftarrows [R\cdot, X, Li, R'\cdot] \rightleftarrows RLi + R'^{*}X$$

This mechanism was weakened, however, by the subsequent failure by ourselves or others to detect CIDNP either in the alkyllithium compounds involved in the exchange or in alkylbromides, like n-butyl bromide which undergoes rapid halogen metal exchange with sec-butyllithium.

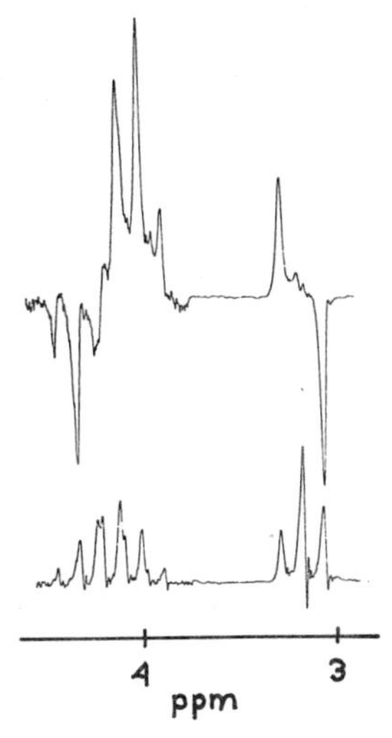

Figure 3. CIDNP from sec-butyl iodide (4.2δ) and n-butyl iodide (3.2δ) during the reaction between n-butyllithium and sec-butyl iodide in benzene. The lower spectrum is from an unpolarized reference sample of the two alkyl iodides.

2.1 Iodine atom transfer

It now appears that the CIDNP observed in alkyl iodides during their reaction with alkyllithiums arises from the previously mentioned geminate pair formation followed by a rapid S_H2 reaction of the radicals with the original alkyl iodide. This forms new radicals which undergo diffusive encounters. An example of this is shown in Figure 3 for the reaction between n-butyllithium and sec-butyl iodide.

n-BuLi + s-BuI

↓ −LiI

n-Bu· ·s-Bu·

↓ s-BuI ↓

(AE) n-Bu*I   s-Bu*I (AE)

+   +

s-Bu·   s-Bu·

DIFFUSIVE ENCOUNTER ↓

2 s-Bu·

↓ s-BuI

2 s-Bu*I (EA)

The EA polarization for s-BuI produced in the diffusive encounter dominates the AE effect produced from the geminate pair because twice as many sec-butyl radicals are involved in the former pairs.

2.2 Magnesium halide transfer

We have recently reported [11] the observation of enhancement in Grignard reagents during their metal-catalyzed reaction with both alkyl iodides and bromides. An example of such polarization is shown in Figure 4 for isobutyl Grignard formed in the iron-catalyzed reaction between isobutyl iodide and ethylmagnesium bromide.

$$i\text{-BuI} \xrightarrow{[Fe]} \underset{\text{ENCOUNTER}}{\overset{\text{DIFFUSIVE}}{- - - \rightarrow}} 2\,i\text{-Bu}\cdot \xrightarrow{EtMgBr} \overset{*}{i\text{-BuMgX}} \quad (EA)$$

We have suggested that this reaction may explain the observation of CIDNP in Grignard reagents during their formation from metallic magnesium [12]. Such enhancements have previously been attributed to the formation of radicals in the reaction between metallic magnesium and organohalides. Our interpretation has been criticized, however, [13] and the matter is far from resolved.

3. OTHER RECENT OBSERVATIONS

It should be noted in conclusion that CIDNP from reactions between organometallics and organohalides or other easily reduced compounds is now being reported rather widely. Alkyllead [14] and halomercury [15] compounds have been reported to produce CIDNP during their reactions with organohalides. CIDNP from reactions between Grignard reagents and ketones [16], peroxides [17], disulfides [18], polyhaloalkanes [19] and acid chlorides [20] has also recently been reported by Savin et al. The surface has barely been scratched, however, in studying the mechanisms of these reactions.

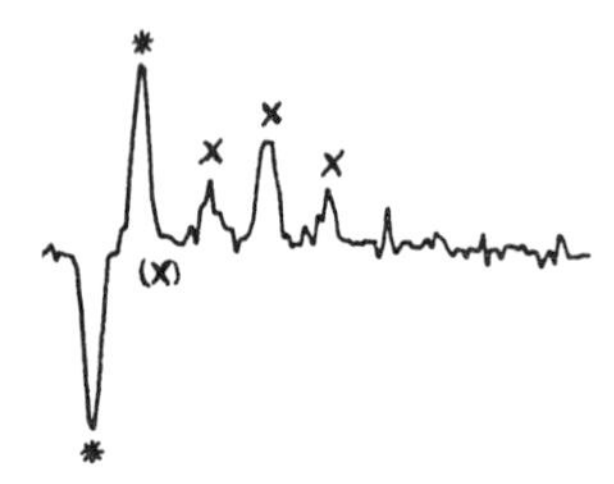

Figure 4. NMR spectrum of the region between 0 and -1 δ obtained 20 seconds after the beginning of the reaction between 0.8 M ethylmagnesium bromide, 1.4 M isobutyl iodide and 10^{-4} M $FeCl_2$ in THF at 37°. The lines arising from isobutylmagnesium halide are indicated with stars (*) and those from ethylmagnesium halide with (x).

REFERENCES

1. H. R. Ward and R. G. Lawler, J. Am. Chem. Soc., 89, 5518 (1967).

2. a. R. Kaptein, P. W. N. M. van Leeuwen and R. Huis, Chem. Comm., 568 (1975); R. Benn, Chem. Phys., 15, 369 (1976); F. J. J. de Kanter, Org. Magn. Reson., 8, 129 (1976); M. Lehnig, Chem. Phys., 8, 419 (1975); M. Lehnig, F. Werner and W. P. Neumann, J. Organometal. Chem., 97, 375 (1975).
b. J. F. Garst, in "Chemically Induced Magnetic Polarization", A. R. Lepley and G. L. Closs, Ed., Wiley, 1973; Chap. 6.

3. D. Bryce-Smith, J. Chem. Soc., 1603 (1956).

4. H. Fischer, J. Phys. Chem., 73, 3834 (1969); G. A. Russell and D. W. Lamson, J. Am. Chem. Soc., 91, 3967 (1969).

5. H. R. Ward, R. G. Lawler and T. A. Marzilli, Tetrahedron Lett., 521 (1970).

6. R. B. Allen, R. G. Lawler and H. R. Ward, ibid., 3303 (1973).

7. R. B. Allen, R. G. Lawler and H. R. Ward, J. Am. Chem. Soc., 95, 1692 (1973).

8. M. Tamura and J. Kochi, J. Organometal. Chem., 31, 289 (1971).

9. A. V. Kramer and J. A. Osborn, J. Am. Chem. Soc., 96, 7832 (1974).

10. H. R. Ward, R. G. Lawler and R. A. Cooper, ibid., 91, 746, (1969).

11. R. G. Lawler and P. Livant, ibid., 98, 3710 (1976).

12. H. W. H. J. Bodewitz, C. Blomberg and F. Bickelhaupt, Tetrahedron Lett., 281 (1972); Tetrahedron, 29, 719 (1973); 31, 1053 (1975).

13. B. J. Schaart, H. W. H. J. Bodewitz, C. Blomberg and F. Bickelhaupt, J. Am. Chem. Soc., 98, 3712 (1976).

14. P. W. N. M. van Leeuwen, R. Kaptein, R. Huis and W. I. Kalisvaart, J. Organometal. Chem. 93, C5 (1975).

15. I. P. Beletskaya, V. B. Volleva, S. V. Rykov, A. L. Buchachenko and A. V. Kessenikh, Div. Chem. Sciences, Bull. Acad. Sciences USSR, 454 (1971).

16. V. I. Savin, I. D. Temichev and F. D. Yambushchev, J. Org. Chem. USSR, 11, 1238 (1975); V. I. Savin and Yu. P. Kitaev, ibid., 2550 (1975).

17. V. I. Savin and Yu. P. Kitaev, J. Gen. Chem. USSR, 46, 385 (1976); V. I. Savin, I. D. Temichev and F. D. Yambushchev, J. Org. Chem. USSR, 12, 275 (1976).

18. V. I. Savin and Yu. P. Kitaev, J. Org. Chem. USSR, 12, 269 (1976).

19. V. I. Savin, A. G. Abulkanov and Yu. P. Kitaev, ibid., 12, 484 (1976).

20. V. I. Savin, ibid., 12, 1857 (1976).

CHAPTER XVI

THE PROBLEM OF PARALLEL RADICAL AND NON-RADICAL MECHANISMS*

R. G. Lawler

Department of Chemistry, Brown University
Providence, R. I. 02912 USA

The most commonly expressed concern about the utility of CIDNP in mechanistic chemistry is a reservation about whether the radical pathway which it reveals is the sole pathway for producing the observed products. The usual reply to this concern is to point out that;

a) Unlike esr detection of free radicals, where spurious signals may have little to do with the main reaction, CIDNP is observed in the reaction products themselves. Except for a few pathological cases where a polarized intermediate may produce the major product by a rapid, non-radical pathway [1], it may be safely assumed that the observation of CIDNP assures that at least some of the product is formed by the direct intermediacy of free radicals.

b) Within (often large) experimental error, CIDNP is produced with the same overall rate constant that characterizes the appearance of product. Thus competing pathways with different rate determining steps seem to be ruled out.

c) Qualitative, or semi-quantitative, comparisons of CIDNP intensities indicate similar magnitudes for products derived from undisputed radical precursors like peroxides or azo compounds and for products formed via questionable pathways.

* This work was supported in part by a grant from the National Science Foundation.

L. T. Muus et al. (eds.), Chemically Induced Magnetic Polarization, 275-281.

In regard to the above, however, the suspicion might be expressed that the only CIDNP effects strong enough to be easily detected are those arising from predominantly radical processes: Mixed pathways may exist but produce CIDNP effects too weak to be detected by present techniques. Indeed, in at least one case [2] it has been suggested that the observed CIDNP intensities are *not* as strong as expected for a purely radical pathway.

Finally, it should be pointed out that not *all* radical mechanisms produce CIDNP since the effect originates in the reactions of *pairs* of radicals. Alas, the most useful of all radical mechanisms, *chain* reactions, generate no CIDNP at all in the product determining propagation steps. CIDNP is produced only by the usually unwanted termination steps or in the equally unpopular "cage effect" which decreases the efficiency of the initiation step. It is, in fact, easily shown that the CIDNP from a product formed in both chain propagation and termination reactions is reduced in magnitude by a factor which is simply the reciprocal of the chain length. Fortunately, chain reactions with long chains are easily detected kinetically through the use of scavengers. The point at which scavengers become less effective, *short* chain lengths, is just that at which the radical-radical reactions producing CIDNP become important. Thus CIDNP should be considered the complement to, rather than a substitute for, scavenging methods for detecting radical reactions.

Unfortunately there are three additional types of observations which may be interpreted as evidence for a parallel non-radical pathway when they are made on samples which also exhibit CIDNP. Each of these will be discussed below.

1. RETENTION OF CONFIGURATION

A high degree (80-90%) of retention of configuration is sometimes observed in products where racemization would be expected intuitively if a bond to the chiral fragment were fully broken during the reaction. The most common CIDNP generating reactions in this category are 1,2-shifts of the Stevens Rearrangement type [3].

$$\text{R-X-}\ddot{\text{Y}} \xrightarrow{\text{RADICAL}} \text{R}\cdot\ \cdot\text{X-}\ddot{\text{Y}} \leftrightarrow \ddot{\text{X}}\text{-Y}\cdot\ (\text{R}\cdot\cdot\text{Я}) \longrightarrow \text{X-Y-R} + \text{X-Y-Я}$$

$$\text{R-X-}\ddot{\text{Y}} \xrightarrow{\text{CONCERTED}} \overset{\text{R}}{\text{X—Y}} \longrightarrow \text{X-Y-R}$$

The contribution of a concerted pathway to these reactions may provide an important test for orbital symmetry rules for reactivity [4].

In terms of the radical pair model for CIDNP we can represent the radical portion of the pathway by the following scheme [5].

$$\text{R-XY} \longrightarrow \overline{\text{R}\cdot\ \text{XY}\cdot} \longrightarrow \overline{\text{R}\cdot \| \text{XY}\cdot} \xrightarrow{F(t)} \overline{\text{R}\cdot\ \text{XY}\cdot}$$

$$f(t) = \tfrac{1}{2} p\, \tau_D^{1/2}\, t^{-3/2} \qquad t > \tau_D$$

Primary branch: λ; $\tfrac{1}{2}(1+r)$ → XY-R, $\tfrac{1}{2}(1-r)$ → XY-Я

Secondary branch: λ; $\tfrac{1}{2}$ → XY-R, $\tfrac{1}{2}$ → XY-Я

PRIMARY RECOMBINATION	SECONDARY RECOMBINATION
PARTIAL RETENTION	COMPLETE RACEMIZATION
NO CIDNP	CIDNP

The symbols are those commonly used [5,6] except for r which is the fraction of retention in the primary recombination step. It is assumed that r = 0 for secondary recombination.

Denoting by P_n and P_n' the probabilities of forming retained and inverted products, respectively, it is easily shown that for a recombination product formed from a singlet-born radical pair

$$P_n = \frac{\lambda}{2}(1+r) + \frac{\lambda}{2}(1-\lambda)p - \frac{\lambda}{2}(1-\lambda)\frac{p}{2}(\pi \tau_D |a_n|)^{1/2}$$

$$P_n' = \frac{\lambda}{2}(1-r) + \frac{\lambda}{2}(1-\lambda)p - \frac{\lambda}{2}(1-\lambda)\frac{p}{2}(\pi \tau_D |a_n|)^{1/2}$$

The first, second and third terms are just the probabilities of product formation by primary recombination, secondary recombination and nuclear spin dependent secondary recombination, respectively. For values of λ and p (the reencounter probability) each of the order of 0.5, τ_D ca. 10^{-11} sec. and a_n ca. 10^9 rad. sec^{-1}, the three terms will have relative magnitudes 50 : 10 : 1. Under these circumstances one may ignore the nuclear spin dependent term. The fraction of retention of configuration, R, then becomes

$$R = \frac{P - P'}{P + P'} = \frac{r}{1 + (1-\lambda)p}$$

Thus for λ and p both 0.5 and r = 1, i.e. no racemization during primary recombination, one predicts 80% retention of configuration. This degree of retention is determined mostly by what happens during primary recombination and depends only weakly on the degree of secondary recombination or, indeed, on the probability, λ , that a reaction will occur during an encounter. The CIDNP intensity and the actual yield of recombination product, $P + P'$, on the other hand, both depend strongly on λ, and the former on p as well.

A corollary of this model, which has been suggested repeatedly, [5,6] and experimentally verified [7], is that the CIDNP enhancement factor should be the same for both the retained and inverted stereoisomers of the recombination product. Furthermore, there is some evidence that r is at least not zero for primary recombination because net retention has been detected in the radical recombination products obtained from optically active azoalkanes [8]. r should be even larger for a pair of radicals formed with no intervening N_2 molecule.

2. LARGE CAGE EFFECT (?)

The traditional methods of detecting the presence of free radicals in reactions which could also proceed in a concerted fashion have involved in one way or another the interception of the intermediate. This could take the form of a) isolation of products derived from scavenged radicals, b) detection of "cross-over" in experiments with mixtures of reactants, or c) observation that the yield of recombination-type products decreases when radical scavengers are added.

For the special case of CIDNP the following points may be made about each of the above methods.

a) If one looks hard enough at the product mixture derived from reactions which exhibit CIDNP it is always possible to find dimers or products derived from atom abstraction reactions; i.e. any reaction producing CIDNP should also yield chemical evidence of the intermediacy of free radicals.

b) The yield of such "cage escape" products is simply 1-P, where P is the yield of recombination product discussed above.

$$1 - P \cong 1 - \lambda[1 + p(1-\lambda)]$$

The first term in brackets is the "cage effect" due to primary recombination and the second term is due to secondary recombination. We thus see that, in our example above, with λ and p each 0.5, a 38% yield of "non-cage" products should still be obtained. Any purely radical mechanism should thus yield at least a few percent of cage escape products. An unusually low yield of such products may well signal competing radical and non-radical pathways.

c) As the reactivity of radical scavengers increases it may be possible to intercept first the radicals undergoing secondary recombination (which also produce CIDNP!) and eventually the primary cage. Both the purely chemical [9,10] and the CIDNP [11,12] consequences of these effects have been explored theoretically [9,11] and experimentally [10,12]. This leads to the phenomenon of "cage substitution" in which a new geminate pair of radicals is produced by reaction of the first formed pair with a scavenging molecule.

3. ISOLATION OF PRODUCTS DERIVED FROM IONS

There has been some speculation [13] that diacyl peroxides undergo ionic dissociation in a reaction paralleling their more common homolytic cleavage.

$$RCO_2O_2CR' \begin{cases} \xrightarrow{\text{RADICAL}} RCO_2\cdot \;\; \cdot O_2CR' \;\; \text{(printed as } RCO_2\cdot \; O_2C \; R'\cdot\text{)} \\ \xrightarrow{\text{IONIC}} RCO_2^- \;\; O_2C \;\; R'^+ \end{cases}$$

We have recently found an example of a reaction in which ions are apparently formed after the homolytic dissociation process [14]. Figure 1 shows the nmr spectrum obtained from trimethylpropionyl benzoyl peroxide in ODCB at 120°. In addition to the expected [15] emission lines from neopentylbenzoate and neopentylbenzene at 3.8 δ and 2.2 δ, respectively, there are three other emission lines labelled I-III. These turn out to be due to the three rearrangement products typically observed during the solvolysis of neopentyl tosylate [16]. Since there is no evidence that the neopentyl radical rearranges at a rate which could compete with diffusive separation of the radical pair, we attribute these products to rearrangement of the carbocation produced from the radical by transfer of

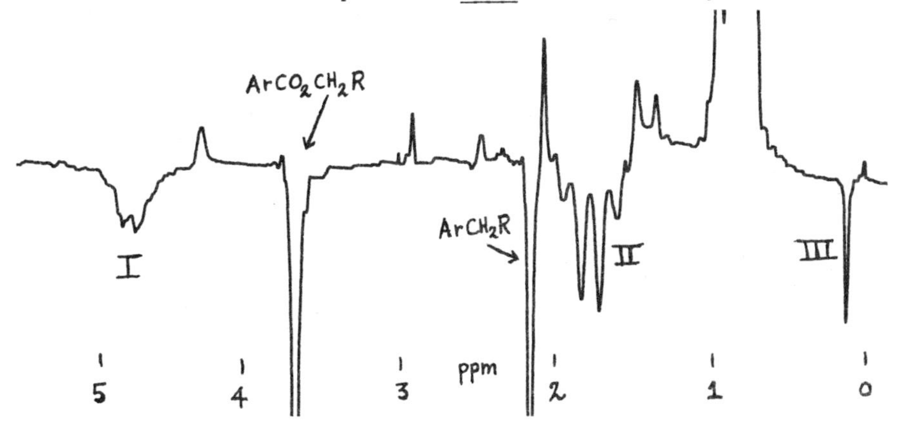

Figure 1. Proton nmr spectrum obtained from trimethylpropionylbenzoyl peroxide in ODCB at 120°.

an electron to the benzoyloxy radical.

$$ArCO_2O_2CCH_2C(CH_3)_3 \longrightarrow ArCO_2\cdot \; \cdot CH_2C(CH_3)_3 \longrightarrow ArCO_2CH_2C(CH_3)_3$$

$$\downarrow -ArCO_2^-$$

$$^+\overset{*}{C}H_2C(CH_3)_3$$

$$\downarrow$$

$$CH_3\overset{*}{C}H_2-\overset{+}{C}(CH_3)_2$$

$$\text{I: } (CH_3)(\overset{*}{H})C{=}C(CH_3)_2 \qquad \text{II: } CH_3\overset{*}{C}H_2-C({=}CH_2)CH_3 \qquad \text{III: } \overset{*}{C}H_2-CH_2-C(CH_3)_2 \text{ (ring)}$$

The observed sign of polarization and its distribution within the rearranged products is in accord with generation of CIDNP in the neopentyl radical followed by rapid rearrangement via the carbocation.

The above results indicate that caution should be exercised in attributing ion-derived products to an ionic process paralleling the production of free radicals. The processes may be well consecutive.

4. CONCLUSION

The observation of a high degree of retention of configuration or the isolation of ionic rearrangement products in reactions exhibiting CIDNP are not necessarily good evidence for competing radical and non-radical pathways. The best evidence for such competition would be based on combined measurements of the degree of retention, the cage effect and the CIDNP intensity.

REFERENCES

1. A. Henne and H. Fischer, Helv. Chim. Acta., 58, 1598 (1975); R. Bausch, H. P. Schuchmann, C. von Sonntag, R. Benn and H. Dreeskamp, Chem. Commun., 418 (1976).

2. U. H. Dolling, G. L. Closs, A. H. Cohen and W. D. Ollis, Chem. Commun., 545 (1975).

3. A. R. Lepley, in "Chemically Induced Magnetic Polarization", A. R. Lepley and G. L. Closs, Eds., Wiley, 1973; Chap. 8.

4. W. D. Ollis, M. Rey, I. O. Sutherland and G. L. Closs, Chem. Commun., 543 (1975).

5. R. G. Lawler, Prog. NMR Spectros., 9, 145 (1975).

6. R. Kaptein, J. Am. Chem. Soc., 94, 6251 (1972).

7. D. Bethell, J. Hollander, EUCHEM Conference on Chemically Induced Magnetic Polarization, St. - Pierre de Chartreuse, France, 21-23 May, 1975.

8. K. R. Kopecky and T. Gillan, Can. J. Chem., 47, 2371 (1969); F. D. Greene, M. A. Berwick and J. C. Stowell, J. Am. Chem. Soc., 92, 867 (1970).

9. R. M. Noyes, Prog. Reaction Kinetics, 1, 129 (1961).

10. T. Koenig and H. Fischer, in "Free Radicals", J. Kochi, Ed., Wiley, 1973; Vol. 1, Chap. 4.

11. R. Kaptein, J. Am. Chem. Soc., 94, 6262 (1972).

12. R. A. Cooper, R. G. Lawler and H. R. Ward, ibid., 94, 552 (1972); R. Kaptein, F. W. Verheus and L. J. Oosterhoff, Chem. Commun., 977 (1971).

13. C. Walling, H. P. Waits, J. Milovanovic and C. G. Pappiaonnou, J. Am. Chem. Soc., 93, 4927 (1970).

14. D. Jacobs and R. G. Lawler, 7th Northeast Regional Meeting of American Chemical Society, Albany, August, 1976.

15. R. E. Schwerzel, R. G. Lawler and G. T. Evans, Chem. Phys. Lett., 29, 106 (1974).

16. G. M. Fraser and H. M. R. Hoffmann, Chem. Commun., 561 (1967).

CHAPTER XVII

EFFECTS OF DIFFUSION ON REACTION RATES

J. Boiden Pedersen

Department of Physics, Odense University
DK-523o, Odense M, Denmark

ABSTRACT. The effects of diffusion on reaction rates are studied by the concentration gradient approach. The treatment covers the range from diffusion controlled to more conventional kinetics, and includes the case of reacting particles that interact through long range forces. A general expression for the rate constant is derived. The concept of reencounters are introduced and the probability density of a reencounter is calculated for free diffusion.

1. INTRODUCTION

The theory of diffusion controlled reactions in its general form has many applications, e.g. the growth of colloidal or aerosol particles, catalysis, ion recombination, charge-transfer reactions, photoinduced charge separation and recombination in chlorophyll assemblies of bacteria and plants, fluorescence quenching, and uv-induced isothermal luminescence.

The basic idea in a diffusion-controlled reaction is that the rate of reaction is equal to the rate at which the reacting molecules approach each other (encounter). This requires an instantaneous reaction when the molecules meet, or identically that the probability of reaction at each encounter approches unity. Obviously the diffusion-controlled rate of reaction is the fastest possible rate. The quenching of fluorescence and reactions of free radicals, are reactions which are close to being diffusion-controlled. The theory of diffusion-controlled reactions was first formulated by Smoluchowski [1], who considered reactant particles with no intermolecular forces. Debye [2] extended the treatment to reactions like those between ions where there are long range

L. T. Muus et al. (eds.), Chemically Induced Magnetic Polarization, 283-296.

forces between reactants. The treament is known as the "concentration gradient" treatment and is given in sections 2 and 3.

Smoluchowski's approach [1] was limited to diffusion controlled reactions due to the choice of boundary condition, the absorption boundary condition. Another boundary condition, known as the radiation boundary condition was given by Collins [3]. Using this boundary condition it is possible to formulate a theory that cover the transition from conventional to diffusion controlled kinetics. This is discussed in section 4.

The concept of reencounters, which was introduced by Noyes [4], is discussed in section 5. This way of considering the mutual motion of two reacting particles has proved useful for the development of the radical pair theory for CIDNP and CIDEP. The probability density of a reencounter plays an important role in these theories and are derived here.

The chapter ends with a discussion of the effect of a finite reaction region. The present Chapter is not intended to be a review of all existing theories; such a review has been given by Noyes [5]. It simply presents the most popular and useful approach, which is based on concentration gradients, in a consistent manner that is convenient for the study of chemical kinetics by CIDNP, CIDEP, and similar techniques. The limitations of the approach is not discussed in details and the reader interested in these aspects are referred to Ref.'s [5] and [6].

2. CONCENTRATION GRADIENTS AND STEADY STATE FLUXES.

The system under consideration consists of three constituents which we denote as fluid, and A and B particles. The fluid serves merely as an inert solvent in which the particles carry out diffusive motion. The A and B particles can react when they are close. On the average the Brownian particles (A and B) are far apart and consequently only two-particle correlations are important. It is further assumed that the potential energy between two particles depends on the relative coordinates only. With these assumptions the motion of the B particles relative to a specific A particle may be described by a Fokker-Planck equation

The idealized model for the reaction of A and B particles is the following. The motion of B particles relative to a specific A particle is given by the diffusion equation.

$$\partial c/\partial t = D\vec{\nabla} \cdot [\vec{\nabla} c + \beta c \vec{\nabla} U] \qquad (1)$$

where $\beta = 1/(kT)$; $c(\vec{r},t)$ is the concentration of B particles at the distance r from the selected A, and U is the potential energy between an A and a B particle. D is the sum of the diffusion coefficients for A and B.

When A and B encounter they may react and after a reaction the B particle disappears from the system whereas the change of state of the A particle is neglected. The A particle acts simply as a reaction sink. A reaction will cause a depletion of B particles near A and this gives rise to a concentration gradient of B towards A. This concentration gradient causes a flux of B towards A which then react etc. In summary the reaction is thought to be due to a concentration gradient which is created by the reaction itself. Normally a steady state is reached within 10^{-7}s and the existence of this steady state flux constitutes the last assumption.

It is convenient to rewrite Eq. (1) as

$$\partial c/\partial t = -\vec{\nabla} \cdot \vec{j}(\vec{r},t) \tag{2a}$$

where the flux $\vec{j}(\vec{r},t)$ is given by

$$\vec{j}(\vec{r},t) = -D(\nabla c + \beta c \nabla U) \quad . \tag{2b}$$

Assuming steady state, $\partial c/\partial t = 0$, gives

$$\vec{\nabla} \cdot \vec{j}(\vec{r},t) = 0 \tag{3a}$$

which, upon assumption of spherical symmetry, may be rewritten to

$$r^{-2}(d/dr)r^{2}j(r) = 0 \tag{3b}$$

where $j(r)$ is the radial flux (particle current density). Eq. (3b) is integrated to

$$4\pi r^{2}j(r) = -J = \text{constant} \tag{3c}$$

which shows that the total current J of B particles towards A is constant and equal to the rate of reaction. Combination of Eqs. (2b) and (3c) leads to

$$-D(dc/dr + \beta c(r)dU/dr) = -J/(4\pi r^{2}) \tag{4}$$

which can be rewritten to

$$D\exp(-\beta U)\left[\frac{d}{dr}c(r)\exp(\beta U)\right] = J/(4\pi r^{2}) \quad . \tag{5}$$

Equation (5) is easily integrated to

$$c(r)\exp(\beta U) - c(\infty) = -J/(4\pi D)\int_{r}^{\infty}\exp(\beta U)r^{-2}dr \tag{6}$$

which gives the steady state distribution of B particles around A in terms of the steady state current J. The quantity $c(\infty)$ is the concentration of B infinitely far from A and may be taken to be the bulk concentration, i.e. the average concentration.

One notes from Eq.(6) that $c(r)$ is equal to the equilibrium value $c(\infty)\exp(-\beta U)$ only for the case of no reaction. But the equilibrium value, as calculated by equilibrium statistical mechanics, is approached as $r \to \infty$. A determination of J or $c(r)$ requires information about the "reactivity" of A with B and is discussed in the two following sections.

3. DIFFUSION CONTROLLED REACTIONS. ABSORBING BOUNDARY

In this section diffusion controlled reactions are considered, i.e. A and B particles react instantaneously when they encounter at $r = d$, the distance of closest approach (the sum of the two radii). Smoluchowski [1] expressed this reactivity by the absorption boundary condition

$$c(d,t) = o \ . \tag{7}$$

If r is set equal to d in Eq. (6) and the boundary condition (7) is used one obtains

$$J/c(\infty) = 4\pi Ddf^{*} \tag{8}$$

where f^* is defined as

$$1/f^* = d\int_d^\infty \exp[\beta U(r)] r^{-2} dr \ . \tag{9}$$

The quantity f^* is a measure of the strengh of the potential energy. One has $f^* = 1$ for $U = o$, $f^* < 1$ repulsive forces, and $f^* > 1$ for attractive forces. Moreover f^* deviates the more from unity the longer the range of $U(r)$ is.

For a coulomb interaction f^* can be given by an analytical expression. But generally f^* must be evaluated by numerical integration, e.g. by a six point Gaussion quadrature. The Debye-Hückel potential is often used for the potential energy between ions in concentrated solutions where the charge-shielding due to the ionic atmosphere becomes important. With that form of $U(r)$ one obtains $f^* \sim 3$ for attraction and $f^* \sim o.1$ for repulsion for an aqueous solution of a divalent ion of concentration $\sim o.o1$ M. For higher concentrations f^* is closer to unity since the ionic shielding makes the potential shorter range.

The current of B particles towards a specific A particle is equal to J, which also equals the rate of encounters of all B's with a specific A. Consequently

$$\text{encounter rate} = 4\pi Ddf^{*}c_{A}(\infty)c_{B}(\infty) \qquad (1o)$$

is the encounter rate of A and B particles. As all encounters lead to reaction Eq.(1o) also gives the rate of reaction. The rate constant for a diffusion controlled reaction is therefore

$$k_2 = 4\pi Ddf^{*} \; . \qquad (11a)$$

This is the maximum obtainable rate constant.

If A and B are identical particles then the rate equation is

$$2dc/dt = -k_2'c^2 \quad , \qquad (11b)$$

with k_2' given by

$$k_2' = 2\pi Ddf^{*} \; . \qquad (11c)$$

4. GENERAL REACTIONS. RADIATION BOUNDARY

The absorption boundary condition is only applicable if every encounter lead to a reaction. If this is not the case another description may be used. The easiest description [6] is to introduce a reaction sink at r = d by changing Eq. (2a) to

$$\partial c/\partial t = -\vec{\nabla}\cdot\vec{j}(r,t) - (K/4\pi d^2)\delta(r-d)c(\vec{r},t) \; . \qquad (12)$$

By integrating Eq. (13) over all space and using Gauss divergence theorem one obtains

$$dn/dt = -\int_{s}\vec{j}(\vec{r},t)\cdot d\vec{s} - Kc(d,t) = -Kc(d,t) \qquad (13)$$

where n(t) is the total number of unreacted B particles at time t. The integral is over the surface of the container (containing the reating system) and equals zero for a closed system, where there is no flux of B particles into or out of the container.

Equation (12) is integrated over the part of space enclosed between a sphere with radius d+ε and the surface of the container. By similar arguments as lead to Eq. (13), one obtains

$$\begin{aligned} dm/dt &= 4\pi(d+\varepsilon)^2 j(d+\varepsilon) \\ &= -4\pi D(d+\varepsilon)^2\left[dc/dr + \beta dU/dr\right]_{r=d+\varepsilon} \end{aligned} \qquad (14)$$

where m(t) is the number of unreacted B particles exterior to a sphere of radius d+ε.

Since the volume of the sphere is small compared to the total volume of the container one may safely equate n and m. By letting $\varepsilon \to o$ and equating Eqs. (13) and (14) one obtains the so called radiation boundary condition

$$Kc(d) = 4\pi Dd^2 [dc/dr + \beta c dU/dr]_{r=d} \ . \tag{15}$$

Of course,-dn/dt is equal to the steady state current J of B towards A, i.e.

$$J = Kc(d) \ . \tag{16}$$

Combination of Eqs. (6) and (16) gives

$$\frac{J}{c(\infty)} = \frac{4\pi dDf^*}{1+4\pi dDf^* \exp[\beta U(d)]/K} \tag{17}$$

and this is the observable rate constant

$$k_{obs} = J/c(\infty) \ , \tag{18}$$

since the rate of reaction per A particle in chemical kinetics is written as

$$dn/dt = k_{obs} c_B(\infty) \ . \tag{19}$$

The numerator in Eq. (17) is equal to the rate constant k_2, Eq. (11), i.e. it is equal to the rate constant for new encounters. If the rate constant is written as

$$k_{obs} = k_2 \Lambda, \tag{20a}$$

with k_2 given by Eq. (11)

$$k_2 = 4\pi Ddf^* \ , \tag{20b}$$

one may interpret the rate of reactions k_{obs} as the rate of new encounters k_2 multiplied by Λ, the probability that a reaction occurs during the collision, which includes all reencounters (cf. the following Section for a definition of a collision). By Eqs. (17), (18), and (2o) one obtains the following expression for Λ

$$\Lambda = k\tau/(1+k\tau) \tag{20c}$$

where

$$k = K/(4\pi d^2 \Delta r) \tag{20d}$$

and an arbitrary Δr has been introduced for dimensional purposes. Consequently τ is given by

$$\tau^{-1} = Df^{*}\exp[\beta U(d)](d\Delta r)\ . \qquad (2oe)$$

One observes that the values of Λ, given by Eq. (2oc), are limited to the range $[o,1]$ as required for a probability. A value of Λ close to one corresponds to a diffusion controlled reaction. The case when $\Lambda << 1$ corresponds to a diffusion independent reaction since

$$k_{obs} \simeq K \exp[-\beta U(d)] \text{ for } \Lambda << 1, \qquad (21)$$

and this is the form appropiate for a slow reaction when the relative diffusion is able to maintain the equilibrium distribution of particles. This result, Eq. (21), can be obtained directly from Eq. (16) by noting that for an equilibrium system $c(r) = \exp[-\beta U(r)]c(\infty)$.

The parameter τ may be interpreted as a characteristic lifetime of the interacting pair, i.e. it is the time spent by the particles in a region where they might react. Alternatively, τ^{-1} is the rate of permanent separation from the "reaction region" (or the dissociation constant for the complex not involving true chemical bonds)and k is the rate of reaction in the region. By denoting an unreacted pair in the reaction volume as A:B and a reacted pair as AB one may write rate equations as

$$(d/dt)[A{:}B] = -(\tau^{-1}+k)[A{:}B]$$

$$(d/dt)[AB] = k[A{:}B].$$

The solution for $[AB](t\to\infty)$ is easily found to be given by Eq. (2oc) when $[A{:}B](o) = 1$, thus justifying the interpretation. The time τ cannot be interpreted as the lifetime for a single encounter; it includes all reencounters and therefore depends also on the long time behaviour of the diffusion. We prefer to denote τ as the collision time.

The present treatment where the reaction process is described by a δ-function sink at r = d and where diffusion inside the sphere with r = d is permitted, is unsatisfactory from a physical point of view. One may exclude diffusion in the sphere by introducing a reflecting boundary condition, $j(d,t)= o$, at r = d. A derivation similar to the one presented here but including the reflecting boundary condition has been given by Schulten and Schulten [8]. This boundary condition ,however, does not affect the result for a δ-function sink [6].

A major problem has been to include a reaction sink of finite extent. This has been solved by Pedersen and Freed [7] which used a numerical technique. Their results differ from those obtained by the earlier approximate theories and may be given in simple analytical expression. They [7] found that Eq.(2oc) is appropriate but the dimensionless product $k\tau$ should be replaced by ($\rightarrow$)

$$k\tau \rightarrow D^{-1} \int_d^\infty r\, k(r)\, dr \tag{22}$$

where $k(r)$ is the rate of reaction for a distance r between the particles. Spherical symmetry and no potential energy has been assumed. For $U \neq 0$ one may calculate $k\tau$ by Eq.(22) and then divide by $f^* \exp(\beta U(d))$. The rate constant $k(r)$ has physical significance; it is the rate constant for the relaxation process that causes a binding at r. If $U=0$ and $k(r)$ equals k in the range $[d,d+\Delta r]$ with $\Delta r \ll d$ then Eq.(22) shows that

$$(k\tau) \rightarrow k\tau \quad \text{and}$$
$$\tau = d\Delta r/D \ . \tag{23}$$

This result, Eq.(23), is identical to Eq.(2oe) obtained using the δ-function sink. Note, however, that in the δ-function approach the parameters k and Δr are completely arbitrary, while all parameters are related to physical quantities in the approach of Pedersen and Freed [7].

In earlier treatments [9] appears a quantity called reaction volume ΔV defined as

$$\Delta V = 4\pi D d f^* \exp[-\beta U(d)]/\tau \tag{24}$$

This reaction volume is usually assumed to equal the total volume swept by the interaction pair, but Eqs.(22) and (23) show that ΔV equals the annular volume of the reaction region.

In the above derivation of expressions for rate constants the fundamental assumptions were: that the reacting particles are present in such low concentrations that only two particle correlations are important; and that a steady state exist. Recently the effect of concentration on the rate of diffusion - controlled reactions has been investigated [1o] with the result that the rate constant increases with concentration. In liquids a steady state is obtained within 10^{-7}s but in that time range the rate constant will be time dependent. By solving the diffusion equation with the assumed initial (non-steady state) distribution of particles the time dependence is obtained [11]. Solutions corresponding to an initial equilibrium distribution is often given [5,11]. The situation where reactive particles are created in pairs and may recombine cannot, of course, be described by an initial equilibrium distribution. A description based on the assumption that the pair is isolated is convenient and is discussed in the following section.

5. ISOLATED PAIRS AND REENCOUNTER PROBABILITIES.

Very often particles are created in pairs, e.g. in a photochemical reaction yielding two radicals with an initial separation of a few Ångstrom. If the concentration of pairs is not too high, an individual pair may be treated as an isolated pair, i.e. the recombination of the pair proceeds without any interference from other pairs. For example, recombination of ion pairs as a function of time has been studied experimentally.

For a description of the motion of one of the partners of an initial pair relative to the other one can again assume that the diffusion equation Eq. (1) is appropriate, and this is the most common description. Noyes [5] used a description based on the random flight model which describes the diffusion process as a series of finite jumps. The long time behaviour, i.e. after ~lo jumps, of these two treatments is identical and the continuous diffusion model Eq.(1) is most convenient for obtaining solutions.

With the intention of simplifying the description of the recombination process Noyes [5] introduced a treatment based on molecular pairs. The main idea is to divide the diffusion process into a series of subsequent events. As will be seen this treatment offers a good deal of physical insight into the diffusion - recombination process.

The physical picture of the process is the following: Two particles which are initially in contact (first encounter) will separate by diffusion if they do not react (recombine). Particles that are closer than $d+\Delta r$, where $\Delta r \leq d$, are said to be in contact. Separated particles (i.e. $r>d+\Delta r$) may diffuse together again with a probability $p<1$, and the particles that meet again are said to reencounter. When the particles are in contact they may react with probability λ. Particles that do not react may reencounter and thus enhance the probability of reaction (relative to a single encounter).

It is convenient to denote as a collision the first encounter and all subsequent reencounters. The probability of reaction per collision Λ is

$$\Lambda = \lambda+\lambda p(1-\lambda)+\lambda p^2(1-\lambda)^2 + \ldots$$

$$= \lambda/[1-p+p\lambda] \tag{25}$$

since the fraction λ reacts at the first encounter; of the remaining $(1-\lambda)$ the fraction p reencounters and of these the fraction λ reacts, i.e. $\lambda p(1-\lambda)$ react during the first reencounter; etc.

The parameters λ and p have no fundamental physical significance but they may be related to parameters appearing in the concentration gradient approach by noting that the experimentally observable rate constant can be written as [5]

$$k_{obs} = 4\pi Ddf^{*}\ \beta' \tag{26}$$

where β' is the probability that two particles separating from a non-reactive encounter will ever react with each other. Now β' is equal to $p\Lambda$ since p is the probability for the first reencounter where the reaction starts. Thus

$$\beta' = \frac{\lambda p/(1-p)}{1+\lambda p/(1-p)} \tag{27}$$

which can be compared with Eq.(2oc).

The probability of a reencounter, p, i.e. the probability that two particles initially separated by a distance r_o will ever attain a separation $d(r_o>d)$ follows by simple arguments for 3-dimensional continuous diffusion. For a steady state system the rate of new encounters at separation d can be calculated as the rate of new encounters at separation $r_o(r_o>d)$ multiplied by $p(r_o\to d)$ the probability of reaching d from an initial separation r_o, i.e.

$$k_2(d) = k_2(r_o)p(r_o\to d). \tag{28}$$

In Eq.(28) it has been used that the rate of new encounters equals the rate of a diffusion controlled reaction, cf. Eq. (1o). By inserting the previously derived expression, Eq.(11), for the rate constant k_2 into Eq.(28) one obtains

$$p(r_o\to d) = \frac{df^{*}(d)}{r_o f^{*}(r_o)} = \frac{\int_{r_o}^{\infty} \exp[\beta U(r)]r^{-2}dr}{\int_{d}^{\infty}\exp[\beta U(r)]r^{-2}dr} . \tag{29}$$

This expression has also been found by numerical solutions to the diffusion equation for non-steady state systems [7]. It is particularly simple for free diffusion (U=o) since then $f^{*} = 1$ and hence $p = d/r_o$. This shows that the probability that the particles will ever reencounter is o.5 when they are initially separated by a distance equal to the molecular diameter. In two dimensions one has p = 1 and the particles will always reencounter, but in three dimensions two particles initially in contact will separate permanently, typically within loo ns, if they do not react.

Another quantity of interest, e.g. for CIDNP and CIDEP, is the probability of first reencounter at time t when the particles separate (i.e. $r=r_o>d$) at time zero. Obviously the intergral of this quantity $f(t,d|r_o)$ over all positive times equals $p(r_o \to d)$.

$$p(r_o \to d) = \int_o^\infty f(t,d|r_o)dt \quad . \tag{3o}$$

More precisely $f(t,d|r_o)$ is the probability density that particles having an initial separation r_o will reach a separation $d < r_o$ for the first time at time t. But this is equal to the rate of reaction at time t for an initial δ-function distribution of particles at separation r_o, if the particles react instantaneously when they reach the separation d,cf. the discussion after Eq.(1o). By these arguments and Eq. (15) one obtains

$$f(t,d|r_o) = -4\pi d^2 D[\partial P(r,t)/\partial r + \beta P(r,t)dU/dr]_{r=d} \tag{31a}$$

where $P(r,t)$ has been used rather than $c(r,t)$ to indicate that P is normalized to unity as required for a probability. The probability distribution $P(r,t)$ in Eq(31a) must satisfy the diffusion equation (1), the boundary condition for instantaneous reaction at $r = d$, Eq (7)

$$P(d,t) = o \tag{31b}$$

and the initial condition

$$P(r,o) = \delta(r-r_o)/(4\pi r^2) \quad . \tag{31c}$$

The diffusion equation (1) has not been solved to yield an analytic expression for $f(t,d|r_o)$ in the presence of forces between the particles. For free diffusion ($U = o$) one can solve the diffusion equation (1) by the Laplace transform method or by eigenfunction expansion and we shall use the latter method.

Since the system is assumed to have spherical symmetry only the radial part of the diffusion eq. (1) is of interest. By averaging Eq.(1) over all angles one obtains

$$\partial P/\partial t = D\Gamma_r P(r,t) = D\frac{1}{r^2}\frac{\partial}{\partial r}r^2\frac{\partial}{\partial r}P(r,t). \tag{32}$$

The diffusion operator Γ_r is hermitian and has a complete set of eigenfunctions $\psi_k(r)$

$$\Gamma_r\psi_k(r) = -k^2\psi_k(r) \tag{33}$$

where k is real. Note that Γ_r is an isotropic operator in the three dimensional space (r,θ,ϕ). If the initial distribution $P(r,o)$ is expanded in eigenfunctions

$$P(r,o)=4\pi\int_o^\infty dk\ k^2\ c(k)\ \psi_k(k) \tag{34a}$$

then by application of Eq's (31), (33) and (34a)

$$P(r,t)=4\pi\int_0^\infty dk\ k^2\ c(k)\ \psi_k(r)\ \exp(-k^2Dt) \tag{34b}$$

Clearly the problem is solved when the eigenfunctions and expansion coefficients are determined. The eigenfunctions satisfy the normalization condition

$$\delta(k-k')/(4\pi k^2)=4\pi\int_0^\infty dr\ r^2\ \psi_k(r)\ \psi_{k'}(r), \tag{35a}$$

and the closure relation

$$\delta(r-r_o)/(4\pi r^2)=4\pi\int_0^\infty dk\ k^2\ \psi_k\ (r_o)\ \psi_k(r). \tag{35b}$$

The closure relation, expressing the completeness of the set of eigenfunctions, gives immediately the expansion coefficient for the δ-function initial distribution Eq.(31c)

$$c(k) = \psi_k(r_o). \tag{36}$$

The eigenfunctions of Γ_F are known, e.g. from the quantum mechanical description of a free particle, to be the spherical Bessel functions

$$\psi_k(r) = A\ j_o(kr)+B\ y_o(kr) \tag{37a}$$

$$j_o(kr) = \sin kr/kr \tag{37b}$$

$$y_o(kr) = \cos kr/kr \tag{37c}$$

The normalized eigenfunctions that satisfies the boundary condition (31b) are

$$\psi_k(r) = \frac{1}{4\pi}\ \frac{2}{\pi}\ \frac{\sin k(r-d)}{k\ r} \tag{37d}$$

and then by Eqs.(34b) and (36)

$$P(r,t)=(2\pi^2 rr_o)^{-1}\int_0^\infty dk\ \sin k(r_o-d)\ \sin k(r-d)\ \exp(-k^2Dt). \tag{38}$$

By application of the trigonometric relation

$$\sin k(r_o-d)\sin k(r-d)=\tfrac{1}{2}\left[\cos k(r-r_o)-\cos k(r+r_o-2d)\right] \tag{39a}$$

and the Fourier transform result

$$\int_{-\infty}^{\infty} e^{ikx}\ \exp(-k^2\tau)dk = (\frac{\pi}{4\tau})\quad \exp(-x^2/4\tau) \tag{39b}$$

one can perform the integration in Eq.(38) with the result

$$P(r,t)=(4\pi^2 rr_o)^{-1}(\frac{\pi}{4Dt})^{\frac{1}{2}}\{\exp[-\frac{(r-r_o)^2}{4Dt}]-\exp[-\frac{(r+r_o-2d)^2}{4Dt}]\}. \tag{4o}$$

The rate of first encounters is now obtained from Eq.(4o) by application of Eqs.(31a) with U=o

$$f(t,d|r_o)) = \frac{d(r_o-d)}{r_o(4\pi Dt^3)^{\frac{1}{2}}} \exp[-\frac{(r_o-d)^2}{4Dt}]. \tag{41}$$

An analytic expression for $f(t,d|r_o)$ has not been obtained for U≠o. One can obtain numerical results by application of the numerical method of Pedersen and Freed [7]. This has been done by Schulten and Schulten [8] which used a finite difference approximation for the time variable. It seems more elegant, however, to diagonalize the "diffusion matrix" as done by Hwang and Freed [12].

APPENDIX. Reaction sink of finite extent.

After the completion of this chapter it became evident that the derivation presented here works equally well for a reaction sink of a finite extent. The full derivation, which is a simple generalization of that given in Sect.4, is given elsewhere [13]. Here we simply give the results for Λ defined in Eq.(2oa) for the free diffusion case (U=o). For a general isotropic reaction sink k(r) defined by

$$\partial c/\partial t = -\vec{\nabla}\cdot\vec{j} - k(r)c(r,t) \tag{A1}$$

one obtains the exact relation

$$\Lambda = \frac{K_2/D}{1+K_1/D} \tag{A2}$$

where K_1 and K_2 are defined as

$$K_1 = \int_d^\infty dr\ r\ k(r) \tag{A3}$$

$$K_2 = d^{-1}\int_d^\infty dr\ r^2\ k(r)\ . \tag{A4}$$

The relations, Eqs.(2oc) and (22), found by Pedersen and Freed by a numerical technique are seen to be appropiate when k(r) is of short range since then $K_1 \simeq K_2$. But for other forms of k(r) one should use the exact relation (A2). Note that Λ may be larger than one since $K_2 \geq K_1$.

REFERENCES.

1. M.v. Smoluchowski, Z.physik.Chem. 92 129 (1917)
2. P. Debye, Trans.Electrochem.Soc. 82, 265 (1942)
3. F.C. Collins and G.E. Kimball, J. Colloid Sci. 4, 425 (1949); F.C. Collins, J. Colloid Sci. 5, 499 (195o)
4. R.M. Noyes, J.Chem.Phys. 22, 1349 (1954); J.Amer.Chem.Soc. 78, 5486 (1956)
5. R.M. Noyes, Prog.React.Kinet. 1, 129 (1961)
6. G.Wilemski and M. Fixman, J.Chem.Phys. 58, 4oo9 (1973)
7. J.B. Pedersen and J.H. Freed, J.Chem.Phys. 61, 1517 (1974); note that Eq.(3.8) should be $\Lambda=k\tau_1/(1+k\tau_1)$.
J.H. Freed and J.B. Pedersen, Adv.Mag.Reson. 8, 1 (1976)
8. Z. Schulten and K. Schulten, J.Chem.Phys. 66, No. 1o (1977)
9. E.g. I Amdur and G.G. Hammes, "Chemical Kinetics", Chapter 2, McGraw-Hill, N.Y., 1966.
1o. B.U. Felderhof and J.M. Deutch, J.Chem.Phys. 64, 4551 (1976)
11. A.T. Bharucha-Reid, Arch.Biochem.Biophys. 43, 416 (1952)
12. L.P. Hwang and J.H. Freed, J.Chem.Phys. 63, 4o17 (1975)
13. J.B. Pedersen, J.Chem.Phys. submitted (1977)

CHAPTER XVIII

GENERAL RELATIONS FOR HIGH FIELD CIDNP

J. Boiden Pedersen

Department of Physics, Odense University
DK-523o Odense M, Denmark

ABSTRACT. General relations for the spin dependent reactivity in high magnetic fields are derived. These relations are independent of the diffusive behaviour of the radicals. Included is a general expression for the singlet-triplet mixing parameter F^*. The results are applied to the continuous diffusion model.

1. INTRODUCTION

This work is based on two very recent articles by Pedersen [1,2]. Several fundamental relations for radical pair CIDNP in high magtic fields are derived, i.e. relations that are independent of the details of the specific diffusion model one may assume for a description of the molecular motion of the radicals. These general results are then applied to the 3-dimensional continuous diffusion model and exact analytical results are given which permit a simple calculation of CIDNP polarizations for all magnitudes of the magnetic interactions and for all diffusional rates. The effect of a finite exchange interaction is included. It is discussed how the involved parameters may be estimated from known data, e.g. bond lengths, viscosity of the solvent, etc. These results will facilitate the interpretation of experimental CIDNP results in terms of quantitative information.

The general ideas of the radical pair mechanism (RPM) are that a chemical reaction of two radicals is spin selective, e.g. only radical pairs whose combined spin state is a singlet state can react; and that the magnetic interactions within the radicals cause a nuclear-spin dependent mixing of the singlet and the triplet states. These general features are common to all RPM-theories, the

L. T. Muus et al. (eds.), Chemically Induced Magnetic Polarization, 297-308.

differences show up in the treatment of the singlet-triplet mixing. In the original works of Closs [3a], Closs and Trifunac [3b], and Kaptein and Oosterhoff [4] an exponential function was used for the probability distribution of the lifetime of the radical pair, i.e. singlet-triplet mixing was only allowed when the radicals were close (in the cage). The spin dependent probability of reaction F was found to depend quadratically on the magnetic interaction. It is remarkable that many of the qualitative features observed for CIDNP were reproduced by this treatment, although quantitatively it was in error.

In liquids two radicals that are initially close together may, if they do not react, separate and later meet again (reencounter). The reencounter mechanism will obviously change the probability of reaction. This effect on the spin-independent raction probabilities has been discussed by Noyes [5]. Adrian [6] and later Kaptein [7] incorporated some reencounter dynamics into the RPM theory, by including a single reencounter, by the use of Noyes' step diffusion model [5]. This treatment gives a square root dependence of F upon the magnetic interaction [8]. These results, however, were of limited value for a quantitative estimation of CIDNP for various reasons: The results depended strongly on a parameter p, the probability of a reencounter, which could only be chosen rather arbitrarily. Futhermore the treatment was limited to small magnetic interactions and non-viscous liquids. Finally it was not at all clear what would be the effect of inclusion of all reencounters. Kaptein [7] gave the reencounter problem a great deal of thought,although his efforts proved unsuccesful. However his work [7], and other works [9,1o] which were also unsuccesful in that respect, inspired us to reexamine the possibility of inclusion of all reencounters by an infinite series.

All reencounters are automatically included in a description based on the Stochastic Liouville Equation (SLE), which was used by Evans et al. [11], and Pedersen and Freed [12, 13]. In fact such a description is in principle exact, the problem being that of obtaining a solution to the SLE. Evans et al. [11] obtained an approximate solution, valid for small magnetic interactions and nonviscous media, which again showed the square root dependence, but here the numerical coefficient did not depend on any parameter p. Since the exchange interaction was approximated by a δ-function it remained unclear what effect a finite exchange interaction would have, but for a negligible exchange interaction and nonviscous liquids these results were reliable.

Pedersen and Freed in their treatment of CIDNP used a numerical technique which was developed for the similar CIDEP problem [13]. Their results are exact for the assumed diffusive models and one may then test the accuracy of the approximations involved in other methods by comparison with these results, cf. [12]. They [12] also give results for viscous liquids, where they find a square root dependence is no longer adequate. The effect of a finite extent of the exchange interaction is studied. Also included are models in which the charge interaction between reacting ionic radicals and their surroundings are accounted for in the Debye-Hückel fashion. One of the results of Pedersen and Freed was that the spin dependent probability of reaction F could be expressed in terms of just two parameters Λ, the spin independent probability of reaction per collision (including all reencounters); and F^* which measures the conversion of triplets to singlets and vice versa. Exact expressions for F were found [12] in terms of Λ and F^*, for any precursor multiplicity. Similar relations have been found by Evans et al.[11], although only valid for small magnetic interactions and in the non-viscous limit. The expressions given by Adrian [4] and Kaptein [5] do not have the correct interrelationship.

By the present approach we have been able to prove that the expressions for F in terms of Λ and F^* found by Pedersen and Freed [12] for specific diffusion models are in fact exact for any diffusion model. Furthermore we have derived a general analytic expression for F^* involving only two parameters which depend on the specific diffusion model. The information needed on the diffusive behaviour is the probability density for first arrival.

2. THEORETICAL APPROACH

We use the spin density matrix in our description of the combined spin system of the two radicals A and B. The rate equation is

$$\frac{d\rho(t)}{dt} = -iH^{\times}(r)\ \rho(t) = -i[H(r),\ \rho(t)] \tag{1}$$

where the Hamiltonian H(r)is given in angular units (rad/s). The Hamiltonian one adopts is conveniently written as

$$H(r) = H^{o}(r) + H' \tag{2}$$

where $H^{o}(r)$ and H', in high magnetic fields, are given by

$$H^{o}(r) = \tfrac{1}{2}(g_a+g_b)\beta_e\hbar^{-1}B_o(S_{az}+S_{bz})+\tfrac{1}{2}[\sum_j^a A_j I_{jz}+\sum_k^b A_k I_{kz}] \times(S_{az}+S_{bz})-J(r)(\tfrac{1}{2}+2\vec{S}_a\cdot\vec{S}_b) \tag{3a}$$

$$H' = \tfrac{1}{2}(g_a-g_b)\beta_e\hbar^{-1}B_o(S_{az}-S_{bz})+\tfrac{1}{2}[\sum_j^a A_j I_{jz}-\sum_k^b A_k I_{kz}](S_{az}-S_{bz}). \tag{3b}$$

In these relations, Eqs. (3), subscript a and b serve to distinguish quantities belonging to the two radicals and the summations extend over all nuclear spin coupled to the unpaired electron spin.

It is convenient to work in the coupled representation, i.e. the singlet-triplet representation. By taking matrix elements of Eq.(1), introducing a simple transformation, and utilizing that $Tr\rho=1$ when there is no reaction, we obtain

$$\frac{d}{dt}\begin{bmatrix}\rho_{ST_o}-\rho_{T_oS}\\ \rho_{ST_o}+\rho_{T_oS}\\ \rho_{SS}-\rho_{T_oT_o}\end{bmatrix}(t) = -i\begin{bmatrix} o & 2J & -2Q\\ 2J & o & o\\ -2Q & o & o\end{bmatrix}\rho(t) \tag{4}$$

where $\rho(t)$ means the column vector at the left hand side of Eq.(4). The quantity Q is given by

$$Q = \tfrac{1}{2}(g_a-g_b)\beta_e\hbar^{-1}B_o+\tfrac{1}{2}(\Sigma^a_j A_jM_j-\Sigma^b_k A_kM_k)\,. \tag{5}$$

Note that Q is called a_n, ω_n or ω_{ab}, by other authors. In writing Eq.(4) we have used that the S, T_o subspace does not couple to the T_+, T_- subspace such that the latter can be left out of the calculation.

Eq. (4) can be solved to

$$\begin{bmatrix}\rho_{ST_o}-\rho_{T_oS}\\ \rho_{ST_o}+\rho_{T_oS}\\ \rho_{SS}-\rho_{T_oT_o}\end{bmatrix}(t) = \begin{bmatrix} c(t) & -i\frac{J}{\omega}s(t) & i\frac{Q}{\omega}s(t)\\ -i\frac{J}{\omega}s(t) & \frac{Q^2}{\omega^2}+\frac{J^2}{\omega^2}c(t) & \frac{QJ}{\omega^2}(1-c(t))\\ i\frac{Q}{\omega}s(t) & \frac{QJ}{\omega^2}(1-c(t)) & \frac{J^2}{\omega^2}+\frac{Q^2}{\omega^2}c(t)\end{bmatrix}\rho(o) \tag{6}$$

where ω, s(t) and c(t) are defined as

$$\begin{aligned}\omega &= (Q^2+J^2)^{\frac{1}{2}}\\ s(t) &= \sin(2\omega t)\\ c(t) &= \cos(2\omega t)\end{aligned} \tag{7}$$

The initial condition will usually be (o,o±1) (or a combination of these) which represents initial singlets (+1) or initial triplets (-1) with no electron polarization ($\rho_{ST_o}+\rho_{T_oS}$) and no initial phase correlation ($\rho_{ST_o}-\rho_{T_oS}$).

The central quantity of interest for CIDNP is the mixing of the singlet and triplet states ($S-T_o$ mixing), i.e. the change of $\rho_{SS}-\rho_{T_oT_o}$ caused by the spin Hamiltonian. By inspection of Eq. (6) one observes that there is no $S-T_o$ mixing for $Q<<J$, i.e. when the radicals are close, more precisely when their relative

distance r is such that $J(r) > Q$. We shall call this region of space for the exchange region and let the outer limit of this region be at $r = r_o$; the inner limit being at $r = d$; the distance of closest approach. Typical values are $d \simeq 4$ Å and $r_o \simeq 6$-8 Å. In the exchange region we approximate the matrix of interaction in Eq. (6) by a unit matrix. The neglect of the sine terms connecting $(\rho ST_o - \rho T_oS)$ and $(\rho ST_o + \rho T_oS)$ is strictly valid only for a short range or low valued exchange interaction such that the average time spend in the exchange region τ_J satisfies $2J\tau_J \ll 1$. We shall return to this point later. Outside the exchange region where $J \ll Q$, we have S-T_o mixing, but note that $\rho ST_o + \rho T_oS$ is constant and decoupled from the other terms, such that we can reduce Eq. (6) to a 2×2 matrix equation by excluding the term $\rho ST_o + \rho T_oS$. By reintroducing $(\rho SS+\rho T_oT_o)(t) = (\rho SS+\rho T_oT_o)(o)$ and making a simple transformation we can rewrite Eq(6) to a convenient form

$$\begin{bmatrix} 2\mathrm{Im}\rho_{ST_o} \\ \rho_{T_oT_o} \\ \rho_{SS} \end{bmatrix}(t) = \begin{bmatrix} c(t) & -qs(t) & qs(t) \\ \frac{1}{2}qs(t) & \frac{1}{2}(1+c(t)) & \frac{1}{2}(1-c(t)) \\ -\frac{1}{2}qs(t) & \frac{1}{2}(1-c(t)) & \frac{1}{2}(1+c(t)) \end{bmatrix} \bar{\rho}(o) \qquad (8)$$

where $q = Q/|Q|$.

We shall be interested in $\bar{\rho}$, the average value of $\bar{\rho}$ (t) when the radicals meet for the first time at $r = d$ starting with an initial separation $r_o > d$ and an initial density matrix $\bar{\rho}'$. From Eq. (8) we see that $\bar{\rho}'$ and $\bar{\rho}$ are connected through

$$\bar{\rho} = \bar{\bar{M}} \cdot \bar{\rho}' \qquad (9)$$

where $\bar{\bar{M}}$ is the matrix obtained by averaging the matrix in Eq(8) with $f(t,d|r_o)$, the probability density of first arrival at $r = d$ at time t for an initial separation r_o. We shall write $\bar{\bar{M}}$ as

$$\bar{\bar{M}} = p \begin{bmatrix} c & -s & s \\ \frac{1}{2}s & \frac{1}{2}(1+c) & \frac{1}{2}(1-c) \\ -\frac{1}{2}s & \frac{1}{2}(1-c) & \frac{1}{2}(1+c) \end{bmatrix} \qquad (1o)$$

where c and s are defined as

$$c = \frac{1}{p}\int_o^\infty f(t,d|r_o)\cos(2Qt)dt \qquad (11)$$

$$s = \frac{1}{p}\int_o^\infty f(t,d|r_o)\sin(2Qt)dt \qquad (12)$$

and $p = \int_o^\infty f(t,d|r_o)dt$ is the probability of at least one reencounter, or alternatively 1-p is the probability that the radicals will never meet.

When the radicals encounter at $r=d$ the singlet pairs react with probability λ. The surviving radicals diffuse apart to $r=r_o$ and start a new reencounter cycle. By defining $\overline{\rho}(n)$ as the average density matrix at the n'th reencounter and $\overline{\rho}(n)'$ as the average value at separation r_o just after the n'th reencounter (i.e. after a possible recombination of the radicals) we obtain from Eq. (9)

$$\overline{\rho}(n) = \overline{\overline{M}}\cdot\overline{\rho}(n-1)' \tag{13}$$

The change of the density matrix due to recombination at the n'th reėncounter is expressed by the relation

$$\overline{\rho}(n)' = (\overline{\overline{1}}-\overline{\overline{\lambda}})\overline{\rho}(n) \tag{14}$$

where, for comparison with [12], we take

$$\overline{\overline{\lambda}} = \begin{bmatrix} o & & \\ & o & \\ & & \lambda \end{bmatrix} . \tag{15}$$

The fraction of radical pairs that recombine at the n'th reencounter is $(o,o,1)\cdot\overline{\overline{\lambda}}\cdot\overline{\rho}(n)$.

The fraction of radical pairs that recombine in a collision (first encounter and all reencounters) can then be written, by Eqs. (13)-(15), as

$$F = (o,o,1)\cdot\overline{\overline{\lambda}}\cdot\sum_{n=o}^{\infty}[\overline{\overline{M}}\cdot(\overline{\overline{1}}-\overline{\overline{\lambda}})]^n\ \overline{\rho}(o) \tag{16}$$

or by summing the series

$$F = (o,o,1)\cdot\overline{\overline{\lambda}}\cdot[\overline{\overline{1}}-\overline{\overline{M}}\cdot(\overline{\overline{1}}-\overline{\overline{\lambda}})]^{-1}\cdot\overline{\rho}(o). \tag{17}$$

3. GENERAL RESULTS

The dimension of the matrices in Eq.(17) is only three and it is therefore possible to carry out the matrix inversion and put Eq. (17) into an analytical form. By Eqs.(1o),(15) and (17) we obtain

$$F = \rho_{T_oT_o}(o)\lambda[p\tfrac{1}{2}(1-c)+p^2\tfrac{1}{2}(1-c)-p^2\tfrac{1}{2}(1^2-s^2-c^2)]/N \tag{18a}$$
$$+ \rho_{SS}(o)\lambda[1-p\tfrac{1}{2}(1+3c)+p^2\tfrac{1}{2}c+p^2\tfrac{1}{2}(s^2+c^2)]/N$$

where

$$N = 1-p\left[1+2c-\tfrac{1}{2}\lambda(1+c)\right]+p^2\left[s^2+c^2+2c-\lambda\tfrac{1}{2}(s^2+c^2+3c)\right]$$
$$- p^3(1-\lambda)(s^2+c^2). \qquad (18b)$$

In an attempt to rewrite Eqs. (18) to a more useful form we shall be guided by the results obtained by Pedersen and Freed [12] i.e. that CIDNP polarizations (i.e. F) are described in terms of just two fundamental parameters: Λ and F^*. The former parameter, Λ, is the spin-independent probability of reaction of singlets for the whole collision, including all reencounters. Note that λ is the corresponding quantity but per encounter. Alternatively, Λ is the probability of forming geminate product for a singlet precursor in the absence of $S\text{-}T_0$ mixing, i.e. $\Lambda = F(S;\ Q = o)$. It follows from Eqs. (11) and (12) that Q=o implies c=1 and s=o, and by Eqs. (18) we then get

$$\Lambda = \lambda/(1-p+p\lambda) \qquad (19)$$

This expression for Λ agrees with Noyes [5] result, cf. [16].

The other parameter, F^*, is defined as $F(T_0;\ \lambda=1)$, i.e. the probability of forming geminate product from a triplet precursor under the condition that singlet radical pairs react with probability one (at r=d). F^* measures the conversion from triplets to singlets, and vice versa, i.e. the effectiveness of the $S\text{-}T_0$ mixing. Eqs. (18) lead to the following expression for F^*

$$F^* = \frac{p(1-c) + p^2\left[(1-c) - (1-s^2-c^2)\right]}{2(1-p)^2 + 3p(1-c) - p^2\left[(1-c) + (1-s^2-c^2)\right]} \; . \qquad (2o)$$

Eq. (2o) is valid for all diffusional models, excluding the spin dependent ones, cf. [12]. The details of the molecular dynamics that determine the reencounter probability $f(t,d|r)$ enter the calculation through s, c and p. We shall later return to the calculation of these quantities.

Finally we note that Eqs. (18) can be rewritten to the relations

$$F(T_o) = \Lambda F^*/\left[1+F^*(1-\Lambda)\right] \qquad (21a)$$

$$F(S)-\Lambda = -\Lambda(1-\Lambda)F^*/\left[1+F^*(1-\Lambda)\right] \qquad (21b)$$

$$F(R.I.)-\frac{\Lambda}{2} = \tfrac{1}{2}\Lambda^2F^*/\left[1+F^*(1-\Lambda)\right] \qquad (21c)$$

where the spin multiplicity of the precursor is written in parentheses. The random initial (R.I.) precursor equals what some authors call F-precursor (free radicals) and results when two radicals, derived from different precursor molecules, encounter. The resulting radical pair have uncorrelated electron spins (R.I.).

The relations Eqs. (21) are independent of the diffusional behaviours and give directly the excess probability of forming geminate product due to the S-T_o mixing, which is the quantity of interest for calculations of CIDNP spectra. Pedersen and Freed [12] found these relations to be exact for 3-dimensional diffusion.

4. CONTINUOUS DIFFUSION MODEL

In order to calculate the quantities p, s, and c, defined in Eq's (11) and (12), we must adopt a diffusion model. We shall assume that the radicals move about independently and that the probability density $P(\vec{r},t)$ for the relative separation of a pair of radicals is determined by the continuous diffusion equation

$$\frac{\partial P(\vec{r},t)}{\partial t} = D\nabla^2 P(\vec{r},t) \tag{22}$$

The diffusion constant D for the relative diffusion is the sum of the diffusion coefficients for the two radicals under consideration. This equation may be solved for $f(t,d|r_o)$, the probability density of first arrival (encounter) at $r=d$, when initially at $r=r_o>d$. The resulting expression is [14-16]

$$f(t,d|r_o) = \frac{d(r_o-d)}{r_o(4\pi Dt^3)^{\frac{1}{2}}} \exp\left[- \frac{(r_o-d)^2}{4Dt} \right] \tag{23}$$

From this expression one easily finds that the probability of at least one reencounter is

$$p = d/r_o \tag{24}$$

or identically; the probability that the radicals will never reencounter is $(r_o-d)/r_o$. The quantities c and s defined in Eqs. (11) and (12), can now be written explicitly as,

$$c = \cos(z)\ \exp(-z) \tag{25a}$$

$$s = \sin(z)\ \exp(-z) \tag{25b}$$

where the parameter z is defined as

$$z = \frac{1-p}{p} \left(\frac{Qd^2}{D}\right)^{\frac{1}{2}}. \tag{25c}$$

The equations (2o), (21), and (25) represent the complete solution to high field CIDNP for the continous diffusion model.

When $z<<1$ the expression for F^* can be simplified. By expansion of c and s in powers of z, substitution in Eq. (2o), and retaining only first order terms one obtains the result,

$$F^* \simeq \tfrac{1}{2}(Qd^2/D)^{\frac{1}{2}} \text{ for } \frac{1-p}{p}(Qd^2/D)^{\frac{1}{2}} << 1 \qquad (26)$$

This asymptotic square root dependence of F^* is the well known result [6,7,11,12] but the numerical coefficient in Eq.(26) and the independence of p is obtained only by approaches that include all reencounters [11,12,1,2]. By comparison with the exact solution Eqs.(2o), (21), and (25) one finds that the approximate expression Eq.(26) is valid for $(Qd^2/D)<0.1$, but cf. [1].

The condition for the validity of Eq.(26) is more restrictive than is generally realized. For the purpose of illustration we note that $(Qd^2/D)^{\frac{1}{2}} = 0.49$ when Q = 15 Gauss, the molecular diameter equals 4 Å, and the viscosity of the solvent equals 13 cp (e.g. ethylene glycol). If a solvent with a viscosity equal to 1 cp (e.g. methanol, ethanol, or isopropanol) is used instead one obtains o.14 for the corresponding value. Quite often one may have to use the complete and exact expression for F^* which, however, represents no problems except that the p-dependence becomes important. The parameter p is related to the extend of the exchange interaction as discussed in the following section.

5. EFFECT OF A FINITE EXCHANGE INTERACTION UPON F^*

In the present approach the exchange interaction is included only via the concept of an exchange volume, i.e. a volume where singlet-triplet ($S-T_0$) mixing is suppressed. All information about the exchange interaction is then placed in the parameter p, the probability of reencounter. In order to test this exchange volume approach, we have compared results derived from the present method with the numerical results of Pedersen and Freed [12]. Note that F^* is independent of p for $(Qd^2/D)<0.1$, cf. Eq. (26).

The probability of a least one reencounter p depends on the "radius" of the exchange interaction r_0 through Eq. (24). By choosing

$$r_0 = 2r^* - d, \qquad (27a)$$

where r^* is determined by

$$|J(r^*)| \simeq |Q|, \qquad (27b)$$

one obtains a perfect agreement with [12]

For all practical purposes the present method can account for the effect of a finite exchange interaction. The accuracy is generally better than 1o% and since a factor of two may result from the finite extent, this is a substantial improvement with respect to a complete neglect of the exchange interaction. Furthermore the accuracy of the method increases with decreasing extent of the exchange interaction and the method becomes exact in the limit of no exchange interaction.

The general effect of a finite exchange interaction is to decrease the magnitude of F^* but the Q-dependence is also changed. It is then, in principle, possible to determine the extent of the exchange interaction (i.e. p) by a CIDNP experiment, cf. the discussion in [1].

6. EFFECT OF A SCAVENGING REACTION UPON F^*

In a scavenging reaction one of the radicals (e.g.A) of the radical pair AB reacts with a molecule (a scavenger) and converts to another radical A'. If such a scavenging reaction takes place during the reencounter cycle it will have two effects: 1) the amount of recombination products is reduced, and 2) the polarization of the recombination products is changed.

An exact expression for F^* including the effect of the scavenging reaction(with first order rate constant k) is given in [1]. The relations Eqs. (21) remain valid, i.e. only F^* is needed. For $k \ll Q$ there is no effect on F^*. For $k \gg Q$ and $(kd^2/D) \ll 1$ the complete expression [1] may be approximated to

$$F^* \simeq \tfrac{1}{2}(kd^2/D)^{\frac{1}{2}}\ Q^2/k^2. \tag{28}$$

This expression shows that the square root dependence of F^* is changed into a quadratic dependence upon Q. Also the magnitude of F^* has been reduced by a factor $(Q/k)^{3/2}$ with respect to the case of no scavenging reaction Eq (26). Note again the independence of p in this limit.

7. ESTIMATION OF PARAMETERS

A time resolved CIDNP experiment may be analyzed to yield a value of F (actually differences between values of F). By comparison of these experimental values of F with theoretical values, calculated by the complete expression for F Eqs. (2o), (21), (24), and (25), one or more parameters entering the expression for F may be determined quantitatively. These parameters are Λ, Q, d, D, p, and k. In order to determine some of these parameters it is necessary that the remaining parameters are known or can be estimated from known data.

The Q-dependence of F is essentially that of F^*, cf. Eqs.(21), and Λ therefore plays the role of a scaling parameter. The parameter Λ can therefore be separated from the other parameters. The need for inclusion of k is indicated by a stronger Q-dependence than the square root dependence.

A reasonable estimate of d, the distance of closest approach, may be obtained from the bond lengths and the geometries of the radicals. Note that the radicals are assumed to be spherical such that orientational averaged values of the radii should be used.

The relative diffusion constant D is the sum of the diffusion constants for the two radicals forming the pair. The single radical diffusion constant D' may be estimated by the Stokes-Einstein relation

$$D' = \frac{kT}{6\pi\eta r} \qquad (29)$$

with the value of the radius obtained above. However, only if the radii of the radicals are much larger than the radius of the solvent molecules can one be sure that Eq.(29) gives a reliable estimate of the diffusion constant. Due to this uncertainty of the accuracy of Eq.(29) for small radicals it is advantageous to determine D' directly by experiment whenever possible, i.e. to perform a relaxation experiment that yields a value for the translational diffusion coefficient.

At present, one cannot estimate the exchange interaction and therefore p for complex radicals. One must then rely on physical intuition. Probably values in the range o.5 to 1 are reasonable. It would, however, be interesting to determine p experimentally.

It may be of interest to determine several parameters by CIDNP, e.g. unknown g-values or hyperfine constants (through Q), p as noted above, and Λ which is essential for a calculation of the yield of the reaction. Furthermore the hydrodynamic radius (r in Eq.(29)) may be determined through D.

It is interesting to note that the experimentally observable second order rate constant also depends on the parameters Λ and D, cf. [16],

$$k_{obs} = 4\pi D d \Lambda \, . \qquad (3o)$$

Equation (3o) is, for example, valid for the reaction of R.I.(F)-radicals when the radicals are uncharged, cf. [16]. Note that the CIDNP polarization F depends on Λ and D as $\Lambda^2/D^{\frac{1}{2}}$ in the low viscosity region, cf.Eqs.(21) and (26). One might then be able to determine D and Λ by measurement of both k_{obs} and F. This method works best for low viscosity systems where the p-dependence of F is negligible.

8. GENERALIZATIONS

In this chapter a recently developed method [1] was used to obtain general and specific relations for CIDNP in high magnetic fields. This method can be generalized [2] such that CIDNP quantities can be calculated "exactly" for any magnetic field strengh and with the effects of scavenging reactions and relaxation included. Such a generalization [2] is of interest because it enables one to treat "complicated systems" with many coupled nuclear spins. The computer requirements are negligible compared to what is needed for the Stochastic Liouville equation approach. This is because the present method has separated the quantum mechanical and the diffusional descriptions. The method is also directly applicable to the magnetic field effect on the recombination process of radical pairs as observed by optical methods.

REFERENCES

1. J.B. Pedersen, J. Chem. Phys., submitted (1977).
2. J.B. Pedersen, Chem. Phys. Lett., submitted (1977).
3(a) G.L. Closs, J.Am. Chem.Soc., 91, 4552 (1969),
(b) G.L. Closs and A.D. Trifunac, J.Am. Chem.Soc., 92, 2183 (197o).
4. R. Kaptein and L.J. Oosterhoff, Chem. Phys. Lett. 4, 195 (1969); ibid p. 214.
5. R.M. Noyes, J. Amer. Chem. Soc., 78, 5486 (1956); "Progress in Reaction Kinetics", 1, 129 (1961).
6. F.J. Adrian, J. Chem. Phys., 53, 3374 (197o); J. Chem. Phys., 54, 3912 (1971).
7. R. Kaptein, J.Am. Chem.Soc., 94, 6521 (1972).
8. Kaptein [7] has J ≠ o in his relations. When J is set equal to zero, as it must be, then the square root dependence follows.
9. J.A. den Hollander, Chem. Phys., 1o, 167 (1975).
1o. J.A. den Hollander and R. Kaptein, Chem. Phys. Lett., 41, 257 (1976).
11. G.T. Evans, P.D. Flemming and R.G. Lawler, J. Chem. Phys. 58, 2o71 (1973).
12. J.B. Pedersen and J.H. Freed, J. Chem. Phys., 61, 1517 (1974).
13. J.B. Pedersen and J.H. Freed, J. Chem. Phys., 57, 1oo4 (1972); Ibid 58, 2746 (1973); ibid 59, 2869 (1973); ibid 62, 179o (1975); J.H. Freed and J.B. Pedersen, Adv. Magn. Reson., 8, 1 (1976).
14. J.M. Deutch, J. Chem. Phys., 56, 6o76 (1972).
15. H.S. Carslaw and J.C. Jaeger, "Conduction of Heat in Solids", Oxford U.P., 1959.
16. J.B. Pedersen, Chapter 17 of this volume.

CHAPTER XIX

NUMERICAL METHODS AND MODEL DEPENDENCE IN CHEMICALLY-INDUCED DYNAMIC SPIN POLARIZATION*

Jack H. Freed

Department of Chemistry, Cornell University,
Ithaca, New York 14853

In this chapter we describe some of the more advanced theoretical methods, which may be usefully employed in the study of chemically-induced spin polarization and related phenomena. The emphasis will be on the finite difference methods developed by Pedersen and Freed [1-6].

1. SOLUTIONS OF THE DIFFUSION EQUATION

1.1 The diffusion equation

Let us consider the diffusion equation:

$$\frac{\partial p(\vec{r},t)}{\partial t} = D\Gamma_{\vec{r}}\, p(\vec{r},t) \tag{1.1}$$

where $p(\vec{r},t)$ is the classical probability or distribution function for motion of a particle (or, in the present problem, the relative motion of a pair of particles), while $D\Gamma_{\vec{r}} = D\nabla_{\vec{r}}^{2}$ is the Markovian operator for the diffusion with diffusion coefficient D. When the problem admits of spherical symmetry (i.e., any potential terms in $D\Gamma_{\vec{r}}$ depend only on $r = |\vec{r}|$ and any boundary conditions depend only on r), then it is most convenient to write $\Gamma_{\vec{r}}$ in spherical polar coordinates in the form:

$$\Gamma_{\vec{r}} = \Gamma_r + (1/r^2)\Gamma_\Omega \tag{1.2a}$$

---

* This work has been supported in part by a grant (CHE 75-00938) from the National Science Foundation.

L. T. Muus et al. (eds.), Chemically Induced Magnetic Polarization, 309-355.

where, for Brownian motion in the absence of potentials, Γ_r, the radial part is given by

$$\Gamma_r = (1/r^2)(\partial/\partial r)r^2(\partial/\partial r) \tag{1.2b}$$

while the angular part Γ_Ω, is:

$$\Gamma_\Omega = (1/\sin\theta)(\partial/\partial\theta)[\sin\theta(\partial/\partial\theta)] + (1/\sin^2\theta)(\partial^2/\partial\theta^2) \tag{1.3}$$

We then can focus on the radial part of the distribution function:

$$p(r,t) = (1/4\pi)\int_0^\pi d\theta\sin\theta \int_0^{2\pi} d\phi\, p(\vec{r},t) \tag{1.4}$$

which obeys the diffusion equation:

$$\partial p(r,t)/\partial t = D\Gamma_r p(r,t) \tag{1.5}$$

1.2 Eigenfunction solutions

The solution to this diffusion equation is well known, and may be expressed as a conditional probability function, or Green's function:

$$p(\vec{R};\vec{r},t) = (4\pi Dt)^{-3/2}\exp\left\{-\frac{(\vec{r}-\vec{R})^2}{4Dt}\right\}$$
$$= \frac{1}{(2\pi)^3}\int \exp\{-i(\vec{r}-\vec{R})\cdot\vec{\rho} - \rho^2 Dt\}d^3\rho \tag{1.6}$$

Then by using the usual spherical Bessel function expansion of $e^{i\vec{\rho}\cdot\vec{r}}$ as well as the addition formula for the Legendre polynomials and their orthonormal properties, one has

$$p(|\vec{r}-\vec{R}|,t) = \frac{1}{2\pi^2}\int_0^\infty e^{-\rho^2 Dt} j_0(\rho R) j_0(\rho r)\rho^2 d\rho \tag{1.7}$$

where $j_0(\rho r)$ is the zero order spherical Bessel function; $j_0(z) \equiv \sin z/z$. This result may be understood as an expansion of the solution in the Hilbert space spanned by the $j_0(\rho r)$ for each different value of r. That is, we may regard this as an expansion in the orthonormal eigenkets:

$$|G_{\ell m}(\vec{\rho},\vec{r})\rangle \equiv |\sqrt{\tfrac{2}{\pi}}\, j_\ell(\rho r) Y_\ell^m(\Omega)\rangle \tag{1.7a}$$

where, because of spherical symmetry, we will only be interested in the $\ell = m = 0$ terms. Thus, we have:

$$p(\vec{R};\vec{r},t) = \int_0^\infty \rho^2 d\rho \sum_{\ell m} |G_{\ell m}(\vec{\rho},\vec{r})> e^{-\rho^2 Dt} < G_{\ell m}(\vec{\rho},\vec{R})| \tag{1.8}$$

and $p|\vec{r}-\vec{R}|,t)$ is obtained from this equation by setting $\ell = m = 0$, to yield the result of eq. 1.7.

Let us now note some useful variations on this formula. If the space is bounded by some outer wall (e.g., the walls of the container) taken to be spherical, then the integral over ρ becomes a sum over ρ corresponding to those discrete values of ρ which satisfy this outer boundary [7]. Of more use is the case of an inner boundary at $r = d$ corresponding to closest contact; thus we have as the range of r: $d \leq r < \infty$. In this case, one replaces the $|G_{\ell m}(\vec{\rho},\vec{r})>$ by the correct eigenfunctions given by Carslaw and Jaeger [7] which include appropriate linear combinations of Bessel functions. Let us call them $|\hat{G}_{\ell m}(\vec{\rho},\vec{r})>$. Now the inner boundary condition will be purely reflective if there is no reaction, but will be at least partially absorptive if there is a finite reactivity. Collins and Kimball [8] considered such a boundary for chemical kinetics. It may be written as:

$$\bar{k}p(d) = D(\partial p/\partial r)_{r=d} \tag{1.9}$$

and it states that the probability flux into $r < d$ [given by $D(\partial \rho/\partial r)_d$] is equal to the rate of reaction at the surface [given by $\bar{k}p(d)$] where $\bar{k}$ plays the role of a reactivity. [In Carslaw and Jaeger's [7] notation $h \equiv \bar{k}/D$.] (We will relate it to a first order rate constant below.) From Collins and Kimball's treatment, one has that the steady-state rate constant, which may be experimentally observed, is

$$k_f = \Lambda 2k_2(d) \tag{1.10}$$

where $2k_2(d) = 4\pi dD$ is the rate of new bimolecular encounters, and we have [6] that Λ is then the fractional probability of reaction. The CK result is:

$$\Lambda = \frac{d\bar{k}/D}{1 + d\bar{k}/D} \tag{1.11}$$

so for $\Lambda << 1$, one has $k_f = 4\pi d^2\bar{k}$, and it is independent of the diffusion, as it should be. (Below, we compare this result with that from finite differences). One has obtained an analytical solution to this problem (cf. [8]). [Another approach is to use the boundary condition of eq. 1.9 but with $h = \bar{k} = 0$, i.e., a pure reflecting wall. Then it is necessary to introduce a reaction at $r = d + \varepsilon$. This is utilized in the finite difference method below].

Now suppose we must generalize the diffusion equation to include, for example, some potential of interaction. That is, we would have a Smoluchowski equation for the relative diffusion:

$$\frac{dp(\vec{r},t)}{dt} = D\nabla_{\vec{r}}^2 p + (D/kT)\vec{\nabla}_{\vec{r}}\cdot[p(\vec{\nabla}_{\vec{r}}U(r))] \equiv D\Gamma_{\vec{r}}p \tag{1.12}$$

where $U(r)$ is the potential energy between particles assumed to depend only on r. The new $p(\vec{r},t)$ will no longer be expressible in terms of the $|G_{\ell m}(\vec{\rho},\vec{r})>$ or $|\hat{G}_{\ell m}(\vec{\rho},\vec{r})>$. Instead, our solution will be expressible in new eigenkets, each of which may be written as an expansion in the $|\hat{G}_{\ell m}(\vec{\rho},\vec{r})>$. The expansion coefficients may then be obtained by perturbation theory, when applicable, or by some other methods. Actually, it is easier to work with a symmetrized form of eq. 1.12. That is, let $p_0(\vec{r})$ be a measure of the equilibrium probability distribution such that $p_0(\vec{r}_i)/p_0(\vec{r}_j)$ is the ratio of probability of finding the radical-pair separated by $\vec{r}_i$ to that for $\vec{r}_j$ (e.g., let $p_0(\vec{r}) = e^{-U(r)/kT}$). Then let

$$\tilde{p}(\vec{r},t) \equiv [p_0(\vec{r})]^{-1/2}p(\vec{r},t)$$

and (1.13)

$$\tilde{\Gamma}_{\vec{r}} \equiv [p_0(\vec{r})]^{-1/2}\Gamma_{\vec{r}}[p_0(\vec{r})]^{1/2}$$

Then, one can show that eq. 1.12 becomes:

$$\frac{d\tilde{p}(\vec{r},t)}{dt} = D\tilde{\Gamma}_{\vec{r}}\tilde{p} = D\nabla_{\vec{r}}^2\tilde{p} + \frac{D\tilde{p}(\nabla_{\vec{r}}^2U(r))}{2kT} + \frac{D|\vec{F}(r)|^2}{(2kT)^2}\tilde{p} \tag{1.14}$$

where

$$\vec{F}(r) \equiv -\vec{\nabla}_{\vec{r}}U(r)$$

which results in $\tilde{\Gamma}_{\vec{r}}$ being a Hermitian operator. One may then, for example, introduce perturbation theory to handle $U(r) \neq 0$. Thus, let us first take the Laplace transform of eq. 1.14 to give

$$\bar{p}(\vec{r},s) = [s - D\Gamma_{\vec{r}}]^{-1}p(\vec{r},0) \tag{1.15}$$

with

$$\bar{p}(\vec{r},s) \equiv \int_0^\infty e^{-st}\tilde{p}(\vec{r},t) \tag{1.15a}$$

Then with

$$\tilde{\Gamma}_{\vec{r}}^{\,0} \equiv \nabla_{\vec{r}}^2$$

while

$$\tilde{\Gamma}_{\vec{r}}^{\,1} \equiv (\nabla_{\vec{r}}^2U(r))/2kT + |\vec{F}(r)|^2/(2kT)^2 \tag{1.16}$$

as the perturbation, one may use a resolvent-type approach [9] to give:

$$[s - D\tilde{\Gamma}_{\vec{r}}]^{-1} = \sum_{n=0}^{\infty} \left[\frac{1}{s - D\tilde{\Gamma}_{\vec{r}}^{0}}(-D\tilde{\Gamma}_{\vec{r}}^{1})\right]^{n} \frac{1}{s - D\tilde{\Gamma}_{\vec{r}}^{0}} \tag{1.17}$$

The conditional probability solution is obtained by letting

$$p(\vec{R},\vec{r},0) = \delta(\vec{r}-\vec{R}) = \int_0^{\infty} \rho^2 d\rho \sum_{\ell,m} |\hat{G}_{\ell m}(\vec{\rho},\vec{r})><\hat{G}^*_{\ell m}(\vec{\rho},\vec{R})| \tag{1.18}$$

Note that the zero-order solution (i.e., n = 0) is just the Laplace transform of eq. 1.8, as it should be. However, the first order correction is given by

$$\bar{p}(\vec{R},\vec{r},s)>^{(1)} = \int_0^{\infty} \rho'^2 d\rho' \sum_{\ell',m'} |\hat{G}_{\ell' m'}(\rho',r')> \times$$

$$\int_0^{\infty} \rho^2 d\rho \sum_{\ell,m} \frac{<\hat{G}_{\ell' m'}(\vec{\rho}',\vec{r})|-D\tilde{\Gamma}_{\vec{r}}^{1}|\hat{G}_{\ell m}(\vec{\rho},\vec{r})>}{s + D\rho'^2} \times$$

$$\frac{1}{s + D\rho^2} <G^*_{\ell m}(\vec{\rho},\vec{R})| \tag{1.19}$$

i.e., one must evaluate the matrix-elements (in $\vec{r}$ space) of $<G_{\ell' m'}(\vec{\rho}',\vec{r})|\tilde{\Gamma}_{\vec{r}}^{1}|\hat{G}_{\ell m}(\vec{\rho},\vec{r})>$. For a spherically symmetric $\tilde{\Gamma}_{\vec{r}}^{1}$, we would get $\ell' = \ell$ and $m' = m$. In a similar manner, one can generate higher-order corrections. A variety of procedures for accomplishing this has been compared by Yoon, Deutch, and Freed [9]. A powerful technique that lends itself to numerical solutions, involves starting with eq. 1.15 and then expanding $\bar{p}(\vec{r},s)$ (this is <u>not</u> the conditional probability) in a complete orthonormal set, e.g., the $\hat{G}_{\ell m}(\vec{\rho},\vec{r})$:

$$|\bar{p}(\vec{r},s)> = \int \rho^2 d\rho \sum_{\ell,m} C_{\ell m}(\vec{\rho},s)|\hat{G}_{\ell m}(\vec{\rho},\vec{r})> \tag{1.20}$$

Then by rewriting eq. 1.15 in the form

$$[s - D\tilde{\Gamma}_{\vec{r}}]|\bar{p}(\vec{r},s)> = |p(\vec{r},0)> \tag{1.15'}$$

and then pre-multiplying by $<\hat{G}_{\ell' m'}(\vec{\rho}',\vec{r})|$ one obtains a set of coupled integral equations for the expansion coefficients $C_{\ell m}(\vec{\rho},s)$:

$$\int \rho^2 d\rho \sum_{\ell,m} \langle \hat{G}_{\ell'm'}(\vec{\rho}',\vec{r}) | (s - D\tilde{\Gamma}_{\vec{r}}) | \hat{G}_{\ell m}(\vec{\rho},\vec{r}) \rangle C_{\ell m}(\vec{\rho},s) = \langle \hat{G}_{\ell'm'}(\vec{\rho}',\vec{r}) | p(\vec{r},0) \rangle \tag{1.21}$$

Note that the coefficient of $C_{\ell m}(\vec{\rho},s)$ in eq. 1.21 may be rewritten as:

$$\int \rho^2 d\rho \sum_{\ell,m} [(s + \rho^2 D)\delta_{\ell\ell'}\,\delta_{mm'}\,\delta(\vec{\rho}-\vec{\rho}') + \langle \hat{G}_{\ell'm'}(\vec{\rho}',\vec{r}) | \tilde{\Gamma}_{\vec{r}}^{(1)} | \hat{G}_{\ell m}(\vec{\rho},\vec{r}) \rangle] \tag{1.22}$$

This then again involves calculating the same matrix elements as above.

Now what are the typical questions asked about $p(\vec{r},t)$ [or $\bar{p}(\vec{r},s)$]? In one case we might want to ask for the total probability over all space $d \le r < \infty$. Suppose, for the moment, we have $U = 0$. Then this corresponds to just the coefficient $C_{0,0}(0,s)$. That is

$$\langle |\bar{p}(\vec{r},s)\rangle_{Av} \equiv \int_d^\infty d^3\vec{r}\; \bar{p}(\vec{r},s) = C_{0,0}(0,s) = \langle \hat{G}_{0,0}(0,\vec{r}) | \bar{p}(\vec{r},s) \rangle \tag{1.23}$$

where the last two equalities follow from the use of eq. 1.20. This quantity will be unity, unless there is a reaction which depletes radicals (e.g., $\bar{k}$ in eq. 1.9 is non-zero). Thus one must calculate in eq. 1.21 (or eq. 1.19) how the coefficient $C_{0,0}(0,s)$ couples to the other coefficients $C_{0,0}(\vec{\rho},s)$ by the perturbation $\tilde{\Gamma}_{\vec{r}}^{(1)}$, which need not be small (but we are assuming it is spherically symmetric). When there is a non-uniform equilibrium distribution, i.e., $p_0(\vec{r}) \neq$ constant, then the averaging prescription becomes:

$$\langle |\bar{p}(\vec{r},s)\rangle_{Av} \equiv \int_d^\infty d^3\vec{r}\,[p_0(\vec{r})]^{1/2}\bar{p}(\vec{r},s) = \langle [p_0(\vec{r})]^{1/2} | \bar{p}(\vec{r},s) \rangle \tag{1.24}$$

In the absence of a chemical reaction, $[p_0(\vec{r})]^{1/2}$ will correspond to the zero eigenvalue solution of $\tilde{\Gamma}_{\vec{r}}$, i.e., it represents the conservation of probability over all space, which is unaffected by the diffusion.

Another question might be the probability of finding the particles separated by $\vec{r}'$ at time t. One then needs

$$\langle\delta(\vec{r}-\vec{r}')|\bar{p}(\vec{r},s)\rangle = \int_0^\infty \rho^2 d\rho \sum_{\ell,m} \hat{G}^*_{\ell m}(\vec{\rho},\vec{r}')C_{\ell m}(\vec{\rho},s) \qquad (1.25)$$

which follows from the form of the Dirac-delta function (eq. 1.18), and it involves a knowledge of <u>all</u> the $C_{\ell m}(\vec{\rho},s)$. Alternatively, we may just ask for a particular radial separation and use $\delta(|r| - |r'|)$. Then, it follows from eq. 1.7, that eq. 1.25 would become:

$$\langle\delta(|r| - |r'|)|\bar{p}(\vec{r},s)\rangle = \int_0^\infty \rho^2 d\rho \hat{G}^*_{00}(\vec{\rho},\vec{r}')C_{0,0}(\vec{\rho},s) \qquad (1.26)$$

1.3 Numerical methods and eigenfunction solutions

An equation like eq. 1.21 may be solved numerically by replacing the integral over ρ by a sum. In fact, this will occur naturally if one introduces a finite outer wall at $r = r_N$ with an appropriate boundary condition, as we have already noted. The resulting coupled algebraic equations can be solved by standard computer techniques for matrix inversion or diagonalization. Such a method is an effective one provided (1) the needed matrix elements (involving integrations over $\vec{r}$) can be conveniently calculated by analytic or numerical methods, and (2) it is not necessary to use too many discrete values of ρ to obtain convergent solutions (i.e., independent of r_N or $\rho_{Min} \equiv 2\pi/r_N$).

The resolvent method should prove more useful for (1) small perturbations or for (2) cases when the integrations of the matrix elements over ρ, are best performed analytically. It could also prove to be a useful method when these integrals are to be done numerically. [It should be noted that the denominators in an expansion like eq. 1.19 can go to zero (for $s = 0$ and ρ or $\rho' \to 0$). This problem can be eliminated by reordering the expansion in a manner shown by Yoon, Deutch, and Freed [9] corresponding to a total-time-ordered cumulant or projection operator procedure.]

Actually, while these methods have proved very useful when applied to problems involving rotational diffusion (e.g., the triplet initial CIDEP mechanism [6,10]), except for perturbation-type approaches, they have yet to be extensively applied to problems involving translational diffusion (e.g., the radical-pair mechanism).

1.4 Finite difference method

We now turn our attention to finite-difference methods, which have been extensively employed in numerical solutions.

The finite difference solution of the diffusion equation amounts to first letting $\hat{p}(r,t) \equiv rp(r,t)$ so that

$$\frac{\partial \hat{p}(r,t)}{\partial t} = D \frac{\partial^2 \hat{p}(r,t)}{\partial r^2} \tag{1.27}$$

and then to let

$$\frac{\partial \hat{p}(r,t)}{\partial r^2} = \frac{1}{\Delta r^2} [\hat{p}(r-\Delta r,t) - 2\hat{p}(r,t) + \hat{p}(r+\Delta r,t)] \tag{1.28}$$

where we have used the standard mean difference form for the second derivative in terms of the radial increment Δr [6]. This application of the finite difference technique is essentially equivalent to transforming the continuous diffusion equation into a discrete Master equation involving a transition-probability matrix $\underline{\underline{W}}$, coupling $p(r,t)$ between discrete values $p(r_0+j\Delta r,t)$ where $j = 0, 1, 2, \ldots, N$. These discrete values form a column vector $\underline{p}$. Thus

$$D \frac{\partial^2 \hat{p}}{\partial r^2} \rightarrow \underline{\underline{W}}\, \hat{\underline{p}} \tag{1.29}$$

and the discrete Master equation becomes:

$$\frac{\partial}{\partial t} \hat{\underline{p}} = \underline{\underline{W}}\, \hat{\underline{p}} \tag{1.29'}$$

with Laplace transform:

$$\hat{\underline{p}}(r,s) = [s - \underline{\underline{W}}]^{-1} \hat{\underline{p}}(r,t=0) \tag{1.30}$$

which can be solved by straightforward matrix diagonalization or inversion methods, once $\underline{\underline{W}}$ is specified [see below]. We now have, by comparison of eqs. 1.28 and 1.29, that:

$$W_{j,j-1} = W_{j-1,j} = D/\Delta r^2 \tag{1.31a}$$

$$W_{j,j} = -2D/\Delta r^2 \tag{1.31b}$$

and this corresponds to a tri-diagonal matrix, which is easily solved. Note that $W_{j,j-1}$ is the transition probability from the j-1 box (at r_{j-1}) to the jth box (at r_j).

We must also be careful to specify boundary conditions. The reflecting-wall boundary condition at d, the distance of closest approach, is just:

$$\partial p(r,t)/\partial r]_{r=d} = 0 \tag{1.32a}$$

or

$$\partial\hat{p}(r,t)/\partial r]_{r=d} = \hat{p}(d,t)/d \tag{1.32b}$$

which, in finite difference notation, becomes:

$$[\hat{p}(d+\Delta r,t) - \hat{p}(d-\Delta r,t)]/2\Delta r = \hat{p}(d,t)/d \tag{1.33}$$

This means, that at the boundary we have

$$W_{0,0} = -2D/\Delta r^2(1+\Delta r/d) \tag{1.34a}$$

$$W_{0,1} = 2D/\Delta r^2 \tag{1.34b}$$

when eq. 1.33 is used to eliminate the $\hat{p}(d-\Delta r,t)$ term that would otherwise appear in eq. 1.28.

Now, in principle, we want solutions over the whole region $d < r < \infty$; but in order to make them tractable, we require a finite outer limit $r_N = d+N\Delta r$. A very useful boundary condition at r_N, which yields convergent solutions (for r_N large enough), even as $t \to \infty$ is the "collecting wall" boundary condition. This amounts to letting $W_{N-1,N} = 0$, so the particles (or radical-pairs) collect at r_N and cannot diffuse back. Then the conservation of probability condition requires that $W_{N,N} = 0$ and $W_{N,N-1} = 2D/\Delta r^2$ as well. This conservation of probability condition may be stated as:

$$\int_d^\infty \frac{\partial\hat{p}(r,t)}{\partial t}\, rdr = D\int_d^\infty \frac{\partial^2\hat{p}}{\partial r^2}\, rdr = 0 \tag{1.35}$$

or in finite difference notation where:

$$\frac{\partial}{\partial t}\,\underline{\hat{p}} = \underline{\underline{W}}\,\underline{\hat{p}} = \underline{0} \tag{1.36}$$

as:

$$\sum_{i=0}^{N} V(i)W_{i,j} = 0 \quad \text{for} \quad j = 0, 1, \ldots, N \tag{1.37}$$

where $rdr \to V(i)$. That is, the weighted sum of elements of $\underline{\underline{W}}$ for each column must be zero. In particular, we have:

$$V(0) = d\Delta r/2 \tag{1.38a}$$

$$V(i) = r_i\Delta r \tag{1.38b}$$

$$V(N) = r_N\Delta r/2 \tag{1.38c}$$

It is often useful to distinguish two regions of space (i) $d \le r < r_M$ and (ii) $r_M < r \le r_N$, such that fine graining in

Δr is required in region (i) to properly account for particle interactions, while much coarser graining in Δr may be utilized in region (ii) corresponding to large separations. One takes Δr in (ii) as f times larger than that of the former region (where $f \sim 10$ to 100). Then eqs. 1.38 become:

$$V(0) = d\Delta r/2 \qquad (1.39a)$$

$$V(i) = r_i \Delta r \qquad \text{for } 0 < i < M \qquad (1.39b)$$

$$V(M) = r_M (1+f)\Delta r/2 \qquad (1.39c)$$

$$V(i) = r_i f \Delta r \qquad \text{for } M < i < N \qquad (1.39d)$$

$$V(N) = r_N f \Delta r/2 \qquad (1.39e)$$

The matrix elements of $\underline{\underline{W}}$ are again given as in eq. 1.31 for $r_i < r_M$. For $r_i > r_M$, they can be obtained from the elements of eq. 1.31 by dividing by f^2. The Mth row is determined by the conservation of probability eq. 1.37 with the V(i)'s of eq. 1.39, giving:

$$W_{M,M-1} = [2/(1+f)](D/\Delta r^2) \qquad (1.40a)$$

$$W_{M,M} = -(2/f)(D/\Delta r^2) \qquad (1.40b)$$

$$W_{M,M+1} = [2/(1+f)f](D/\Delta r^2) \qquad (1.40c)$$

So far, this procedure will merely give numerical solutions to the diffusion equation whose analytic solution is known (e.g., eq. 1.6). We, therefore, consider the new features, such as reactivity and potentials of interaction. Rather than introducing reactivity by an inner boundary condition like eq. 1.9, it is more convenient to explicitly include the effects of a finite reactivity upon radical contact by adding to the diffusion equation a term:

$$\left.\frac{dP}{dt}\right]_{rxn} = -k(r)p(r,t) \qquad (1.41a)$$

where k(r) plays the role of a first-order chemical rate constant. One may use a variety of forms for the functional dependence of k(r) on r. Usually, the simple form:

$$k(r_i) = k\delta_{i,0} \qquad (1.41b)$$

has been used, where k is a constant. This represents a "sphere of influence" for the colliding radicals extending from d to $d+\Delta r_k$ with $\Delta r_k = \Delta r$. The effect of this reaction, then, is explicitly included in an augmented $\underline{\underline{W}}$ matrix. It will be seen below that this form leads to the following result for Λ, the fractional probability of reaction:

$$\Lambda = k\tau_1/(1 + k\tau_1) \tag{1.42}$$

with

$$\tau_1 = d\Delta r_k/D \tag{1.42'}$$

the characteristic "lifetime" of the interacting pair. One may rewrite $\tau_1^{-1} \equiv 4\pi Dd/\Delta V$ where ΔV is the "reaction volume." A comparison of this result with that from [8] yields $\bar{k} \to k\Delta r_k$.

We now consider the inclusion of the interactive potential U(r) in the form of eq. 1.12. It is rather easy to deal with this new term in the finite difference method. We first define a function F(r) by

$$F(r)\hat{e}_{\vec{r}} \equiv (1/kT)\vec{\nabla}U(r) = (1/kT)[\partial U(r)/\partial r]\hat{e}_{\vec{r}} \tag{1.43}$$

where F(r) plays the role of the force (but in units of inverse length) in the radial direction represented by unit vector $\hat{e}_{\vec{r}}$. Equation 1.27 now becomes:

$$\frac{\partial \hat{p}(r,t)}{\partial t} = D\,\frac{\partial^2 \hat{p}(r,t)}{\partial r^2} + \frac{D}{r}\frac{\partial}{\partial r}\,[\hat{F}(r)\hat{p}(r,t)] \tag{1.44}$$

with $\hat{F}(r) \equiv rF(r)$. The effect of this force is thus to introduce new terms into the $\underline{\underline{W}}$ matrix when the finite difference approach is employed, i.e.,

$$\frac{\partial}{\partial r}\,[\hat{F}(r)\hat{p}(r,t)] \to \frac{\hat{F}(r-\Delta r)\hat{p}(r-\Delta r,t) - \hat{F}(r+\Delta r)\hat{p}(r+\Delta r,t)}{2\Delta r} \tag{1.45}$$

We summarize the additional contributions to $\underline{\underline{W}}$:

$$D^{-1}W^F_{0,0} = r_1 F(1)/\Delta r d \tag{1.46a}$$

$$D^{-1}W^F_{0,1} = F(0)/\Delta r \tag{1.46b}$$

$$D^{-1}W^F_{j,j-1} = -F(j)/2\Delta r \qquad 0 < j < M \tag{1.46c}$$

$$D^{-1}W^F_{j,j} = (2\Delta r)^{-1}[F(j+1r_{j+1}/r_j - F(j-1)r_{j-1}/r_j] \qquad 0 < j < M \tag{1.46d}$$

$$D^{-1}W^F_{j,j+1} = F(j)/2\Delta r \qquad 0 < j < M \tag{1.46e}$$

$$D^{-1}W^F_{M,M-1} = -F(M)/\Delta r(1+f) \tag{1.46f}$$

$$D^{-1}W^F_{M,M} = -F(M-1)r_{M-1}/r_M\Delta r(1+f) \tag{1.46g}$$

$$D^{-1}W^F_{M,M+1} = 0 \tag{1.46h}$$

when one chooses r_M such that for $r > r_M$, $U(r) = 0$. Note that these equations are obtained by applying the conservation of probability condition eq. 1.37 wherever necessary. (Note that now $W_{i,j}$ in eq. 1.37 includes the additional terms due to the forces [6,11].) It should be clear, from the concept of $\underline{\underline{W}}$ as a transition-probability matrix, that we require Δr to be small enough so the off-diagonal elements of $\underline{\underline{W}}$ are nonnegative, while the diagonal elements of $\underline{\underline{W}}$ must be nonpositive. An inspection of these elements leads to the sufficiency condition that:

$$\Delta r < |2/F(j)| \tag{1.47}$$

where, usually $|F(0)|$ is the largest of the $F(j)$'s. The solutions may now be obtained by matrix inversion or by diagonalization. It is possible to symmetrize this matrix by the transformation:

$$\underline{\underline{\tilde{W}}} \equiv \underline{\underline{S}}\ \underline{\underline{W}}\ \underline{\underline{S}}^{-1} \tag{1.48}$$

where $\underline{\underline{S}}$ is the diagonal matrix whose matrix elements are:

$$S_{00} = 1 \tag{1.49a}$$

$$S_{ii} = S_{i-1,i-1}\sqrt{W_{i-1,i}/W_{i,i-1}} \qquad i > 0 \tag{1.49b}$$

$$(S^{-1})_{ii} = (S_{ii})^{-1} \tag{1.49c}$$

It is, thus, necessary to have $W_{i-1,i}/W_{i,i-1} > 0$, so one must let $W_{N-1,N} > 0$, but a value of $W_{N-1,N}/W_{N,N-1} < 10^{-10}$ is sufficient [11].

One may then diagonalize $\tilde{W}$ with a real orthogonal transformation represented by the matrix $\underline{\underline{O}}$ such that [11]:

$$\underline{\underline{w}} = \underline{\underline{O}}\ \underline{\underline{\tilde{W}}}\ \underline{\underline{O}}^{tr} = \underline{\underline{O}}\ \underline{\underline{S}}\ \underline{\underline{W}}\ \underline{\underline{S}}^{-1}\underline{\underline{O}}^{-1} = \underline{\underline{T}}\ \underline{\underline{W}}\ \underline{\underline{T}}^{-1} \tag{1.50}$$

where $\underline{\underline{w}}$ is the diagonal form consisting of real, nonpositive eigenvalues [12]. Then eq. 1.30 becomes:

$$\underline{\hat{p}}(r,s) = \underline{\underline{T}}^{-1}[s-\underline{\underline{w}}]^{-1}\underline{\underline{T}}\hat{p}(r,0) \tag{1.51a}$$

or

$$\hat{p}(r_i,s) = \sum_{j,k=0}^{N} \frac{(T^{-1})_{ij}(T)_{jk}}{s-w_{jj}}\hat{p}(r_k,0) \tag{1.51b}$$

so

$$\hat{p}(r_i,t) = \sum_{j,k=0}^{N} (T^{-1})_{ij}e^{+w_{jj}t}(T)_{jk}\hat{p}(r_k,0) \tag{1.52}$$

Then

$$p(t) \equiv \int_d^\infty r dr \hat{p}(r,t) \to \sum_{i=0}^{N} V(i)\hat{p}(r_i,t) \qquad (1.53)$$

is readily obtained as a sum of exponentially decaying quantities. In the absence of any chemical reactions, $p(t) = 1$ is the conservation of probability condition. It arises because there must then be one eigenvalue $w_0 = 0$ and this corresponds to the sum: $\sum_{i,k=0}^{N} V(i)(T^{-1})_{i0}T_{0k}\hat{p}(r_k,0)$ which must equal $\sum_{k=0}^{N} V(k)\hat{p}(r_{k,0})$. Thus, we must have

$$V(k) = \sum_i V(i)(T^{-1})_{i0}T_{0k} \quad \text{or} \quad \frac{V(k)}{T_{0k}} = c \qquad (1.54a)$$

independent of k from which it follows (using $T^{-1}T = \underline{\underline{1}}$) that

$$\sum_i V(i)T^{-1}_{ij} = c\delta_{0,j} \qquad (1.54b)$$

Now let us consider the limit $t \to \infty$. Then for

$$w_{00} = 0 \text{ and } w_{jj} < 0 \text{ for } j \neq 0$$

we get

$$\lim_{t\to\infty} \hat{p}(r_i,t) = \sum_{k=0}^{N} (T^{-1})_{i0}T_{0k}\hat{p}(r_k,0) \qquad (1.55)$$

But by eqs. 1.51 and 1.52, this is easily seen to be equivalent to taking $\lim_{s\to 0} s\hat{p}(r_i,s)$, since, in this limit, only the terms involving the zero eigenvalues of $\underline{w}$ persist. In general then, for any matrix $\underline{\underline{A}}$ (which is time independent but may be complex) and which has an eigenvalue spectrum a_j, such that $\mathrm{Re}\,a_j \lesssim 0$, it will follow, given $\dot{\underline{f}}(t) = \underline{\underline{A}}\underline{f}(0)$, that

$$\lim_{s\to 0} s\underline{g}(s) = \lim_{t\to\infty} \underline{f}(t) \qquad (1.56a)$$

where

$$\underline{g}(s) = \int_0^\infty e^{-st}\underline{f}(t)dt \qquad (1.56b)$$

Now when there is a chemical reaction occurring at $r = d$, there is no longer complete conservation of probability (unless there is an inner collecting box). This is then replaced by conservation of all particles (or probability) that reach the collecting wall at r_N at $t \to \infty$, so that we may write

$$\lim_{t\to\infty} p(t) = \lim_{t\to\infty} V(N)\hat{p}(r_N,t) = \lim_{s\to 0} sV(N)\hat{p}(r_N,s) \tag{1.57}$$

It is interesting to ask about the relationship between solutions such as eqs. 1.51 or 1.52 and eigenfunction expansion methods. The diagonalization of $\underline{\underline{W}}$ will yield eigenfunctions corresponding to linear combinations of the basis vectors: $|r_i\rangle$ representing the finite grid points in the finite difference method. Each linear combination must correspond to one of the Bessel-function-type eigenfunction chosen for the appropriate outer and inner boundary conditions (including the reactivities). In particular, a finite outer wall at r_N will mean discrete values of ρ such that $\rho_{min} \sim 2\pi/r_N$. But because of the discreteness of the space we now also have a $\rho_{max} \sim \pi/\Delta r$ (i.e., an upper cutoff to the allowed wave-vectors).

1.5 Interaction potentials

A) Ionic interactions. A convenient way to represent spin-independent Coulombic forces between charged radicals in ionic solution is to use the usual Debye formulas. Thus, one may write:

$$U(r) = (e^2 Z_a Z_b/\varepsilon r)[e^{-\kappa(r-d)}/(1+\kappa d)] \tag{1.58}$$

where eZ_a and eZ_b are the charges on the radicals, and where κ, the reciprocal thickness of the ionic layer, obeys:

$$\kappa^2 = (4\pi e^2/\varepsilon kT) \sum_i n_i Z_i^2 \tag{1.59}$$

where ε is the dielectric constant and n_i is the number density of the ith type of particle of charge Z_i.

One finds [6] that the effect on the reactivity Λ is merely to require

$$\tau_1^{-1} = \tau_{1,u}^{-1} f^* \exp[U(d)/kT] \tag{1.60a}$$

where

$$(f^*)^{-1} = d \int_d^\infty \exp[U(r)/kT]dr/r^2 \tag{1.60b}$$

and $\tau_{1,u}$ is the value for the uncharged radicals given by eq. 1.42.

B) Pair correlation functions. One can show that U(r) is the potential of averaged forces between the spin-bearing molecules. Thus, we may obtain U(r) from the pair-correlation function g(r), i.e.,

$$\ln g(r) \equiv -U(r)/kT \tag{1.61}$$

so that one has from eq. 1.43

$$F(r) = -\frac{\partial \ln g(r)}{\partial r} \tag{1.62}$$

When these expressions are incorporated into eq. 1.12, then in the limit $t \to \infty$, $p(\vec{R};\vec{r},t)$ will yield the equilibrium $g(r)$, while, for finite times, $-F(r)\hat{e}_{\vec{r}}$ is the driving force acting to restore this equilibrium. The Debye-Hückel formula, eq. 1.58, which includes charge-shielding by the solvent, is only one example. One is able to use pair-correlation functions $g(r)$ obtained from theories of equilibrium statistical mechanics or even from computer dynamics calculations, since for the finite difference method, it is sufficient to have numerical solutions to $g(r)$. An analysis of $p(\vec{R};\vec{r},t)$ for hard-sphere potentials appropriate for liquid ethane is given by Hwang and Freed [11] utilizing the finite difference approach.

2. THE STOCHASTIC LIOUVILLE EQUATION: CIDNP AND CIDEP

2.1 Stochastic Liouville equation

The most general form of the stochastic Liouville equation appropriate for the relative diffusion of two spin-bearing molecules, which includes the spin dynamics, has been derived by Hwang and Freed [13] and is given in terms of a position-dependent, spin density matrix $\rho(\vec{r},t)$:

$$\frac{\partial \rho(\vec{r},t)}{\partial t} = -i\mathcal{H}^{\times}(\vec{r})\rho(\vec{r},t) + D\Gamma_{\vec{r}}\rho(\vec{r},t) + DT_{\vec{r}}\rho(\vec{r},t) + \mathcal{K}_{\vec{r}}\rho(\vec{r},t) \tag{2.1}$$

Here $\mathcal{H}^{\times}(\vec{r})$ is the Liouville operator associated with the spin Hamiltonian $\mathcal{H}(\vec{r})$ (i.e., for any two operators A and B, $A^{\times}B \equiv [A,B]$). Also, $\Gamma_{\vec{r}}$ is the diffusion operator of the previous section. The term $T_{\vec{r}}$ is given by:

$$T_{\vec{r}} \equiv \frac{1}{2kT}\,\vec{\nabla}_{\vec{r}}\cdot[(\vec{\nabla}_{\vec{r}}\mathcal{H}^{+}(\vec{r}))\rho(\vec{r},t)] \tag{2.2}$$

where $\mathcal{H}^{+}(\vec{r})$ is the anti-commutator form (i.e., $A^{+}B \equiv [A,B]_{+} = AB + BA$). This term gives an effective spin-dependent force which is to be included into the diffusion. It represents the back-reaction of the spins, whose Hamiltonian depends on $\vec{r}$, onto the diffusional process. In the high-temperature limit (i.e., $|\mathcal{H}|/kT \ll 1$) it is shown to be associated with relaxation of the spins to thermal equilibrium. Equation 2.1 may be derived by first writing the quantum mechanical Hamiltonian equation of motion for the complete liquid and then passing to the semi-classical limit, where the nuclear motions become classical while the spin systems remain quantum mechanical. Then the relative nuclear motions are assumed to obey simple Brownian motion. [In fact, more generally, eq. 2.1 may be interpreted such that the electronic spin and orbital states (as well as nuclear spin states) are treated in terms of a quantum-mechanical density matrix, while the nuclear motions are described by a classical probability distribution function.] The density matrix $\rho(\vec{r},t)$ then includes both the properties of a spin-density matrix and the classical $p(\vec{r},t)$ for the relative diffusion. A more complete discussion of this is given by Hwang and Freed [13].

In eq. 2.1, the operator K is introduced phenomenologically, when needed, to represent reactivities for the radical pair. Note that when the interacting molecules have no spin, then eq. 2.1 reduces to the diffusion equation for $p(\vec{r},t)$ discussed in Sect. 1 (cf. eq. 1.1 or 1.12).

When we deal with CIDEP and CIDNP, the important quantities we need are: (1) the total probability function:

$$P(\vec{r},t) = \mathrm{Tr}\rho(\vec{r},t) \tag{2.3a}$$

or

$$P(t) = \int d^3\vec{r}P(\vec{r},t) \tag{2.3b}$$

Also

$$\rho(t) = \int d^3\vec{r}\rho(\vec{r},t) \tag{2.3c}$$

[where $P(\vec{r},t) = p(\vec{r},t)$] and Tr implies a trace over spin states, and the time-dependent polarization of radical a, given by:

$$P_a(\vec{r},t) \equiv -2\mathrm{Tr}\{\rho(\vec{r},t)S_{az}\} \tag{2.4a}$$

or

$$P_a(t) \equiv \int d^3\vec{r}P_a(\vec{r},t) \tag{2.4b}$$

i.e., the difference in populations between spin up and spin down (the sign convention yields positive equilibrium polarizations P_{eq}). The quantity $P_a(t)$ is of fundamental importance for CIDEP, while the quantity:

$$F(t) \equiv 1 - P(t) \tag{2.5}$$

is of fundamental importance for CIDNP.

We will, as in Sect. 1, assume only r-dependent terms in $U(r)$ and $H(r)$, and let $\hat{\rho}(r,t) \equiv r\rho(r,t)$. Then, again using Laplace transforms, we have:

$$s\hat{\rho}(r,s) - \hat{\rho}(r,0) = -iH^{\times}(r)\hat{\rho}(r,s) + D\left[\frac{\partial^2}{\partial r^2} + \frac{1}{r}\frac{\partial}{\partial r}\hat{F}(r)^T\right]\hat{\rho}(r,s) + K_r\hat{\rho}(r,s) \tag{2.6}$$

where

$$\hat{F}(r)^T \equiv \hat{F}(r) + \frac{1}{2kT}\frac{\partial H^+(r)}{\partial r} \tag{2.7}$$

We need only solve this equation by straightforward generalizations of the methods employed in Section 1, i.e., eigenfunction expansion procedures or finite differences. We follow [6] and employ finite difference below in discussing the radical pair mechanism. Also, for simplicity, we initially let $\hat{F}(r)^T = \hat{F}(r)$, so, if not for the spin Hamiltonian term $H^{\times}(r)$, we would just have the diffusion problem of Section 1.

2.2 Finite difference method

How do we now introduce the effects of $H^{\times}(r)$? Recall that for each finite-difference value r_i we would have a value of $\hat{p}(r_i,s) = \mathrm{Tr}\hat{\rho}(r_i,s)$. Thus, for each r_i we have a matrix of values of $\hat{\rho}(r_i,s)$. Let $\hat{\rho}(r_i,t)$ be represented by an L dimensional matrix, so there are L^2 matrix elements. Then the vector space in which $\underline{\hat{\rho}}(s)$ is defined will be an $L^2\times(N+1)$ dimensional space formed from the direct product of the L^2 spin-superspace and the N+1 dimensional space spanned by the r_i for $0 \le i \le N$. The complete solution will now become a matrix equation in this space:

$$[s\underline{\underline{1}} - \underline{\underline{K}}' - \underline{\underline{W}}' + i\underline{\underline{\Omega}}]\underline{\hat{\rho}}(s) = \underline{\hat{\rho}}(0) \tag{2.8}$$

The $\underline{\underline{W}}'$ matrix is just the $\underline{\underline{W}}$ matrix of Section 1, but with each element replaced by the product of that element and an $L^2\times L^2$ unit matrix, since $D\Gamma_r$ is independent of spin. The $\underline{\underline{\Omega}}$ matrix represents the matrix elements of $H^{\times}(r)$. It consists of blocks of $L^2\times L^2$ matrices for each value of r_i; i.e., it has a block diagonal form. The $\underline{\underline{K}}'$ matrix includes the effects of the reactivities.

One solves the matrix eq. 2.8 for the elements of $\underline{\hat{\rho}}(s)$ or $\hat{\rho}(r_i,s)$. This could be performed by diagonalization techniques as discussed in Section 1. Instead, let us consider the long time limit so that it is only necessary to consider the $s\to 0$ case, i.e., we solve for

$$P \equiv \lim_{t\to\infty} P(t) = \lim_{s\to 0} sP(s) \tag{2.9}$$

and

$$P_a^\infty \equiv \lim_{t\to\infty} P_a(t) = \lim_{s\to 0} sP_a(s) \qquad (2.10)$$

This is accomplished by solving eq. 2.8 for a value of s, small enough that $\hat{\underline{p}}(s)$ has converted to its limiting value. (Note that the eigenvalues of the matrix $\underline{\underline{W}}' + \underline{\underline{K}}' - i\underline{\underline{\Omega}}$ will be complex: the real nonpositive contributions coming from $\underline{\underline{W}}' + \underline{\underline{K}}'$ and the imaginary parts from $-i\underline{\underline{\Omega}}$.)

2.3 Density matrix elements and spin Hamiltonian

Let us now consider the particular matrix elements of ρ that are required. We first express them in terms of the standard singlet and triplet states of the radical pair, using S, T_0, and $T_\pm$, while ρ_{AB} refers to the ABth matrix element of ρ.

Then

$$P(t) = \rho_{SS}(t) + \rho_{T_0T_0}(t) + \rho_{T_+T_+}(t) + \rho_{T_-T_-}(t) \qquad (2.11)$$

Now we rewrite $2S_{az} = (S_{az} - S_{bz}) + (S_{az} + S_{bz})$. Then we easily find that

$$P_a(t) = -[\rho_{ST_0}(t) + \rho_{T_0S}(t)] + [\rho_{T_-T_-}(t) - \rho_{T_+T_+}(t)] \qquad (2.12a)$$

and

$$P_b(t) = -2\mathrm{Tr}\{\rho(t)S_{bz}\} = [\rho_{ST_0}(t) + \rho_{T_0S}(t)] + [\rho_{T_-T_-}(t) - \rho_{T_+T_+}(t)] \qquad (2.12b)$$

Let us now consider the spin-Hamiltonian $H(r)$ for the interaction of radical pair A-B.

We write this as:

$$H(r) = H^0(r) + H' \qquad (2.13)$$

where $H^0(r)$ is diagonal in the singlet-triplet representation, while the off-diagonal part, H' of $H(r)$ is independent of r. Then we have

$$H^0(r) = \frac{1}{2}(g_a + g_b)\beta_e\hbar^{-1}B_0(S_{az} + S_{bz}) + \frac{1}{2}\left(\sum_j^a A_j\vec{I}_j + \sum_j^b A_k\vec{I}_k\right) \times \cdot(\vec{S}_a + \vec{S}_b) - J(r)\left(\frac{1}{2} + 2\vec{S}_a\cdot\vec{S}_b\right) \qquad (2.14a)$$

$$H' = \frac{1}{2}(g_a - g_b)\beta_e \hbar^{-1} B_0 (S_{az} - S_{bz}) + \frac{1}{2}\Big(\sum_j^a A_j \vec{I}_j - \sum_j^b A_k \vec{I}_k \Big) \times$$

$$\cdot (\vec{S}_a + \vec{S}_b) \qquad (2.14b)$$

Thus H' only includes differences in g-values and hyperfine energies between the two interacting radicals as discussed in previous chapters. J(r) is the exchange interaction between radicals which depends explicitly on r.

We will now simplify the analysis to high-field experiments. Thus we only need consider the $A_j I_{jz} S_{az}$-type hyperfine components. As a result of this, only the S and T_0 states are found to couple to give induced polarizations (except when the initial triplet polarization is operative [6,10]. Thus, for the high-field radical-pair mechanism, we can neglect the $\rho_{T_\pm T_\pm}(t)$ in eqs. 2.12. We can now use the eqs. 2.14 to find the matrix elements of $H(r)^\times \rho$ in the subspace defined by the S and T_0 levels. One finds that

$$\begin{pmatrix} [H^\times(r)\rho]_{SS} \\ [H^\times(r)\rho]_{ST_0} \\ [H^\times(r)\rho]_{T_0S} \\ [H^\times(r)\rho]_{T_0T_0} \end{pmatrix} = \begin{pmatrix} 0 & -Q & Q & 0 \\ -Q & 2J(r) & 0 & Q \\ Q & 0 & -2J(r) & -Q \\ 0 & Q & -Q & 0 \end{pmatrix} \begin{pmatrix} \rho_{SS} \\ \rho_{ST_0} \\ \rho_{T_0S} \\ \rho_{T_0T_0} \end{pmatrix} \qquad (2.15)$$

Here:

$$2Q \equiv (g_a - g_b)\beta_e \hbar^{-1} B_0 + \Big(\sum_j^a A_j^a M_j^a - \sum_k^b A_k^b M_k^b \Big) \qquad (2.16)$$

so 2Q is the difference in ESR resonant frequencies between separated radicals A and B. Equation 2.15 thus defines the 4×4 block form of $\underline{\Omega}$ of eq. 2.8 for each value of r_i. The only r dependence is in $J(\bar{r})$. One expects an exponentially decaying exchange interaction, which we write as:

$$J(r) = J_0 \exp[-\lambda(r-d)] \qquad (2.17)$$

One also finds that $[H^\times \rho]_{T_+T_+} = [H^+ \rho]_{T_-T_-} = 0$, thus confirming the fact that the $T_\pm$ states do not contribute to the polarization process in the high field limit.

2.4 Spin-dependent reactivity

We now consider the superoperator $K(r)$ and its associated matrix $\underline{\underline{K}}'$. In particular, we wish to consider a spin selective reaction between the radical pair when in contact. For definiteness, we assume that only radical pairs in the singlet state may react. Thus, we may introduce equations like eqs. 1.41a and b, but only for singlet states. Since $K(r)$ is phenomenologically introduced, we need to use it in a fashion that is consistent with the properties of the density matrix. It is well known that a decay of diagonal density matrix elements [e.g., $\rho_{SS}(d,t)$], will lead to a lifetime-uncertainty-broadening for associated off-diagonal elements (e.g., $\rho_{ST_0}(d,t)$ and $\rho_{T_0S}(d,t)$ are broadened by the mean of the decay rates of S and T_0 states). These two effects are well represented by writing [2]:

$$K\rho = [-k(r)/2][|S\rangle\langle S|\rho + \rho|S\rangle\langle S|] \equiv [-k(r)/2]|S\rangle\langle S|^{+}\rho \quad (2.18)$$

If we use eq. 1.41b for k(r), then $\underline{\underline{K}}'$ is completely defined. [Note that, in general, off-diagonal elements ρ_{ij} are not completely independent of the diagonal elements ρ_{ii} and ρ_{jj}, due to the inequality $\mathrm{Tr}(\rho^N)^2 \leq 1$ where $\rho^N \equiv \rho/\mathrm{Tr}\rho$ (which follows from the fact that in diagonal form all elements of ρ^N must be real and positive [14]). This formal requirement is what should be associated with the Heisenberg uncertainty-in-lifetime effect (usually written as $\Delta\omega\Delta t \geq 1$), on ρ_{ij}, $i \neq j$, with eq. 2.18 representing the equality.]

However, it is not necessary to be satisfied simply with a phenomenological treatment of reactivity. This is because we have in eqs. 2.6-2.7 a spin-dependent potential and associated force $(1/2kT)[\partial H^{+}(r)/\partial r] = (1/2kT)(\frac{1}{2} + 2\vec{S}_a\cdot\vec{S}_b)^{+}[\partial J(r)/\partial r]$, which, for singlet reactivity, will lead to bonding attraction of the singlet state but anti-bonding repulsion of the triplet state (cf. Fig. 1). This tells us that eq. 2.6 already includes chemical reactivity for the radical-pair even with the phenomenological $K_r = 0$. This is as it should be, since the spin-dependent Smoluchowski eq. 2.6 has been derived from the complete semi-classical many-body problem including all the interactions. It is then only approximate in (1) its treatment of the surrounding molecules as a simple diffusive background and (2) its coarse-graining-in-time neglect of momentum of the radical-pair, which is equivalent to letting the momentum relax instantaneously. The former ignores solvent structure, while the latter is really inadequate in dealing with strong interactions, although the more complete spin-dependent Fokker-Planck theory including momentum still contains the same spin-dependent potential. Thus, either in the Smoluchowski or Fokker-Planck forms, this theory supplies the spin-dependent, but adiabatic, reactive trajectories, which are determined by Coulomb and exchange interactions. This is referred to as a self-consistent (SC) method, since it requires that the effects of $H(r)$ must appear <u>both</u> in the commutator $H^{\times}(r)$ representing the purely dynamical motions of the spins in eq. 2.6 <u>and</u> the total force $\hat{F}(r)^T$ in eq. 2.7. Also note that the anti-commutator $[H^{+}(r)/2kT]$ plays a role formally analogous to the phenomenological $k/2\ |S\rangle\langle S|^{+}$ in eq. 2.18.

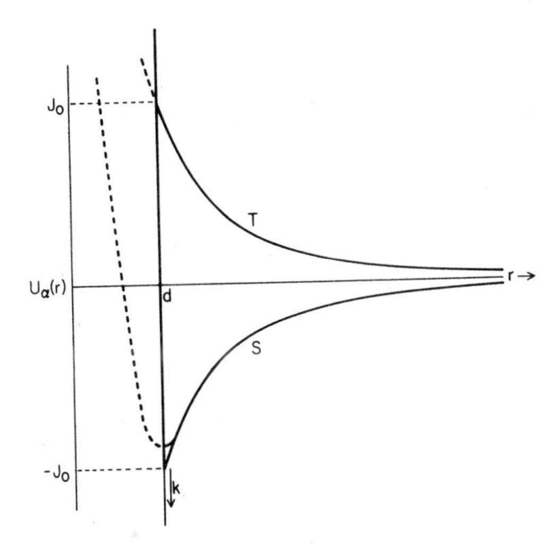

Figure 1. Spin Dependent Potentials $U_\alpha(r)$ as a function of r the interradical separation. The solid curves labeled S and T represent the exchange potentials utilized for singlet and triplet states respectively, with a reflecting wall at r = d. The dashed lines represent the usual continuation of the potentials in the absence of a reflecting wall. (By permission from Pedersen and Freed [3].)

We will, in our discussion of specific models below, find that we can approximate the particular SC model based upon Fig. 1 (referred to in [6] as the EFP model), by replacing the $\mathcal{H}^+(r)/2kT$ term in eq. 2.6 with a phenomenological

$$K\rho = -k(r)|S\rangle\rho_{SS}\langle S| \tag{2.19}$$

but with $\mathcal{H}(r)$ still in the commutator of eq. 2.6. We will call this the ASC model (in [6] it is referred to as the EFA model). We will favor the ASC model in our discussion (partly also for its greater simplicity), while the fully phenomenological form of eq. 2.18 is employed in the original work of ref. [2].

2.5 Initial conditions and transfer factors

It now only remains to specify initial conditions in order to be able to solve eq. 2.6. We will usually consider a radical-pair at some initial separation, r_I. Thus $\rho(r,0) = \rho_0\delta(r-r_I)/r_I^2$. Thus in the case of pure singlet at r_I = d one has in finite-difference form:

$$\hat{\rho}_{S,S}(r_i,0) = \delta_{i,0}/V(0) \tag{2.20}$$

It follows from the linear homogeneous form of eq. 2.1 (or 2.6) that we can superpose solutions obtained for different initial conditions, to obtain the solution to the problem involving an initial condition, which is itself a linear combination of already solved initial conditions. We refer to this as the superposition principle.

In problems involving random collisions (i.e., RI) in which there are forces between radicals, the diffusion affects the rate of the initial encounter. Thus it is necessary to start the radical-pair (for RI) at a position r_I such that $\hat{F}(r_I)^T \simeq 0$. Then, to get around the arbitrariness of such an initial state, we define a transferred polarization, etc., by:

$$P_a^\infty(d_t) \equiv [k_2(r_I)/k_2(d)]P_a^\infty(r_I) = (r_I/d)P_a^\infty(r_I) \qquad (2.21)$$

As a result of the effects of the forces between radical-pairs, $P_a^\infty(d_t)$ is not the same as that for a true initial condition of $r_I = d$, since it includes the effects of the forces on the initial approach before contact of the radical-pair. In the case of a finite absorbing wall at $r = r_N$, eq. 2.21 must be rewritten as:

$$P_a^\infty(d_t) \xrightarrow[r_N]{\text{finite}} d^{-1}(r_I^{-1} - r_N^{-1})^{-1}(1 - d/r_N)P_a^\infty(r_I) \qquad (2.22)$$

This expression comes about as follows. One has from eqs. 1.10 that $k_2(r_I) \propto r_I/d = t_f(r_I)^{-1}$ (or inversely with the probability of encounter) for diffusional encounters from an infinite medium. The probability that particles separated at r_N will encounter is, however, d/r_N, and since there is an absorbing (or collecting wall at r_N), one must subtract this amount, i.e., $k_2(r_I) \propto 1/[t_f(r_I) - t_f(r_N)] = d^{-1}(r_I^{-1} - r_N^{-1})$. Similarly for $k_2(d_t)$ (neglecting forces) $\propto [1 - t_f(r_N)]^{-1} = (1 - d/r_N)^{-1}$. (See below for a definition of t_f.)

3. RESULTS AND MODEL DEPENDENCE

As we have previously noted, it is possible by numerical methods to obtain solutions for a wide range of models which may not be accessible to analytic solutions. One of the significant findings in the original work [1-6] is that the numerical results (where at $t = 0$ the particles are in contact) may be summarized by a series of relatively simple expressions, and that most of these expressions are model <u>independent</u>.

3.1 CIDNP

Let us first consider CIDNP. We note that Λ, the fractional probability of reaction (for $Q = 0$) of singlet (for singlet reactivity) may be defined as

$$\Lambda \equiv F_0(S) \quad (3.1)$$

(Note, in this notation the subscript refers to $Q = 0$ and $F(i)$ means i initially.) This fractional probability of reaction includes all re-encounters of the radical-pair, which we shall refer to as the complete collision. We then define:

$$F^* \equiv \lim_{\Lambda \to 1} F(T_0) = \lim_{\Lambda \to 1} [F(T_0) - F_0(T_0)] \quad (3.2)$$

since $F_0(T_0) = 0$, so that F^* measures the conversion from triplets to singlets for the whole collision. A set of results that are exact and model independent are:

$$-[F(S) - F_0(S)] = -[F(S) - \Lambda] = (1 - \Lambda)F(T_0) \quad (3.3)$$

i.e., the net decrease in reaction for pure singlets per collision due to $Q \neq 0$ is just the probability a singlet does not react for $Q = 0$ (i.e., $1-\Lambda$) times the probability pure triplets do ultimately react because of mixing of triplets and singlets (Q-mixing). The superposition principle may then be applied to obtain the result for random initial condition (i.e., equal amounts of singlet and triplet):

$$F(RI) = \frac{1}{2} [F(S) + F(T_0)] \quad (3.4)$$

then eq. 3.3 may be rewritten as:

$$[F(RI) - F_0(RI)]/F_0(RI) = F(T_0) \quad (3.5)$$

where $F_0(RI) = \frac{1}{2}\Lambda$. The role of the parameter F^* appears in the exact relation:

$$F(T_0) = \Lambda F^*[1 + F^*(1-\Lambda)]^{-1} \quad (3.6)$$

so that eqs. 3.3 and 3.5 become:

$$F(S) = \Lambda[1 + F^*(1-\Lambda)]^{-1} \quad (3.7)$$

and

$$F(RI) = \frac{1}{2} \Lambda(1 + F^*)[1 + F^*(1-\Lambda)]^{-1} \quad (3.8)$$

[Note that these expressions are easily converted for T_0 reactivity by interconverting $F(S)$ and $F(T_0)$ in all the definitions and expressions above. The $T_\pm$ states are still treated separately and independently in the high field case.] The triplet or singlet initial cases are of interest for CIDNP due to recombination of a geminate radical pair which were originally formed in the triplet or singlet state. The random initial case is appropriate for a radical-pair which experience a random encounter. One may combine the result of eq. 3.8 with the usual steady-state bimolecular rate of encounters, eq. 1.10, to obtain the effective bimolecular spin-dependent reaction rate (for singlets) of:

$$k_f^S = \mathcal{F}(RI)k_2(d) \tag{3.9}$$

Again, it is interesting to note that these above results are all model independent and exact as demonstrated by the numerical results. It is only in the precise forms of Λ and $\mathcal{F}^*$ that there is model dependence. Note that typically Λ obeys the form of eq. 1.42 but with τ_1 given by eq. 1.42a for no interaction potentials, while it is given by eq. 1.60 for interaction potentials (e.g., Coulombic interactions). We have found that for a model, which corresponds to an apparent space-dependent diffusion-coefficient [i.e., the replacement of D by $D(r) = D(1 - a/r)$] and based on Oseen's tensor, one obtains eq. 1.60, but with f* replaced by $\tilde{f}^*$:

$$(\tilde{f}^*)^{-1} = d\int_d^\infty \exp[U(r)/kT][r^2(1 - a/r)]^{-1}dr \tag{3.10}$$

Equation 1.60 also applies for models involving the spin-dependent exchange interaction, provided $|\hbar J_0/kT|$ is not much greater than unity (see below).

We now consider $\mathcal{F}^*$. In the case of small values of $J_0 d/\lambda D$ it is found to be independent of J(r) and a function only of the dimensionless variable Qd^2/D. (See [6] for a discussion of the solutions to eq. 2.6 in terms of dimensionless variables.) In the interaction-free case, it is then possible to obtain good agreement with the numerical results with the analytic form:

$$\mathcal{F}^* \simeq \frac{\frac{1}{2}\{1+\frac{1}{2}\ln[1+(Qd^2/D)^{\frac{1}{2}}]\}(Qd^2/D)^{\frac{1}{2}}}{1+\frac{1}{2}\{1+\frac{1}{2}\ln[1+(Qd^2/D)^{\frac{1}{2}}]\}(Qd^2/D)^{\frac{1}{2}}} \tag{3.11}$$

as demonstrated in Table 1. The simple form $\mathcal{F}^* \sim \frac{1}{2}(Qd^2/D)^{\frac{1}{2}}$, valid for $(Qd^2/D)^{\frac{1}{2}} << 1$, confirms the important role of the re-encounter mechanism as discussed in the previous chapters. A likely interpretation of the logarithmic term in eq. 3.11 in terms of the "initial encounter" mechanism is presented below. Note that in the asymptotic limit: $\lim_{(Qd^2/D)\to\infty} \mathcal{F}^* = 1$, i.e., with infinitely rapid Q-mixing all the initially triplet spins are completely converted to singlet states and react (cf. eq. 3.2). [Note a simpler form of eq. 3.11 (Table 1) was slightly misprinted in [6].]

This, then, is an example of how numerical results can be converted to useful approximate analytic forms. In this, it is

Table 1. Dependence of F^* on (Qd^2/D) for small values of J(r)

Qd^2/D	F^{*a} (numerical)	F^{*b} (theory)	F^{*c} (simple)
1.6×10^{-3}	0.019863	0.01998	0.019608
1.6×10^{-2}	0.067974	0.06281	0.059484
1.6×10^{-1}	0.20099	0.18940	0.16667
1.6	0.49620	0.47117	0.38743
1.6×10^{1}	0.78435	0.78305	0.66667
1.6×10^{2}	0.93588	0.93585	0.86347
1.6×10^{5}	1.0000	0.99875	0.99503

a) Results computed by finite differences.
b) Results predicted by eq. 3.11
c) Results predicted by simple form $F^* = \frac{1}{2}(Qd^2/D)^{\frac{1}{2}}/[1+\frac{1}{2}(Qd^2/D)^{\frac{1}{2}}]$.

useful to be guided by (1) analytic solutions to simpler models; and (2) consideration of proper limiting cases.

The effects of Debye-Hückel-type Coulombic forces on F^* may be approximately represented by letting

$$F^* \simeq F_u^* f^{*(1+\delta)} \tag{3.12}$$

where F_u^* are the results obtained for uncharged radicals and $\delta \simeq \frac{1}{4}$ for attraction and $0 \le \delta \le \frac{1}{4}$ for repulsion (with $Q \sim 10^8$ sec^{-1}, $D \sim 10^{-5}$ cm^2/sec), but more generally δ is somewhat sensitive to Q and κd. They are given in graphs in [6]. Note that from the simple form of $F^* \approx \frac{1}{2}(Qd^2/D)^{\frac{1}{2}}$ valid for $(Qd^2/D) \ll 1$ [cf. eq. 3.11], one might expect to replace the d by an effective interaction distance, f^*d (cf. eq. 1.60b), so that $F^* \simeq f^* F_u^*$. Thus the small $\delta \ne 0$ in eq. 3.12 reflects the long-range effect on the relative diffusive motion affecting the re-encounter dynamics and Q-mixing. This illustrates an important difference between CIDNP and the usual models of liquid-state reaction kinetics.

When Oseen's tensor is introduced, the effects of F^* are somewhat more complex, but are discussed in detail in [5]. In general the ratio F_{OS}^*/F_N^* (OS = Oseen, N = without Oseen tensor), range from ca. 1/2 to 1.3. Again, to a first approximation, one should replace d in eq. 3.11 by $\tilde{f}^*d$, so for small Qd^2/D one expects $F_{OS}^*/F_N^* \approx \tilde{f}^*/f^* \approx \tau_{1,N}/\tau_{1,OS}$.

The above discussion is appropriate for $J_0d/\lambda D \ll 1$. When, however, $J_0d/\lambda D \gtrsim 1$, the results for F^* become weakly dependent upon F^*, tending to reduce it in magnitude. This effect is roughly expressed by:

$$F^*_{J=0}/F^* \approx 1 + \frac{5}{3}(\lambda d)^{-1}\ln[1 + (J_0 d^2/D)(\lambda d)^{-1}]\ln[1 + (Qd^2/D)^{0.45}] \quad (3.13)$$

The physical picture here is that of an "exchange-volume," i.e., a region extending beyond the contact distance d and satisfying $J(r) > Q$, so that Q-mixing is suppressed. Since it is found that for small Qd^2/D the effects of re-encounters after longer separations are needed for Q-mixing, this excluded volume has less of an effect for small Qd^2/D. The effects of Oseen's tensor and shielded Coulombic attractions are also modified by appreciable values of $J_0 d/\lambda D$. Again one finds enhanced effects for larger Qd^2/D (cf. [5]). Also, one should note how, by numerical methods, it is possible to calculate the simultaneous effects of these various interactions. It is then frequently possible to interpret them in terms of one's understanding of the physical models.

Another important property of the CIDNP solutions is given by the transfer factor t_f, defined as:

$$t_f \equiv \Delta F(RI, r_I)/\Delta F(RI, d) \quad (3.14)$$

where

$$\Delta F(RI, r_I) = F(RI) - F_0(RI) \quad (3.15)$$

with an initial separation of $r_I \geq d$. For a simple diffusive model,

$$t_f = d/r_I \quad (3.16)$$

Also, t_f is the probability that two particles initially separated by r_I will encounter at least once at $r = d$. That eq. 3.14 is equivalent to this simple definition, follows from the fact that for random initial condition, the CIDNP process only starts upon initial encounter. Equation 3.14 may be usefully employed with eq. 3.8. (The results for singlet and triplet initial have been discussed in [6].) When (shielded) Coulombic forces are introduced, then one finds quite a good fit to:

$$t_f = 1 - f^*/f^*(r_I) \quad (3.17a)$$

where

$$f^*(r_I)^{-1} = d\int_d^{r_I} \exp[U(r)/kT](dr/r^2) \quad (3.17b)$$

so that

$$f^* = \lim_{r_I \to \infty} f^*(r_I) \quad (3.18)$$

Now one way of looking at $t_f(r_I)$ is that it must equal the ratio of the rates of new bimolecular encounters at separations of r_I versus d. That is, $t_f(r_I) = k_2(d)/k_2(r_I)$ which is easily shown to be in agreement with eq. 3.17. Similarly, with Oseen's tensor one expects $t_f = 1-\tilde{f}^*/\tilde{f}^*(r_I)$ by analogy with eq. 3.7 and using eq. 3.10.

3.2 CIDEP polarizations

For CIDEP we again obtain a series of exact expressions:

$$P_a^\infty(RI)/F(RI) = -P_{a,k=0}^\infty(S) = P_{a,k=0}^\infty(T_0) \tag{3.19}$$

where the subscript k = 0 indicates no chemical reaction. This is easily rearranged with the use of eqs. 3.3 and 3.4 to give

$$P_a^\infty(RI) = \frac{1}{2}\Lambda[1+F(T_0)]P_{a,k=0}^\infty(T_0) \tag{3.20}$$

Also one finds

$$P_a^\infty(T_0) = (1+F^*)[1+(1-\Lambda)F^*]^{-1}P_{a,k=0}^\infty(T_0) \tag{3.21}$$

and

$$P_a^\infty(S) = (\Lambda-1)P_{a,k=0}^\infty(T_0) \tag{3.22}$$

Thus, we see that all the CIDEP polarizations are obtained from a knowledge of Λ, F^*, and $P_{a,k=0}^\infty(T_0)$. Again these exact relations are independent of the model. We, therefore, only have to discuss $P_{a,k=0}^\infty(T_0)$ and its model dependences. [Note that these results are based upon using eq. 2.19 for $K(r)$. The results for eq. 2.18 are discussed in [2,6]. They require some modification of the above exact expressions for $k \neq 0$ for smaller values of J_0. However, small values of J_0, according to the self-consistent approach, should mean that the spin-dependent reactivity is negligibly small (see also below).]

The numerical results can be incorporated into the approximate form:

$$|P_{a,k=0}^\infty(T_0)| \approx \left(\frac{Qd^2}{D}\right)^\varepsilon \frac{2|J_0|\tau_1(\lambda) + [\frac{3}{2}(\lambda d)^{\varepsilon'}][2J_0\tau_1(\lambda)]^2}{1+[2J_0\tau_1(\lambda)]^2} \tag{3.23}$$

where

$$\tau_1(\lambda) \approx (d/D\lambda)[1+(\lambda d)^{-1}] \tag{3.24}$$

and $\varepsilon \sim \frac{1}{2}$ for $(Qd^2/D) \leq 0.016$ while $\varepsilon' \sim 1$ for $\lambda d >> 1$ and $(Qd^2/D) \leq 0.016$, but ε and ε' become smaller as these inequalities are violated. Also, the sign of $P^{\infty}_{a,k=0}(T_0)$ is determined as follows:

$$\mathrm{Sign}[P^{\infty}_{a}] = [\mathrm{Sign}\ Q]\times[\mathrm{Sign}\ J] \tag{3.25}$$

Note from eq. 3.23 that for large J_0, P^{∞}_{a} takes on an asymptotic value:

$$\lim_{J_0\to\infty} |P^{\infty}_{a,k=0}(T_0)| \approx \frac{3}{2(\lambda d)^{\varepsilon'}}(Qd^2/D)^{\varepsilon} \tag{3.26}$$

while the maximum value of P^{∞}_{a} occurs at

$$J_0(\max) \approx [2\tau_1(\lambda)]^{-1} \tag{3.27}$$

A special case of the exponential decay model, referred to as the contact exchange model, for which $J(r_j) = J_0\delta_{j,0}$, may be obtained by letting $\lambda d \to \infty$, while $\lambda^{-1} \to \Delta r_J$. This yields:

$$|P^{\infty}_{a,k=0}(T_0)| \approx (Qd^2/D)^{\varepsilon} \frac{2|J_0|\tau_{1,J}}{1+(2J_0\tau_{1,J})^2} \tag{3.28}$$

with

$$\tau_{1,J} \equiv d\Delta r_J/D \tag{3.28a}$$

where Δr_J is the very small extent of the exchange region. The $(Qd^2/D)^{\frac{1}{2}}$ dependence in eqs. 3.26-3.28, for smaller values is indicative of the re-encounter mechanism discussed in the previous chapters. However, for larger values of (Qd^2/D) (e.g., viscous solvents), the required $S\text{-}T_0$ mixing can occur extensively, so one expects the effects of the "complete collision" to be dominated by the initial encounters. This is shown by the expression, valid for small J_0, which we have found to represent the Q dependence for large Qd^2/D:

$$|P^{\infty}_{a,k=0}(T_0)| \approx 1.18(Qd^2/D)^{\frac{1}{2}}[2J_0\tau_1(\lambda)]\left\{1 - \frac{1.56(Qd^2/D)^{\frac{1}{2}}}{1+1.56(Qd^2/D)^{\frac{1}{2}}} - \frac{0.8\ \ln[1+\frac{1}{2}(Qd^2/D)^{\frac{1}{2}}]}{1+1.6(Qd^2/D)^{(0.48+\delta)}\ln[1+\frac{1}{2}(Qd^2/D)^{\frac{1}{2}}]}\right\} \tag{3.29}$$

for $|2J_0\tau_1(\lambda)| << 1$.

Here δ increases slowly from zero as Qd^2/D becomes $\gtrsim 16$ and is larger for smaller λ. This fits the data very well, cf. Table 2. Note that the correction terms include (Qd^2/D) to the first power, which should represent the effects of the initial encounter [see below]. The results for large $J(r)$ are somewhat more complex,

Table 2. Dependence of P_a^∞ on (Qd^2/D) for small values of $J(r)$.[a]

Qd^2/D	$r_{ex} = 2$ Å[b]	$r_{ex} = 4$ Å[c]	$r_{ex} = 8$ Å[d]
1.6×10^{-4}	21.4 (21.1)	42.0 (44.0)	97.1 (95.0)
1.6×10^{-3}	64.3 (63.3)	127 (132)	292 (285)
1.6×10^{-2}	169 (170)	331 (355)	761 (766)
1.6×10^{-1}	326 (332)	637 (693)	1436 (1496)
1.6	323 (300)	617 (625)	1290 (1350)
16	163 (159)	280 (332)	466 (716)

a) Polarizations given as $10^3\times P_a^\infty$. The first number is the numerical solution, the number in parentheses is from eq. 3.29 but with $\delta = 0$. Also $r_{ex} \equiv \lambda^{-1} 5 \ln 10$.
b) $2J_0\tau_1(\lambda) = 1.450$.
c) $2J_0\tau_1(\lambda) = 3.021$.
d) $2J_0\tau_1(\lambda) = 6.525$.

so we do not reproduce them here. In all cases, however, one finds that

$$\lim_{Q\to\infty} P_a^\infty = 0 \tag{3.29a}$$

The asymptotic dependence of P_a^∞ with $J_0\tau_1(\lambda) >> 1$ may be understood in terms of an effective region of polarization. The inner region, where $J(r) > J_0(\max)$, is primarily effective in quenching any polarization by a Heisenberg exchange mechanism; while the desirable or polarization-effective range in $J(r)$, i.e., $J(r) \sim J_0(\max)$, merely moves out farther from r_0. (Note also that one may consider different functional forms for $J(r)$, but the above results are not very sensitive to such variations [6].)

We may attempt to predict the effects of Coulombic forces on CIDEP polarizations as follows. For $2J_0\tau_1 < 1$ and $(Qd^2/D) << 1$, then $P_a^\infty \approx (Qd^2/D)^{\frac{1}{2}} 2J_0\tau_1(\lambda)$. Then, to a first approximation we would let $\tau_1(\lambda)$ have the $f^{*-1}\exp[-U(d)/kT]$ dependence that appears in eq. 1.60. [This will not be such a good approximation if λ is small, i.e., a long-range $J(r)$.] Also we would let $d \to f^*d$. This simple analysis would lead to:

$$P_a^\infty/P_{a,u}^\infty \approx \exp[-U(d)/kT] \tag{3.30}$$

which is in accord with the actually calculated trends, although it is not quantitative. In particular [with $r_{ex}/d = 1$ with $r_{ex} \equiv \lambda^{-1} 5 \ln 10$], for attractive forces the calculated result is about 80% of the simple prediction of eq. 3.30 while for repulsive forces it is about 160-180%. This is as though one should use a $U(r)$ with r a little larger than d in eq. 3.30. As (Qd^2/D) becomes larger, the trends are affected by the changed dependence on this variable. A similar analysis for large $J_0\tau_1(\lambda)$ utilizing eq. 3.26 suggests that for (Qd^2/D) small, P_a^∞ should only depend weakly upon Coulomb forces. The repulsive forces yield an enhancement (typically less than 50%), while attractive forces give reductions by factors up to 3. These effects may again be understood by recognizing that for higher J_0-values the region of effective polarization moves out to $r > d$ where the Coulomb forces are reduced.

When Oseen's tensor is included, one sees effects which can be rationalized similarly. Thus for small J_0, $(P_a^\infty)_{OS}/P_a^\infty \approx \tau_{1OS}/\tau_1 \times F_{OS}^*/F^* \sim 1$, while for much larger Qd^2/D, this ratio $\rightarrow f^*/\tilde{f}^*$. The results for high J_0, show only a small effect, again because the effective region is $r > d$, where Oseen tensor effects are reduced.

3.3 Convergence of the solutions

The above results are based upon obtaining convergence with the finite difference method. One can summarize these conditions as: (1) Δr must be small enough; (2) r_N and r_M must be large enough; and (3) s be small enough that the limit $s \rightarrow 0$ has been achieved. Condition (1) is found to be satisfied if $J(r_i)/J(r_i+\Delta r) = e^{\lambda\Delta r} \lesssim 5$, while (2) for r_M requires that $J(r_M) << Q$. Note that a small s implies a large t. That is, consider eqs. 1.51 and 1.52. The limit on $\hat{p}(r_i,s)$ is achieved when $s << |w_{min}|$, where w_{min} is the smallest non-zero eigenvalue of $\underline{\underline{W}}$; while for $\hat{p}(r_i,t)$ it is achieved when $t^{-1} << |w_{min}|$. Thus, we see that the s and t needed for convergence should be approximately related as $s_{conv} \sim t_{conv}^{-1}$. In general, one finds that $s_{conv} \propto Q$ (provided r_N is sufficiently large). In particular, for CIDEP one has

$$s_{conv} \lesssim Q/3 \tag{3.31a}$$

while for CIDEP one has:

$$s_{conv} \lesssim Q/80 \tag{3.31b}$$

(where s_{conv} means that the associated P^∞ or F^* is within 10% of its limiting value) e.g., for $Q \sim 10^8$ sec^{-1}, P^∞ is generated in about 3×10^{-8} sec and F^* takes about 0.8×10^{-6} sec to develop.

This large difference between CIDEP and CIDNP convergence may be understood in terms of the basic re-encounter mechanism. Thus CIDEP involves the two-step path, e.g.,

$$\rho_{T_0} \xrightleftharpoons{Q} \rho_{S,T_0} - \rho_{T_0,S} \xrightleftharpoons{J(r)} \rho_{S,T_0} + \rho_{T_0,S},$$

only the first of which involves Q-mixing and re-encounters. On the other hand, CIDNP involves two successive Q-mising steps, e.g.,

$$\rho_{T_0} \overset{Q}{\rightleftarrows} \rho_{S,T_0} - \rho_{T_0,S} \overset{Q}{\rightleftarrows} \rho_S - \rho_{T_0} .$$

Thus re-encounters after longer periods of separation are needed for CIDNP. Also, as the polarization effects take longer to develop, other processes such as individual radical T_1's and T_2's (ca. 10^{-6} sec) and scavenging reactions, begin to affect the polarization process (cf. [6]). Note that a position and spin-independent "radical-pair"-scavenging reaction can be introduced by adding a term $-k\hat{\underline{\rho}}$ to eq. 1.27 (or eq. 1.29). This does not affect the diagonalization of $\underline{\underline{W}}$. Hence the new eigenvalues become $w_{jj} \to (w_{jj}-k)$. This is equivalent to letting $s \to s+k$ in eq. 1.51, so instead of taking the limit $s \to 0$, one may now take the limit $s \to k$. But now $\lim_{s\to 0} s\hat{p}(r_i,s+k) = 0$ for finite k. However, if we look at the polarization of the radical pairs, as they are scavenged, then we want to collect the polarization contributions expressed as the time integral:

$$P^{\infty} \equiv \int_0^{\infty} P_k(t)k\,dt = kP_k(s=0) = kP(s-k) \qquad (3.32)$$

where $P_k(t)$, etc., implies that we have included $-k\hat{\underline{\rho}}$ in the density matrix equation of motion.

Similarly, we have for F^*:

$$F^* \equiv \int_0^{\infty} F_k^*(t)k\,dt = kF_k^*(s=0) = kF^*(s=k) \qquad (3.33)$$

Equations 3.32 and 3.33 thus replace eqs. 2.9 and 2.10. When $k \to 0$ these two approaches become equivalent. But when $k \gtrsim s_{conv}$, the polarization generating mechanism will be interfered with by the rate process. In particular for s (or k) > 3Q one finds that instead of the $(Qd^2/D)^{\frac{1}{2}}$ dependence of P_a^{∞} and F^*, one now obtains $P^{\infty} \propto (Qd^2/D)$ and $F^* \propto (Qd^2/D)^2$. This is due to the fact that the polarization processes are quenched before the reencounter mechanism can be effective, and it is only the effect of the initial encounter which can be observed. The linear and quadratic dependences of P^{∞} and F^* on (Qd^2/D) reflect the respective one and two step Q-mixing already noted. The initial encounter mechanism is discussed in more detail in the previous chapters. Note that the logarithmic correction used in eq. 3.11 and Table 1 for F^* may well be reflecting contributions from the initial encounter mechanism as Qd^2/D gets large. See also eq. 3.29 and Table 2 for P_a^{∞}, where the role of the initial encounter mechanism is even clearer.

The convergence of the solutions with r_N (for small enough s) may be related to that for s just discussed. Thus, if we use the diffusion expression

$$D = \langle\Delta r^2\rangle/6t \approx (2r_N)^2/6t \tag{3.24}$$

(where we have assumed that the radical-pair must diffuse apart to r_N and then return for a re-encounter after maximum separation), then we have:

$$r_N/d \approx (\frac{3}{2}\, t_{conv} D/d^2)^{\frac{1}{2}} \approx (\frac{3}{2})^{\frac{1}{2}}(D/s_{conv} d^2)^{\frac{1}{2}} \propto (D/Q)^{\frac{1}{2}} \tag{3.35}$$

Thus, as Q decreases (or D increases), reencounters after longer distances of separation are needed to provide effective Q-mixing and a larger r_N is required.

3.4 Self-consistent method

In the self-consistent method, one sets $K = 0$ in eq. 2.6, and one uses the complete form of $\hat{F}(r)^T$ in eq. 2.7. As a result, the diffusion becomes spin-dependent due to the spin-dependent forces, and the reactivity is explicitly included for $|\hbar J(r)/kT| \gg 1$. The results of such a model have been summarized in [6], and we only touch some of the salient points here. In particular, Fig. 1 corresponds to $U_{SS}(r) \approx -U_{TT}(r) \approx -\hbar J(r)$, $r > d$ and $U_{ST_0}(r) = U_{T_0S}(r) \approx 0$. This results from recognizing that from hydrogen-atom-pair potential surfaces

$$U_{SS}(r) \simeq (H_0 + H_1)/(1 + S) \tag{3.36a}$$

$$U_{TT}(r) \simeq (H_0 - H_1)/(1 + S) \tag{3.36b}$$

where H_0 is the "Coulomb integral," $H_1 = -\hbar J(r)$ the exchange integral, and S is the overlap integral. Our simplified model is based upon the fact that $H_1 \gg H_0$ for $r > d$ and H_1 is the main source of the attractive forces, while S tends to be small. The strong Coulomb repulsive forces for $r \leq d$ are approximated by a reflecting wall at $r = d$ corresponding to $U_{SS}(r) = U_{TT}(d)$ for $r < d$. It is this model, for which $U_{ST_0}(r) \simeq 0$, so that $\rho_{ST_0}(r)$ experiences no net forces, that is approximated by the use of eq. 2.19 as discussed above. [Suppose, however, we were to employ a model for which $U_{SS}(r) = -U(r)$ and $U_{TT}(r) = 0$ for $r > d$, corresponding to an attractive and binding potential for singlets but no potential for triplets (i.e., $H_0 \sim H_1$). Then $U_{ST_0}(r) = -\frac{1}{2}U(r)$, and the resulting self-consistent model would bear a close relation to the phenomenological approach based upon eq. 2.18.]

The main results of the SC model are to demonstrate that (1) since substantial CIDNP polarizations require non-negligible values of Λ, and large values $\hbar J_0/kT \gg 1$ are required for reactivity, then the finite range and magnitude of J(r) must be considered in a complete treatment of CIDNP; (2) the effect of the spin-selective reaction of singlets is well approximated by the ASC method (cf. eq. 2.19) in which only the diagonal density matrix elements for singlets react; (3) to a large extent (for RI), the CIDEP polarizations are independent of the details of the spin-selective chemical reaction and they are just linearly dependent on F, i.e., one may just as well use the ASC method as the SC method to calculate P^∞/F (although P^∞ and F are significantly altered); however, (4) since the reaction region is around $r \sim d$, the CIDEP polarization is developed in a region for which $r > d$ such that $\hbar J(r)/kT < 1$.

Note that τ_1 and F^* are easily calculated by eqs. 1.60 and 3.12, respectively, with $U(r) = \hbar J(r)$, and with $\delta = 0$, as expected for very short range interactions. When $\Lambda < 10^{-3}$, then one begins to see CIDEP polarization effects due to the differences in diffusion rates for singlets versus triplets in the polarization region. This leads to a slight excess of singlets generating polarizations. Such effects might possibly be of importance for faster diffusion where non-negligible values of P^∞/F are predicted, but F itself is negligibly small [3].

3.5 Nonspherical radicals

While all current theories are for spherically symmetric exchange interactions, most interacting radicals are expected to display anisotropic features in their exchange interactions and their ability to react. On the basis of our discussion of the SC method it is possible to make some qualitative comments about effects from nonspherical features. First we note that for CIDNP, one expects the primary effect is to lead to a reduction in Λ compared to that predicted for a spherically-symmetric J(r). This is because only that fraction of reencounters for which $\hbar J(\vec{r})/kT > 1$ is important. The effect on CIDEP will, however, be different. It is illustrated in Fig. 2 for a spherical radical (e.g., H atom) interacting with a nonspherical one. Only in Region 1 is $\hbar|J_0|/kT > 1$, as required for a reaction to occur. When a reaction can occur, net triplet character (symbolized by F) is created. Then subsequent reencounters will occur at different regions in Fig. 2 each with its own characteristic range of values of J. Thus while geometric factors will substantially reduce F (via the reduction in Λ), one may still anticipate that P^∞/F (which is independent of F for spherical radicals) need not be very significantly altered. This is because a wide range of J values experienced in reencounters can still lead to comparable polarizations, largely due to the asymptotic dependence of P^∞/F on J_0 (cf. Fig. 2b).

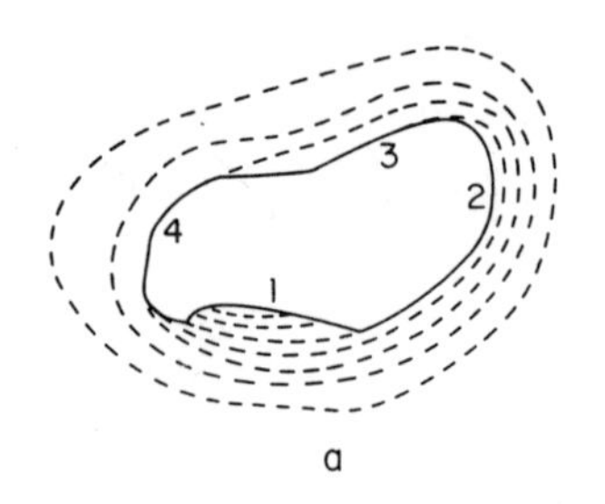

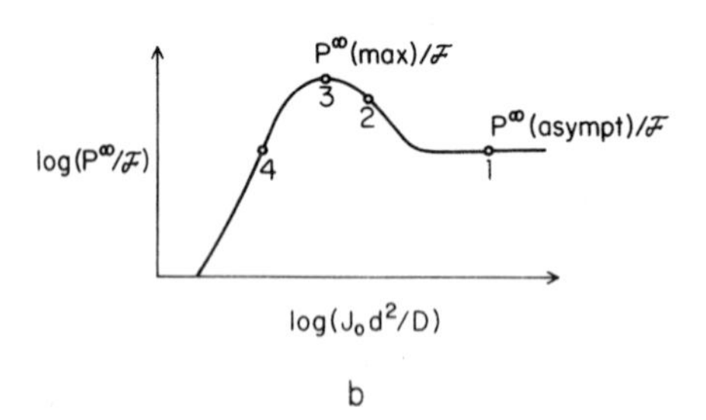

Figure 2. Nonspherical radicals. (a) Suggested contours of constant J value about a nonspherical radical interacting with a spherical radical (e.g., J varies by a factor of 10 between adjacent curves). Spin-selective chemical reaction may occur only at Region 1. (b) Typical variation of P^{∞}/F with J_0d^2/D for spherical radicals showing suggested equivalent points corresponding to Regions 1-4 in (a). (By permission from Pedersen and Freed [3].)

4. SPECIAL TOPICS

4.1 Heisenberg spin exchange

We have already discussed the negative role that Heisenberg spin exchange plays in the CIDEP polarization process. It is possible to explicitly calculate the effects of spin depolarization due to Heisenberg spin-exchange. Also, it must be included in a complete treatment of signal intensities [6] (cf. Chapter by Pedersen). Furthermore, such calculations are useful for spin-relaxation studies involving spin exchange [15]. One obtains such results by the finite difference approach by selecting as the initial case: $P_a(t=0) = -1$, or more precisely:

$$2\mathrm{Re}\rho_{S,T_0}(r) = \delta(r-r_I)/r_I^2 \qquad (4.1)$$

while $\rho_{SS} = \rho_{TT} = \mathrm{Im}\rho_{S,T_0} = 0$. The calculation then leads to

$$-\Delta P(r_I) = P_a^\infty(r_I, t\to\infty) - 1 \tag{4.2}$$

where $P_a^\infty(r_I, t\to\infty)$ is the polarization which remains at the end of the collision, after having started with the initial conditions given by eq. 4.1. Thus $\Delta P(r_I)$ measures the change in polarization at the end of the collision. Typical results have been obtained for the transferred $\Delta P(d_t)$ (cf. eq. 2.22). The results for a contact exchange model $[J(r_j) = J_0\delta_{r_j,0}]$ bear a simple relation to the well-known analytic result:

$$\Delta P(d_t) \approx \frac{(2J_0\tau_{1,J})^2}{1 + 4(J_0 + Q^2)\tau_{1,J}^2}[1 - H] \tag{4.3}$$

where $H(Qd^2/D, J_0d^2/D)$ is a small correction due to the effect of successive reencounters, which tend to generate new polarizations [6] and $\tau_{1,J}$ is given by eq. 3.28a. When one includes the finite range of the exchange, then for small $J_0 < J_0(\max)$ one has a very similar result:

$$\Delta P(d_t) \approx (2J_0\tau_1(\lambda))^2[1 - H'] \tag{4.4}$$

with $H' \approx H$ and $\tau_1(\lambda)$ is given by eq. 3.24. However, for $J_0 >> J_0(\max)$, one finds that $\Delta P(d_t) > 1$, representing the fact that the depolarization can occur for $r > d$. This effect is approximated by:

$$\Delta P(d_t) = g(J_0d^2/D, \lambda d) \quad \text{for } (|J_0|d^2/D) >> 1 \tag{4.5}$$

where

$$g(J_0d^2/D, \lambda d) \approx 1 + (\lambda d)^{-1}\ln[1 + (J_0d^2/D)(\lambda d)^{-1}] \tag{4.6}$$

The Heisenberg exchange frequency ω_{HE} is then given by [15]:

$$\omega_{HE} = k_2(d')\mathcal{N}p(d_t) \tag{4.7}$$

where we have let

$$k_2(d') \equiv 2\pi Dd' \tag{4.8a}$$

(cf. eq. 1.10) and

$$d'/d \equiv g(J_0d^2/D, \lambda d) \tag{4.8b}$$

while $p(d_t) \equiv \Delta P(d_t)/g(J_0d^2/D, \lambda d)$. In this notation $k_2(d')$ is the "effective" rate of bimolecular collisions, while $p(d_t)$ (ranging from 0 to 1) is the probability of exchange per collision.

Recently the combined effect of the finite range of J(r) and of ionic effects has been studied [16]. These results may be summarized by eqs. 4.7, 4.8 and the approximate forms:

$$p(d_t) \approx \frac{(2J_0\tau_1)^2}{1 + (2J_0\tau_1)^2}\left[1 - \frac{h(Qd^2/D)}{1 + 4(J_0^2 + Q^2)\tau_1^2}\right] \tag{4.9}$$

where

$$\tau_1 \approx [d^2/D\lambda d][1 + (\lambda d)^{-1}]f^{*-1}\exp[-U(d)/kT] \tag{4.9a}$$

$$h(Qd^2/D) \sim (Qd^2/D)^{\frac{1}{2}}f^* \quad \text{for } Qd^2/D \lesssim 0.16 \tag{4.9b}$$

and

$$g(J_0d^2/D), \lambda d) \simeq \left\{f^* + (\lambda d)^{-1}\ln\left[\left(\frac{J_0d^2}{D}\right)(\lambda d)^{-1}f^* + 1\right]\right\} \tag{4.10}$$

The first term in eq. 4.10 is just the usual Debye-Hückel effect on the collision diameter, and the second term approximates the correction due to the finite range of $J(r)$ as well as the Coulombic forces. More detailed results will be given elsewhere [16].

4.2 Spin polarization in two dimensions

The possibility of observing spin polarization in two dimensions is interesting from the point of view of problems in surface catalysis and membrane biophysics. It is no less interesting theoretically, since it affords us a chance to explore the spin polarization mechanism from a different point of view with perhaps new insights. Here we summarize some of the interesting observations one is able to make, in particular from a finite-difference point of view [17]. One can utilize the theory outlined in Sects. 1 and 2 for three dimensions with very little change. The main points are that the diffusion operator given by eqs. 1.2 and 1.3 are first written in cylindrical coordinates; one integrates over the single angle; and one assumes D_z, the diffusion parallel to the principal cylinder axis is zero, so particles remain on the surface. Then the finite-difference solution in two dimensions requires that we now let $\hat{p}(r,t) \equiv p(r,t)$ and the matrix elements of $\underline{\underline{W}}$ given by eqs. 1.31 now become:

$$W_{j,j\pm 1} = (D/\Delta r^2)[(1 \pm \Delta r/2r_j)] \tag{4.11a}$$

and

$$W_{j,j} = -2D/\Delta r^2 \tag{4.11b}$$

Note, here, that $W_{j\leftarrow j+1}/W_{j\leftarrow j-1} = (1+\Delta r/2r_j)/(1-\Delta r/2r_j)$, which says that the rate of transition from larger to smaller values of r is greater than the reverse rate. In three dimensions, one sees by eqs. 1.31, that they are equal. This "inward diffusion" effect, in two dimensions, which is purely geometrical, has a very important consequence, viz., $t_f(r_I)$ the reencounter probability for initial separation r_I, equals unity independent of r_I. Thus, no matter how far a radical pair confined to an infinite surface is initially separated, if one were to wait a long enough time, then they would ultimately encounter (provided only they are not scavenged or destroyed in some other manner). This has to have important consequences in the spin polarization process, since it means that the radical-pair "collision" is never complete as $t \to \infty$, unless other processes, such as radical scavenging, radical T_1, or radicals leaving the surface succeed in terminating the process.

We have presented a preliminary summary in ref. [18] of S-T_o results for finite r_N but as $t \to \infty$. The effect of an outer collecting wall at r_N is to terminate the process, but our results show they tend toward their limiting values for $r_N \to \infty$ as a function of $\ln(r_N/d)$. Here we summarize some of our S-T_o results [18] from the other point of view of large r_N but finite t, or more precisely finite $s > 0$. In order to do so, it is useful to define a quantity L given by:

$$L(s,c) \equiv \ln[1 + c(D/sd^2)^{\frac{1}{2}}] \tag{4.12}$$

so $\lim_{s \to 0} L(s,c) \to \infty$. Then for $L(s,2) > \ln(r_I/d)$ one has:

$$t_f(r_I) \approx 1 - \frac{\ln(r_I/d)}{1 + L(s,2)} \tag{4.13}$$

which approaches unity as $s \to 0$. Similarly, one has for $\Lambda(s)$:

$$\Lambda(s) \approx \frac{k\tau_1 L(s,\ 4/3)}{1 + k\tau_1 L(s,\ 4/3)} \tag{4.14}$$

so the reactivity also approaches unity as $s \to 0$. This makes sense, since for finite $k > 0$, the radical-pair will continue to reencounter until they finally react. We may therefore think of τ_1, the effective duration of the collision as becoming $\tau_1 L(s,\ 4/3)$. Now, for CIDNP:

$$F^*(s) \simeq \frac{(Q/2s)^2}{1+(Q/2s)^2(Qd^2/D)^{-0.2}[L(s,2^{-\frac{1}{2}})]^{-1}[1+(Qd^2/D)^{0.2}L(s,2^{-\frac{1}{2}})]} \tag{4.15}$$

which for small sd^2/D becomes:

$$F^* \simeq \frac{L(s,2^{-\frac{1}{2}})(Qd^2/D)^{0.2}}{1 + L(s,2^{-\frac{1}{2}})(Qd^2/D)^{0.2}} \tag{4.16}$$

which increases as $s \to 0$ to its maximum value of unity. Equation 4.15, in its essentially quadratic dependence upon Q, is thus very different than the $Q^{\frac{1}{2}}$ behavior of eq. 2.32 which is appropriate for three dimensions. In fact, it is more nearly the behavior associated with an "initial encounter" mechanism (cf. discussion below eq. 3.33). On this basis, one might venture to suggest that in two dimensions, the role of the initial encounter, in which Q mixing occurs during the encounter, is the dominant process. The role of the reencounter is then essentially just to begin anew the "initial encounter" process.

Why is there such a change from the three-dimensional mechanism? It appears likely that this is due to the "inward diffusion" effect in two dimensions (cf. discussion of eqs. 4.11), which will tend to keep the radical-pair closer together for longer periods of time. Nevertheless, it is interesting to note that all of the model-independent, exact relations eqs. 2.24-2.29 still hold true.

We see a very similar effect in the results for CIDEP. Thus for contact exchange [and $(4/7)(D/sd^2)^{\frac{1}{2}} > 1$], one has:

$$P_a(s) \simeq \frac{\frac{5}{6}(Q/s)}{1+b(Q/s)(Qd^2/D)^{\varepsilon}} \left(\frac{2J_0\tau_{1,J}}{1+\frac{13}{4}(2J_0\tau_{1,J})^2 L(s,\frac{4}{7})}\right) \tag{4.17}$$

with $b = 5/2$ and $\varepsilon = 0.2$ for $(2J_0\tau_{1,J})^2(13/4)L(s, 4/7) \ll 1$ and $b = 3/4$ and $\varepsilon = -0.15$ for $(2J_0\tau_{1,J})^2(13/4)L(s, 4/7) \gtrsim 1$. Equation 4.17 shows a linear dependence on Q, which is again just what is expected for the "initial encounter" mechanism. When we introduce a finite λ, then the results are approximated by:

$$P_a(s) \approx \frac{(8/7)(Q/s)}{1+\frac{8}{3}(Q/s)(Qd^2/D)^{0.25}} 2J_0\tau_1(\lambda)$$

$$\text{for } (2J_0\tau_1(\lambda))^2 \frac{8}{3} L(s, \tfrac{4}{7}) \ll 1 \tag{4.18a}$$

with

$$\tau_1(\lambda) \approx \frac{7}{8}\frac{d}{\lambda D}\left[1+\frac{1}{2}\frac{1}{\lambda d}\right] \tag{4.18b}$$

and by:

$$P_a(s) \approx \frac{\frac{3}{5}(D/20sd^2)^{3/2}(Qd^2/D)[2J_0\tau_1(\lambda)]^{-0.02}}{1+\frac{13}{2}[L(s,\frac{4}{7})]^{2.75}(D/20sd^2)^{3/2}(Qd^2/D)^{0.85}(d^2/\tau_1(\lambda)D)^{\frac{1}{2}}} \tag{4.19}$$

for $(2J_0\tau_1(\lambda))^2 \frac{8}{3} L(s,\frac{4}{7}) > 1$.

On the other hand, the results for Heisenberg spin exchange do not suggest a fundamental change in mechanism, except for the fact that $t_f = 1$. Thus for contact exchange one has:

$$\Delta P(s,d_t) \approx \frac{[L(s,\frac{3}{10})]^2(2J_0\tau_1)^2}{1 + [L(s,\frac{3}{10})]^2 4(J_0^2 + Q^2)\tau_1^2} [1 - f] \tag{4.20}$$

where $f(Qd^2/D) \approx (Qd^2/D)^{1/2}$ for $Qd^2/D < 1$, and $f(Qd^2/D) = 1$ for $Qd^2/D \gtrsim 1.6$. One can recover the three-dimensional result of eq. 4.3 merely by letting $L(s, 3/10) \to 1$ in eq. 4.20 [except for $f(Qd^2/D)$].

Again, for finite λ we have:

$$\Delta P(s,d_t) \approx [L(s,\tfrac{15}{8})]^2(2J_0\tau_1(\lambda))^2[1-f]$$
$$\text{for } (2J_0\tau_1(\lambda))^2[L(s,\tfrac{15}{8})]^2 \ll 1 \tag{4.21}$$

and

$$P(s,d_t) \sim 1 + \frac{\ln(1 + J_0 d^2/D)}{\frac{\lambda d}{\sqrt{3}}[L(s,\frac{15}{8})]^2} \quad \text{for } (2J_0\tau_1(\lambda))^2 L(s,\tfrac{15}{8})^2 \gtrsim 1 \tag{4.22}$$

The above results can be approximately related to solutions summarized in [18] for Λ, F^*, and P_a as a function of r_N/d by letting

$$s^{-1} \to [(r_N)^2/4D][\ln(r_N/d) - \tfrac{1}{2}] \tag{4.23}$$

Another very interesting feature of the two-dimensional theory compared to three dimensions is the fact that the convergence conditions on r_M, and r_N are significantly altered. These changes occur because of the $(1 \pm \Delta r/2r_j)$ factor in eq. 4.11 for $W_{j,j\pm1}$, and because of the new role of the reencounter process. One finds that the "inward diffusion" effect is considerably amplified by introducing the factor $f \gg 1$ at r_M (cf. eqs. 1.40), and this interferes with the convergence of the solutions. That is, in $\vec{\rho}$ space, one is introducing an additional reflected "wave" component at r_M. Such an effect is found to be unimportant in three dimensions but of considerable importance in two dimensions. Thus, it was deemed desirable to avoid altogether any change in the finite-difference element (i.e., let $f = 1$). This means that it is not conveniently possible to use values of $r_N/d \sim 10^3$ as was found necessary for convergence in three dimensions. However, because of the changed role of the reencounter mechanism in two dimensions, and the relatively simple asymptotic dependence on $\ln(r_N/d)$ [17,18], it becomes only necessary to use values of r_N/d such that the inequality

$$(r_N/d)^2[\ln(r_N/d) - \tfrac{1}{2}] > 4D/d^2 s \sim 4Dt/d^2 \tag{4.24}$$

(cf. eq. 4.23) is obeyed; i.e., r_N is large enough to converge to the correct solution for finite t (or s). Once this is satisfied, the solution is obtained as a function of s anyway! In this manner, satisfactorily convergent solutions are obtained for $r_N \lesssim 25$.

The use of these quantities $\Lambda(s)$, $F^*(s)$, $P_a(s)$, and $\Delta P(d_t,s)$ in the time-evolution expressions of actual observables (cf. [6] and Pedersen's chapter for three dimensions), will be given elsewhere [17]. However, we may simply note here that if k represents a first order rate constant for interrupting the process (e.g., rate of desorption from a surface), then one is interested in values for $s \sim k$.

Finally we note these two-dimensional results are a good example of how, by a combination of approximate fits of the numerical results to relatively simple analytical forms, one is able to obtain a great deal of insight into what is physically a very different result than in three dimensions.

4.3 Low-field spin polarization

Low-field spin polarization is discussed in Adrian's chapter. One finds that only CIDNP effects are important, while CIDEP effects are too small to be of interest. From our point of view, in low fields it is necessary to include the $\rho_{T_\pm T_\pm}$ terms in eq. 2.11 to determine the CIDNP effects. This is because of the significant role played by $S-T_{\pm 1}$ mixing by the hyperfine interactions. That is, suppose one of the interacting radicals has no nuclear spins, while the other has a nuclear spin I. We class the combined electronic and nuclear spin states by $|i,M_I\rangle$ where $i = S, T_0$, or $T_\pm$, while M_I is the nuclear-spin quantum number. Then we can write for the Hamiltonian $\mathcal{H}(r)$ [cf. eq. 2.13] in this basis:

$\lvert S,M_I\rangle$	$\lvert T_0,M_I\rangle$	$\lvert T_+,M_I-1\rangle$	$\lvert T_-,M_I+1\rangle$
$2J(r)$	A	$-B$	B^+
A	0	B^-	B^+
B^-	B^-	C^-	0
B^+	B^+	0	$-C^+$

(4.25)

where:

$$A \equiv \frac{1}{2}(g_a-g_b)\beta_e\hbar^{-1}B_0 + \frac{1}{2}AM_I \tag{4.26a}$$

$$B^\pm \equiv \frac{1}{\sqrt{8}}A[I(I+1) - M_I(M_I \pm 1)]^{\frac{1}{2}} \tag{4.26b}$$

$$C^\pm = \frac{1}{2}(g_a+g_b)\beta_e\hbar^{-1}B_0 + \frac{1}{2}A(M_I \pm 1) \tag{4.26c}$$

Thus, in low fields, one now has a 16×16 matrix representation of $H^{\times}$ for each value of r (i.e., $\underline{\underline{\Omega}}$ is still block diagonal, but each block is 16×16). However, this leads to matrices of huge size to be solved. Thus, for typical high-field solutions one has $r_N \sim 5\times10^3$ Å, or N ∼ 300 (with M ∼ 100, Δr = ¼ Å and f ∼ 100). In low field, the complete super matrix will be of order 16×300 = 4,800 with a bandwidth of 33. Thus considering just the banded portion of such a matrix, one needs to store ∼ 160 K elements as double precision complex numbers (16 bytes/matrix element). This means that the matrix storage alone requires about 2.4 Megabytes of core, in order to employ a standard matrix inversion routine, such as DGELB (cf. Table A #4), which is otherwise a very reliable method. Because of such enormous core requirements, we have been examining other computer algorithms which would permit the extensive use of storage devices, because at a given step in the solution they need only operate on a portion of the complete supermatrix [19]. Simpler Gaussian elimination methods (cf. Table A #5) can be used in such a manner. But, without employing Gaussian elimination with complete pivoting as does DGELB, one encounters difficulties, because the CIDNP super-matrices are ill-conditioned. For our purposes, this means that for small s, the off-diagonal elements of $\underline{\underline{\Omega}}$ are not small compared to the diagonal elements.

Diagonalization methods, which do not require complete matrix storage in core, and which are not subject to the ill-conditioned problem, may well be the most fruitful, (e.g., Table A #2). Other approaches are based on a direct study of time-dependent solutions as opposed to solutions in Laplace space (cf. below).

4.4 Time-dependent solutions

We have already noted that the stochastic-Liouville equation in the form of eq. 2.8 could be solved by standard matrix diagonalization procedures operating on $A = \underline{\underline{W}}' - \underline{\underline{K}}' + i\underline{\underline{\Omega}}$ (cf. Table A #1) by analogy to the method outlined by eqs. 1.50–1.53 for diagonalizing $\underline{\underline{W}}'$ as employed in [11]. This would yield time-dependent solutions when desired (cf. eq. 1.52). Such methods would require the storage in core of A as well as the complex orthogonal matrix $\underline{\underline{O}}$ (provided $\underline{\underline{W}}' - \underline{\underline{K}}'$ has been previously symmetrized by a matrix like $\underline{\underline{S}}$ of eq. 1.48).

For problems involving very large matrices (e.g., low-field CIDNP), one may propose an alternative procedure [19]. First, one may explicitly solve the finite difference equations in t space [20]. That is, one solves eq. 2.1 (cf. eq. 2.8) as:

$$\hat{\rho}(t) = [-i\underline{\underline{\Omega}} + \underline{\underline{W}}' + \underline{\underline{K}}']\Delta t\hat{\rho}(t-\Delta t) + \hat{\rho}(t-\Delta t) \qquad (4.27)$$

Actually, it is found [19] that eq. 4.27 does not lead to a stable finite difference method, because it does not exactly preserve conservation of probability (for $\underline{\underline{K}}' = 0$). This problem can be resolved as follows. We first rewrite eq. 4.27 in a manner that is correct to lowest order in Δt (as is eq. 4.27 itself). That is

$$\hat{\underline{\rho}}(t) \approx (\Delta t\underline{\underline{W}}' + \underline{\underline{1}})(\underline{\underline{1}} - i\underline{\underline{\Omega}}\Delta t + \underline{\underline{K}}'\Delta t)\hat{\underline{\rho}}(t-\Delta t) \quad (4.28a)$$

Then:

$$\hat{\underline{\rho}}(t) \approx (\Delta t\underline{\underline{W}}' + \underline{\underline{1}})[\exp(-i\underline{\underline{\Omega}}\Delta t + \underline{\underline{K}}'\Delta t)]\hat{\underline{\rho}}(t-\Delta t) \quad (4.28b)$$

Equation 4.28b is the basis of an effective finite difference method [19]. One then solves $\hat{\underline{\rho}}$ for each time t in terms of the previously obtained solution at the previous time $t-\Delta t$. The stability condition on $\underline{\underline{W}}'$ for such a method is

$$\frac{2\Delta t}{(\Delta r)^2} D \leq 1 \quad (4.29)$$

so that the diagonal elements of $\underline{\underline{W}}'\Delta t$ are smaller than unity. (A convenient value of $\Delta t \sim 10^{-12}$ sec for $D \sim 10^{-5}$ cm^2/sec, $\Delta r \sim \frac{1}{4}$ Å). However, such a method still requires that $\underline{\underline{A}}$ be stored in core. We can improve on this by taking further advantage of the separation of the solution into the two parts relating to $\underline{\underline{K}}'-i\underline{\underline{\Omega}}$ and $\underline{\underline{W}}'$, and recognizing that if $\underline{\underline{W}}'$ is spin-independent, then we can separately diagonalize each $L^2 \times L^2$ block of $\underline{\underline{K}}'-i\underline{\underline{\Omega}}$ as needed. Now the $\underline{\underline{W}}'$ matrix has simple elements (cf. eqs. 1.31). Thus, eq. 4.28b is easily solved for each value of r_i utilizing only a small portion of the total $L^2 \times N$ dimensional space [19]. The ensuing iterative process can efficiently employ an external storage device with a minimum core requirement (cf. Table A #3).

In this way, one can contemplate newer computer algorithms to solve more challenging CIDNP/CIDEP problems.

4.5 The triplet mechanism for CIDEP

This topic is discussed in the chapter by Atkins from the point of view of analytical or perturbation methods. The main point to be made here is that the triplet mechanism involves rotational diffusion which modulates the zero-field triplet tensor. Thus an analytical solution of the density matrix analogous to eqs. 1.17-1.19 given for translational diffusion is appropriate. The eigenfunctions for rotational diffusion are the generalized spherical harmonics. Such a point of view was utilized by Pedersen and Freed [6,10]. When, however, the tumbling motion slows down sufficiently and/or the zero-field splitting increases, then the lowest order perturbation approach breaks down, and one must use a numerical approach analogous to eqs. 1.21 and 1.22 to solve for the coupled algebraic equations resulting from the eigenfunction expansions in generalized spherical harmonics. A comparison of these numerical solutions has shown that perturbation-type approaches are adequate for

$$D^2 \lesssim \frac{1}{2}\,[\omega_0^2 + \tau_R^{-2}] \tag{4.30}$$

where D is the zero-field splitting constant, ω_0 is the ESR Larmour frequency, and τ_R is the rotational correlation time. When this inequality is not satisfied, then the numerical solutions are required. They are discussed further in [6] and [10].

ACKNOWLEDGMENT

I wish to thank Mr. Gary P. Zientara for his extensive help in preparing these lectures.

REFERENCES

1. J.B. Pedersen and J.H. Freed, J. Chem. Phys. 57, 1004 (1972).
2. J.B. Pedersen and J.H. Freed, J. Chem. Phys. 58, 2746 (1973).
3. J.B. Pedersen and J.H. Freed, J. Chem. Phys. 59, 2869 (1973).
4. J.B. Pedersen and J.H. Freed, J. Chem. Phys. 61, 1517 (1974).
5. J.B. Pedersen and J.H. Freed, J. Chem. Phys. 62, 1790 (1975).
6. J.H. Freed and J.B. Pedersen, Adv. Mag. Res. 8, 1 (1976).
7. H.S. Carslaw and J.C. Jaeger, Conduction of Heat in Solids, Oxford Univ. Press, London and New York, 1959, p. 382.
8. F.C. Collins and G.E. Kimball, J. Colloid Sci. 28, 425 (1949).
9. B. Yoon, J.M. Deutch, and J.H. Freed, J. Chem. Phys. 62, 4687 (1975).
10. J.B. Pedersen and J.H. Freed, J. Chem. Phys. 62, 1706 (1975).
11. L.P. Hwang and J.H. Freed, J. Chem. Phys. 63, 4017 (1975).
12. J.B. Pedersen, in Electron-Spin Relaxation in Liquids (L.T. Muns and P.W. Atkins eds.), Ch. III, Plenum, New York, 1972.
13. L.P. Hwang and J.H. Freed, J. Chem. Phys. 63, 118 (1975).
14. U. Fano. Rev. Mod. Phys. 29, 74 (1957).
15. M.P. Eastman, R.G. Kooser, M.R. Das, and J.H. Freed. J. Chem. Phys. 51, 2690 (1969).
16. S.A. Goldman, J.B. Pedersen, and J.H. Freed (to be published).
17. G.P. Zientara and J.H. Freed (to be published).
18. J.H. Freed, ACS Symp. Issue 34, Ch. 1 (1976).
19. G.P. Zientara and J.H. Freed (unpublished).
20. Z. Schulten and K. Schulten, J. Chem. Phys. (in press).

Appendix: Table A. Typical Numerical Methods in the Solution of CIDNP/CIDEP Problems[a]

General Method	Equation Solved[b]	Specific Method	Typical Numerical Algorithm(s)[d]
(1) Matrix Diagonalization (for time dependent solutions)	Solve for $\underline{\underline{a}}_d = \underline{\underline{T}}\,\underline{\underline{A}}\,\underline{\underline{T}}^{-1}$ where $\underline{\underline{a}}_d$ is diagonal	(1) Diagonalization of the entire matrix	Complex Jacobi Rotations
(2)		(2) Diagonalization with use of storage devices	Complex Rotations via a modified Givens method
(3)		(3) Analytic diagonalization (each r-value) with iterative diffusive steps	Explicit forward difference or Crank-Nicolson schemes
(4) Direct Solution via matrix inversion (solution in Laplace space)	Solve $\underline{\hat{\rho}}(r,s) = (s\underline{\underline{1}} - \underline{\underline{A}})^{-1}\underline{\hat{\rho}}_0(r)$	(4) Inversion of the entire matrix	Gaussian elimination with complete pivoting
(5)		(5) Inversion via locally active algorithms	Simple Gaussian elimination in combination with other techniques
(6) Direct Solution via iterative matrix inversion	Solve equation as for (4) and (5) iteratively[c]	(6) Inversion row by row by successive approximations	Gauss-Seidel or Successive Over-relaxation iterative methods

Table A (continued)

Major Mathematical Constraint(s)	Major Computing Requirements	Core Requirements[f]	External Storage Permitted	Useful for Mini-Computer	Status[i]
(1) None	$\underline{\underline{A}}$ (banded), $\underline{\underline{T}}$ (full) must fit in core	$(2L^2+1)(N\times L^2) + (N\times L^2)$	No	No	WM
(2) None	Sophisticated use of storage devices	Minimal-approx. $2(2L^2+1)$	Yes	Yes	WIP
(3) Eq. 4.29	Minimal core requisites, and as (2)	Minimal-approx. $3(L^2\times L^2)$	Yes	Yes	WIP
(4) $(s\underline{\underline{I}} - \underline{\underline{A}})$ must be nonsingular (as it is for $s \neq 0$)	$\underline{\underline{A}}$ (banded) must fit in core	$(2L^2+1)(N\times L^2)$	No	No[g]	WM[j]
(5) $(s\underline{\underline{I}} - \underline{\underline{A}})$ must be diagonally dominant	Solution collapses if mathematical condition not satisfied due to round-off errors	Approx. $(2L^2+1)(L^2\times 20)$	No	Yes[h]	WIP
(6) Same as for (5)[e]	Solution will not converge if mathematical condition not satisfied	Minimal-approx. $(2L^2+1)$	Yes	Yes	No future utility seen in CIDNP/CIDEP

Table A (continued)

Speed of Solution	Utility	References
(1) Fastest time resolved (TR) method	Small N, high field for TR studies	Gordon and Messenger in Muus' Electron Spin Relaxation in Liquids, Plenum, New York (1972), Ch. 13.
(2) Moderate TR method	Low field, large N for TR studies (large matrices)	Numerical Analysis of Symmetric Matrices, by H. Schwarz, Prentice-Hall, Englewood Cliffs, N.J. (1973); and Mathematical Methods for Digital Computers, by Ralston and Wilf, Wiley, New York (1960), Vols. I, II; among others.
(3) Slowest TR method	Low field, moderate N for TR studies (large matrices)	Difference Methods for Initial Value Problems, by Richtmyer and Morton, Wiley-Interscience, New York (1967), pp. 17, 189.
(4) Fastest method (for given s value)	High field in 2 and 3 dimensions	Subroutine DGELB, from the IBM Scientific Subroutine Package.
(5) Moderate V.E.	Low field, large N not feasible for 3D $s \to 0$ limit	Same as (2).
(6) Very slow	Unfavorable as CIDNP/CIDEP matrices do not suit mathematical requirements	Matrix Iterative Analysis, R. Varga, Prentice-Hall, Englewood Cliffs, N.J. (1962), p. 58.

Footnotes to Table A

a) Table prepared by G.P. Zientara.
b) $\underline{\underline{A}} = \{\underline{\underline{W}}' - i\underline{\underline{\Omega}} - \underline{\underline{K}}'\}$.
c) If we let $\underline{\underline{A}} = (\underline{\underline{L}} + \underline{\underline{D}} + \underline{\underline{U}})$ where $\underline{\underline{L}}$, $\underline{\underline{D}}$, $\underline{\underline{U}}$ are matrices composed of the elements of $\underline{\underline{A}}$ below the diagonal, the diagonal elements, and the elements above the diagonal respectively. Using the Gauss-Seidel method one would solve for the kth approximation to $\hat{\underline{\rho}}(r,s)$ from:

$$\underline{\rho}^{(k)}(r,s) = (s\underline{\underline{1}} - \underline{\underline{D}})^{-1}[\hat{\underline{\rho}}_0 + (\underline{\underline{L}} + \underline{\underline{U}})\underline{\rho}^{(k-1)}(r,s)].$$

d) Many more algorithms (some favorable, some not) exist than are mentioned here.
e) By diagonally dominant is meant,

$$(s - A_{ii}) > \sum_{\text{all } j \neq i} |A_{ij}| \quad \text{for all } i.$$

This condition suffices for our problems as more rigorous mathematical treatments lead to similar conditions [cf. James and Riha, SIAM J. Numer. Anal. 12, 137 (1975)].
f) Given as the number of matrix elements needed in core for solution (1 matrix element = 16 bytes). This is based on an L×L Hamiltonian matrix and N locations used in finite difference r-space.
g) If this is implemented with a computer algorithm utilizing storage devices, it is extremely time consuming and therefore unfavorable.
h) For CIDNP/CIDEP in the contact exchange cases only.
i) WM = working method, WIP = work in progress on the numerical method.
j) The method used by Pedersen and Freed in their studies.

CHAPTER XX

CHEMICALLY INDUCED MAGNETIC POLARIZATION IN SYSTEMS OF BIOCHEMICAL INTEREST*

G. L. Closs

Department of Chemistry, The University of Chicago,
Chicago, Illinois 60637, USA

ABSTRACT. Donor-acceptor reactions on chlorophyll and its derivatives are being discussed as model systems for photochemical charge transfer reactions. The CIDNP observed can be explained by the radical pair theory. The polarized triplet esr spectrum of bacteriochlorophyll _in vivo_ is given as a possible example of the radical pair mechanism operating in intact biological systems.

INTRODUCTION. A survey of the literature shows that the phenomenon of CIDNP so far has had only marginal applications to biochemistry. As a matter of fact at the time of this writing the only reaction in a biological system which shows some electron polarization related to the CIDEP mechanism has been uncovered in the field of photosynthesis. However, there exist a few applications of CIDNP to _in vitro_ reactions of molecules which are indirectly related to biochemical problems. Most of the reactions are of the charge-transfer type and this short chapter is meant to summarize reported work as well as to point out possible applications.

1. PHOTOCHEMICAL DONOR-ACCEPTOR REACTIONS OF PHOTOSYNTHETIC PIGMENTS STUDIED BY CIDNP

Charge transfer processes play a key role in the mechanism of photosynthesis. Model systems involving chlorophyll and its derivatives have been studied by esr for some time, and more recently CIDNP has been employed to gain information on such processes. The first re-

---

* This work has been supported by the National Science Foundation.

L. T. Muus et al. (eds.), Chemically Induced Magnetic Polarization, 357-367.

port of experiments on photochemically generated CIDNP involving a chlorophyll derivative was made by Tomkiewicz and Klein who illuminated chlorophyll-a in the presence of quinone and observed a CIDNP signal due to the protons of the quinone.[1] This work was repeated by Roth and Lamola who did a more careful study, the outcome of which did not support the findings or conclusions of the earlier workers.[2] We will discuss in this section some work carried out in the author's laboratory on a very similar system.[3]

The system to be discussed can be summarized in the scheme shown in (1.1). A solution containing an acceptor pigment (A) and an electron donor (D) is irradiated with light of a wavelength which only excites A to its first excited singlet state (*A).

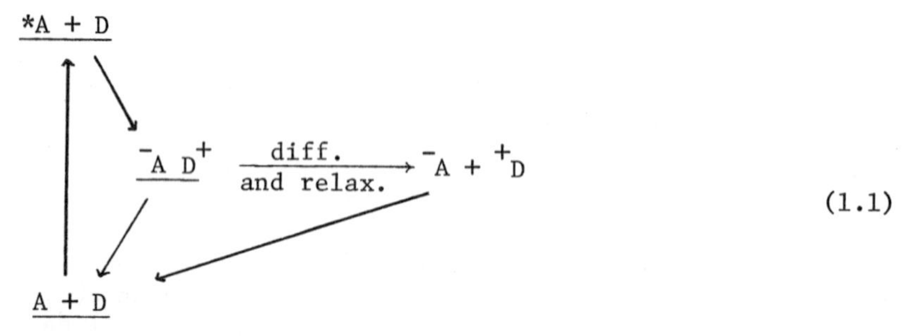

(1.1)

Charge transfer may take place forming a geminate radical ion pair which, in a good dielectric medium, can separate into free radical ions. Geminate electron back-transfer can produce polarization in the reactant if the escaping ions live long enough to lose their polarization by relaxation. Eventually the uncorrelated ions are also reconverted to reactants. Two points are left open at this time; it is not excluded that electron transfer takes place from the triplet level and the ions may be converted to neutral radicals by either capture or loss of a proton. In our experiments the acceptor is chlorophyll a (Chl) or one of its derivatives. The donors are hydroquinone, catechol and derivatives of hydroquinone. Roth and Lamola carried out essentially the "mirror image" experiment in which the chlorophyll was oxidized by quinone acceptor.

What we hoped to learn from these experiments was the multiplicity of the state of the chlorophyll molecule from which electron donation took place, and to gain information on the charge distribution in the negative ion (Chl^-).

If a well-degassed solution of chlorophyll a ($< 10^{-3}$ M) in methanol-d_4-chloroform-d containing 0.02 molar hydroquinone is irradiated with light of wavelength > 600 nm inside a 90 MHz nmr spectrometer, one observes a very strong emission from the aromatic protons of the hydroquinone (Figure 1). This emission can be observed for hours and the solution, when analyzed after the experiment, reveals no new products. We will address ourselves to the question why no polarizations are seen for the chlorophyll later and first focus attention on the donor polarization.

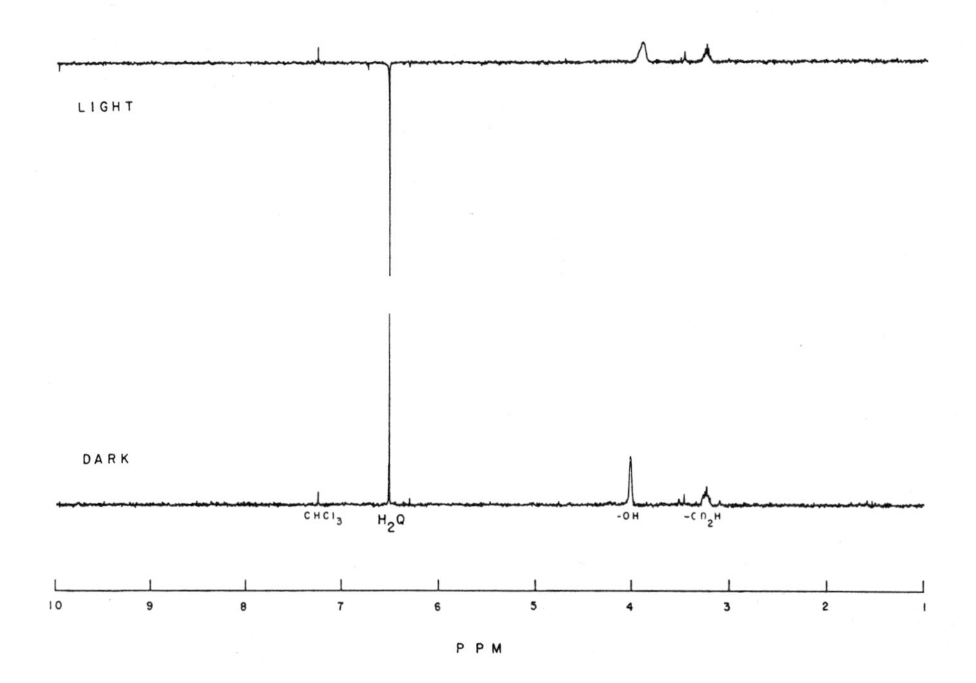

Fig. 1. Light and dark spectrum of a solution of chlorophyll a and hydroquinone in $CDCl_3$ (see text for concentrations).

If we interpret this result within Scheme 1.1, application of the net effect rule implies interception of the triplet state (Δg: +; a_i: −; ε: +). However one line does not yield too much information and other donors were tried in similar experiments. As shown in Figure 2A catechol gives equivalent results (emission) with many resolved lines.

An important experiment involves p-methoxyphenol as a donor (Figure 2B). The fact that the ortho protons are in emission and the m-protons give weak enhanced absorption rules out the triplet mechanism as a source of polarization (it is not likely that the two types of ring protons depend on two different mechanism for cross-relaxation). The second information we gain is that in the geminate pair the hydroxy proton has been lost to the solvent forming the neutral radical. This follows from the positive hyperfine interaction at the meta position.

Using the known hyperfine interactions for the neutral o-semiquinone and for p-methoxyphenoxy radicals and their g-factors and the g-factor of $Chl^{(-)}$, it was possible to calculate the expected polarization. The results are shown in Figure 3. The agreement is excellent considering that the hyperfine interactions in $Chl^{(-)}$

were not taken into account.

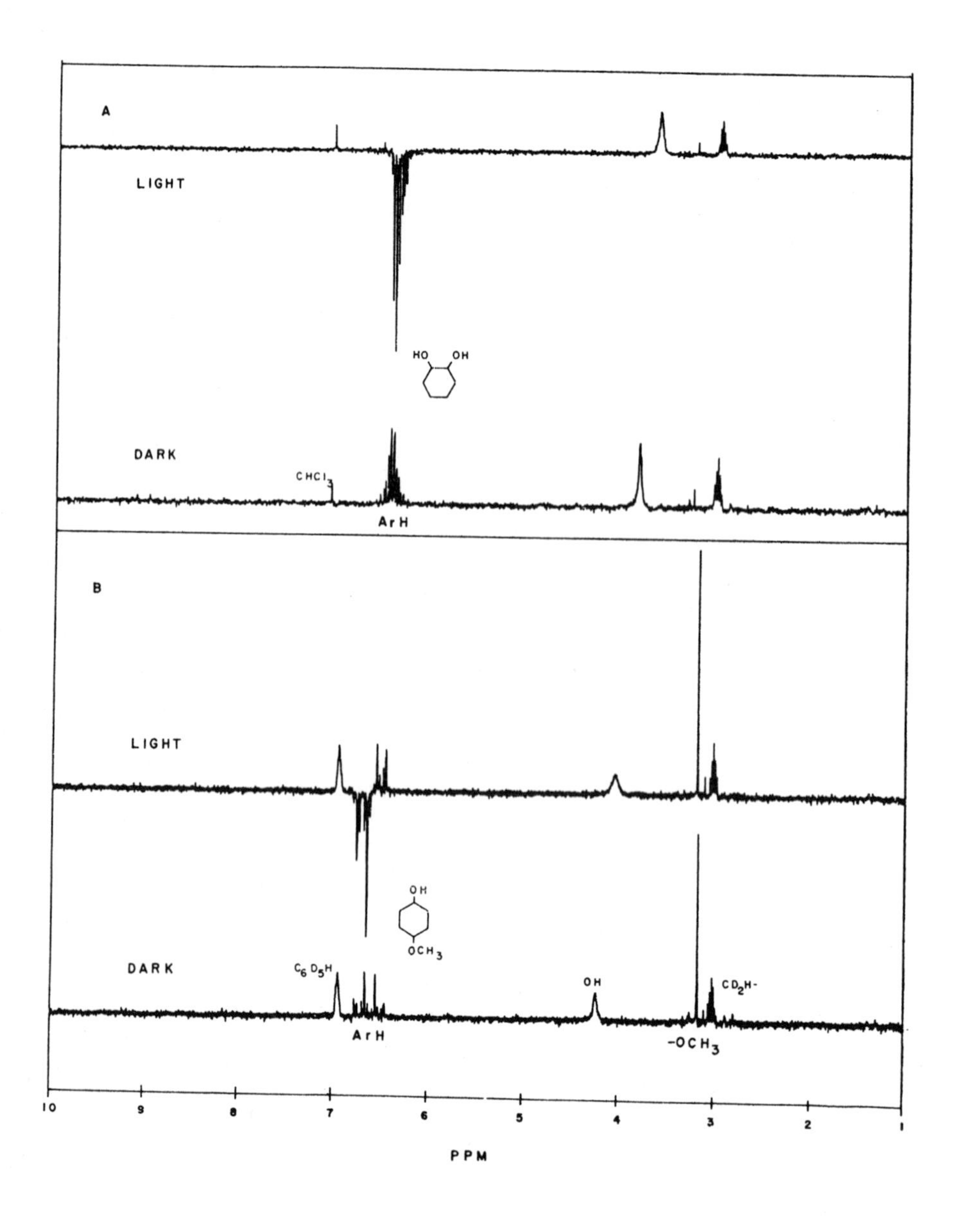

Fig. 2. Light and dark spectra of a solution of chlorophyll a in $CDCl_3$ and A) catechol, B) p-methoxyphenol.

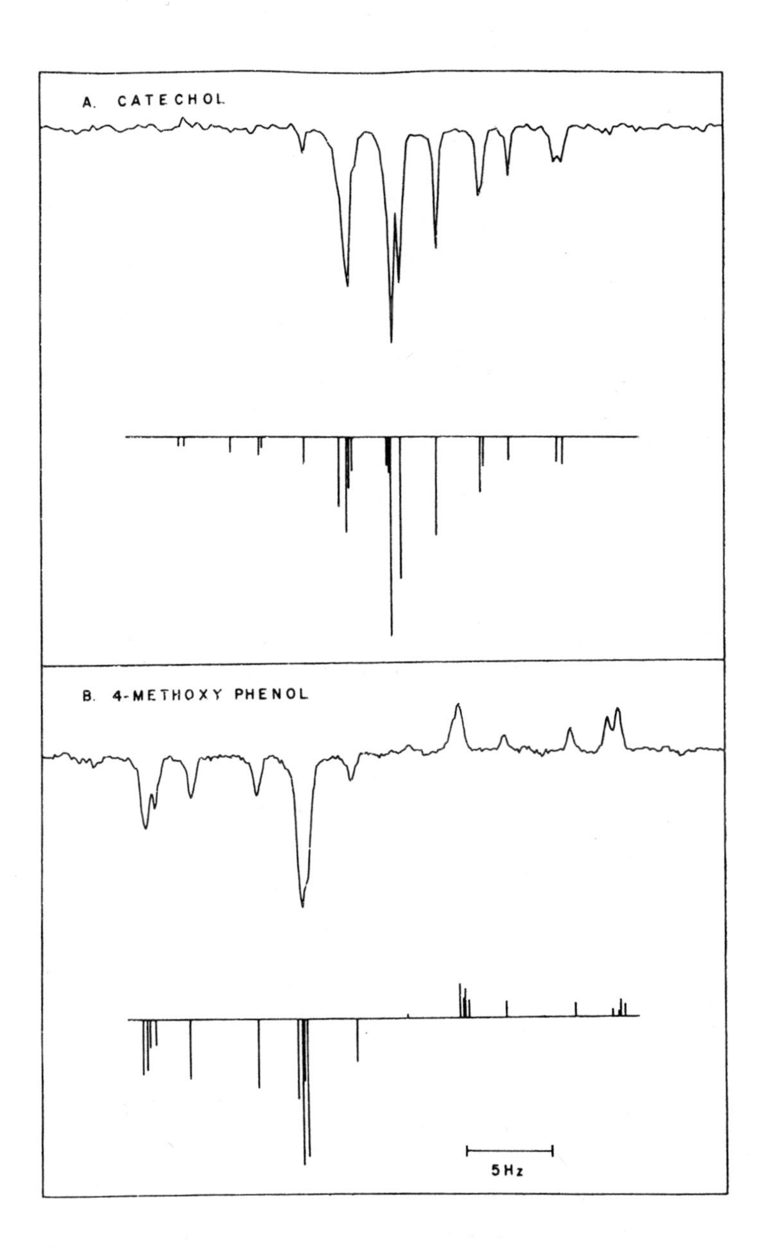

Fig. 3. Expanded aromatic region of Figure 2 and calculated transition intensities (see text for parameters).

It appears at this stage that Scheme 1.1 can be amended as shown in (1.2)

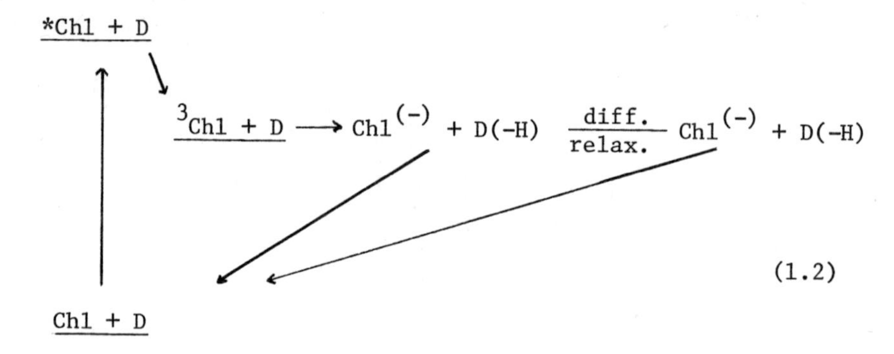

At this stage it is unresolved whether the Chl$^{(-)}$ takes up a proton to give ChlH. Although it is quite likely, there is no evidence for or against it. It is worthwhile mentioning that there is no polarization with p-dimethoxy benzene as a donor, nor if the solvent is changed to acetonitrile.

Next, we turn to the absence of any noticeable polarization on Chl. At first sight, this may be attributed to two causes. First, T_1 in Chl may be too short and therefore the signals may be too weak to detect. Second, the hyperfine interactions may be too small in Chl$^{(-)}$. None of these are likely to be the cause, because measurement of the T_1's show them to be much shorter than in the donor but not short enough to wipe out the effect. The magnitude of the hyperfine interactions, although not known for Chl$^{(-)}$ can be estimated from related models and should be large enough for noticeable polarization to develop.

The answer came from experiments on pyro-methylpheophorbide (PMPh), a close analog to chlorophyll in which the long phytol chain has been replaced by a methyl ester, the magnesium has been replaced by two hydrogens, and the carbomethoxy group at position 10 has been replaced by a hydrogen. This compound gives the identical results in the polarization experiments as Chl.

Signal averaging in the dark for several hundred scans produced all the important signals of PMPh as shown in Figure 4. However, when the light was turned on most of the signals disappeared. There are three exceptions. These are the signals of the three methyl groups which are isolated from the π-system by one or more carbon atoms (the methyl ester at 3.4 ppm and the two methyl groups at positions 4 and 8 at 1.6 ppm). These are the protons which do not carry any significant amount of hyperfine interaction in any paramagnetic form of PMPh. Unless one wants to assume the unlikely event that CIDNP polarization for all the other peaks is negative and exactly cancels the dark signals, one has to conclude that the signals have broadened to the point where they cannot be recognized.

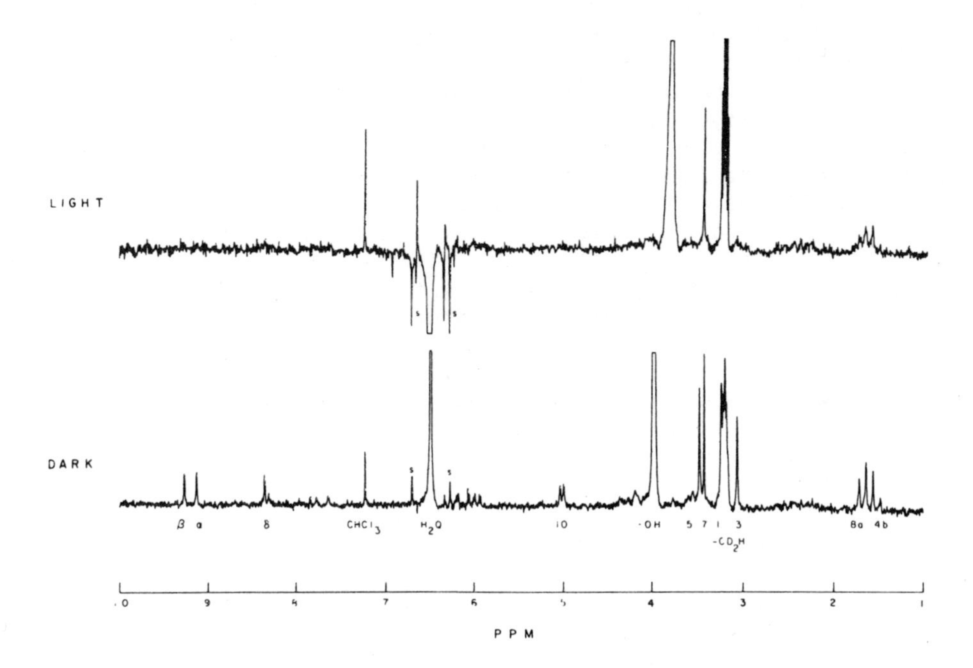

Fig. 4. Time averaged dark and light spectrum of PMPh and hydroquinone in $CDCl_3$.

There are two potential mechanisms for broadening of signals from the chlorophyll derivatives, both involving exchange with a paramagnetic species. The first is degenerate triplet-groundstate exchange of the type (1.3).

$$\text{Chl} + {}^3\text{Chl} \rightleftharpoons {}^3\text{Chl} + \text{Chl} \quad . \qquad (1.3)$$

That this exchange is indeed fast enough to cause broadening is shown by the fact that broadening occurs in the absence of a donor.[4] In that case the only paramagnetic species in solution is the triplet state and equation(1.3) accounts for the broadening. Another source for the broadening is the electron exchange between the ion and the neutral molecule. Such processes have been observed before

$$\text{Chl}^- + \text{Chl} \rightleftharpoons \text{Chl} + \text{Chl}^{(-)} \qquad (1.4)$$

in other electron transfer reactions.[5] In the system under discussion it is likely that both processes are going on simultaneously.

Electron and triplet energy exchange are processes which shorten T_2 in the groundstate molecule. Their effect on T_1 is relatively minor if the paramagnetic species are kept in very low concentrations. Therefore the exchange processes (1.3) and (1.4) do not account for the absence of CIDNP. They only make it unobservable by broadening the lines until the signals are buried in noise. If this is correct, a slight modification in the experiment should make the CIDNP signals come out of the noise. This modification is based on the fact that all paramagnetic species will disappear on the time scale of milliseconds when the light is turned off, while the nuclear T_1's are much longer. The experiment then involves interfacing a light shutter with the computer controlling the nmr spectrometer and irradiating the sample for a few seconds to build up the polarization to a steady state. Next, the light is turned off and within a few milliseconds a rf-pulse is applied. Fourier transform of the collected free induction decay should give a sharp spectrum containing almost all the steady state polarization. That this indeed works is shown in Figure 5 which shows the time average of a number of such light-dark-pulse sequences. The reaction is the photoreduction of PMPh with hydroquinone. The most prominent signal is the emission signal of the methyl group in position 5. Equally revealing is the lack of polarization of

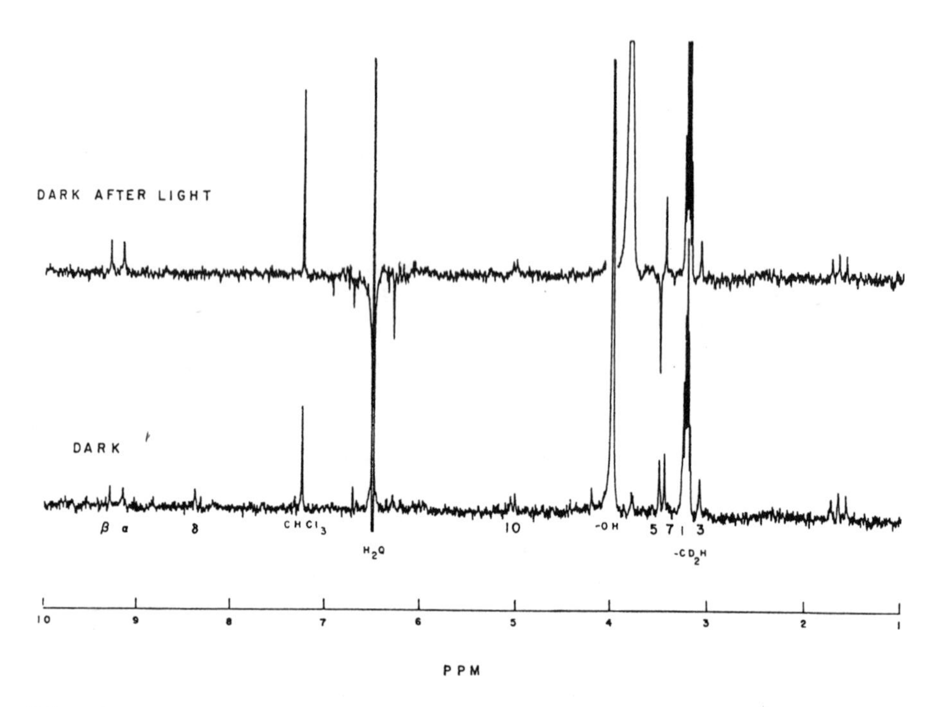

Fig. 5. Time averaged spectrum of PMPh and hydroquinone in $CDCl_3$. The upper trace is an accumulation of spectra obtained by the light-dark-rf pulse sequence described in the text.

the methyl group at position 3. Protons α and β are slightly enhanced and δ is weakly in emission, cancelling its dark signal. A more quantitative evaluation of these spectra is underway and should yield the spin densities in the negative ion.

Similar light pulsed experiments can be carried out on the pigment cations produced by the experiments of Roth and Lamola.[2] These results, together with the line broadening experiments resulting from triplet-singlet exchange should give the spin-density distributions of the three most important paramagnetic states of chlorophyll and its derivatives.

2. ELECTRON SPIN POLARIZATION IN PHOTOSYNTHETIC BACTERIA

Electron spin polarization in excited triplet states in the solid state at low temperature is a well understood phenomenon with many examples. It arises from the fact that the spin-orbit coupling tensor connecting the first excited singlet state with the triplet state is anisotropic. Therefore the three zero-field splitting states of the triplet state receive population from the singlet state with different rates. Similarly, the tensor connecting the triplet levels with the ground state is anisotropic and the depopulation rates of the triplet levels are different. This leads in steady state to a polarization different from Boltzmann distribution. This effect is the basis for the triplet mechanism of CIDEP in doublet states.

Chlorophyll derivatives are no exception when studied in a matrix at low temperature _in vitro_. The important feature of that polarization is that it is determined by the molecular axis system. However, when photosynthetic bacteria, whose electron acceptor has been reduced with dithionate, are frozen in a matrix and irradiated with red light, a triplet state esr spectrum can be observed and attributed to the triplet state of bacterio-chlorophyll.[6] The polarization of this spectrum does not follow the usual pattern and differs from the _in vitro_ polarization of bacterio-chlorophyll. The salient feature of the _in vivo_ spectrum is its independence of molecular axis orientation relative to the magnetic field. Instead the spectrum is polarized entirely by the field. That is, what in the high-field approximation becomes T_+, T_o, T_- states carry the same relative populations regardless with which zero-field states the high-field states correlate. In the bacterio system the T_o state carries the excess population.

This type of triplet polarization is highly unusual and must find its origin in a mechanism that is independent of the molecular axis system. One possible explanation is as follows:[7] In the photosystem the first excited singlet state of bacterio chlorophyll transfers an electron to the primary acceptor, which in turn transfers it to the next acceptor until a reduced acceptor is encountered in the transfer chain. There must be such reduced acceptors because the system has been chemically reduced with dithionite. At this point the radical ion pairs are separated far enough to make singlet and triplet degenerate.

The wavefunction will evolve under the influence of Δg-effects and hyperfine interactions to acquire triplet character. From perturbation arguments it follows that T_0 of the ion pair state receives most of the population if it is degenerate with the singlet state. Since further electron transfer is blocked, the system can do nothing but decay back to lower lying states, which are the ground state of bacterio-chlorophyll and its triplet state. If spin is conserved in this decay process, T_0 of the bacterio-chlorophyll will be carrying excess population in steady state.

The mechanism, while plausible, is at the point of this writing not confirmed nor ruled out by additional experiments. It is however sufficiently interesting to deserve inclusion in this discussion.

Recently the observation of CIDEP in bacterio-chlorophyll in vitro and in green plant chloroplasts has been reported by several groups.[8,9] It was suggested that the polarized radicals are formed via the triplet mechanism rather than via the radical pair mechanism. The relationship of these experiments to the above mentioned triplet state polarization is at present unclear.

3. OUTLOOK

One of the reasons not more systems of interest to biochemists have shown CIMP effects may be that people have not looked for it. In electron transport phenomena, which after all comprise a major part of biochemistry, the potential exists for both CIDEP and CIDNP. The latter suffers from the effect large molecular weight has on nuclear relaxation by broadening lines to the point where they can no longer be resolved. Also transition metals are frequently involved in the electron transport chain and with their large g-factor anisotropies they can be expected to wipe out the small effects caused by hyperfine interactions. Nevertheless it is safe to predict that there should be other cases of CIMP, particularly CIDEP.

REFERENCES

1. M. Tomkiewicz and M.P. Klein, Proc. Nat. Acad. Sci. USA, 70, 143 (1973).
2. A.A. Lamola, M.L. Manion, H.D. Roth and G. Tollin, Proc. Nat. Acad. Sci. USA, 72, 3265 (1975).
3. S.G. Boxer, Ph.D. Thesis, The University of Chicago, 1976.
4. S.G. Boxer and G.L. Closs, J. Am. Chem. Soc., 97, 3268 (1975).
5. H.D. Roth and A.A. Lamola, J. Am. Chem. Soc., 96, 6270 (1974).

6. P.L. Dutton, J.S. Leigh and M. Seibert, Biochem. Biophys. Res. Commun., **46**, 406 (1972); P.L. Dutton, J.S. Leigh and D.W. Reed, Biochim. Biophys. Acta, **292**, 654 (1973).
7. M.C. Thurnauer, J.J. Katz and J.R. Norris, Proc. Natl. Acad. Sci. USA, **72**, 3270 (1975).
8. J.R. Harbour and G. Tollin, Photochem. Photobiol., **191**, 163 (1974).
9. R. Blankenship, R.A. McGuire and K. Sauer, Proc. Natl. Acad. Sci., USA, **72**, 4943 (1975).

CHAPTER XXI

TRIPLET OVERHAUSER MECHANISM OF CIDNP*

F. J. Adrian

The Johns Hopkins University Applied Physics Laboratory, Laurel, Maryland, USA

1. INTRODUCTION

Soon after the discovery of CIDNP it was proposed that the observed nuclear spin polarizations resulted from a small deviation from equilibrium in the electron spin states of the chemically generated radicals (such as all states equally populated), which was converted by electron-nuclear cross relaxation into a large deviation of the nuclear spin states from thermal equilibrium [1,2]. This model is often called an Overhauser mechanism, because of its similarity to the Overhauser effect in which microwave pumping of the electron spin states of a paramagnetic species in a magnetic field leads to large nuclear spin polarizations [3]. This model failed to account for many features of CIDNP, however, and it was superseded by the radical pair mechanism [4,5]. This did not prove the Overhauser mechanism is nonexistent, however. It could be that it is usually much less efficient than the radical pair mechanism, which therefore dominates the polarization process and hides the smaller effects of the Overhauser mechanism. If so, the Overhauser mechanism could be important in cases where the radical pair mechanism is inoperative, or if there is a very large initial electron spin polarization.

Thus, interest in the Overhauser mechanism was restimulated somewhat by the discovery that large net electron polarizations are produced during the photolysis of a number of carbonyl compounds such as quinones, aromatic ketones, etc. [6], by a triplet mechanism in which the photoexcited molecule undergoes a

*This work has been supported by the U. S. Naval Sea Systems Command under Contract N00017-72-C-4401.

L. T. Muus et al. (eds.), Chemically Induced Magnetic Polarization, 369-381.

spin-selective intersystem crossing to a triplet which reacts to yield a pair of electron spin polarized radicals [7]. This raised the question of whether nuclear spin polarization could be produced by this triplet mechanism combined with transfer of the electron polarization to the nuclei. Several observations of nuclear spin polarization in photochemical reactions involving quinones have lent some support to this mechanism [8]. Of particular interest was the observation of a polarized light dependence of the ^{19}F polarization during photolysis of fluoranil in chloroform [8b], because it had been shown that electron spin polarization by the triplet mechanism can depend on the orientation of the electric vector of a polarized light source with respect to the external field [6c,d,7c].

In this chapter we shall discuss the Overhauser mechanism of CIDNP, its limitations and, in a few cases, its advantages, with emphasis on cases where the initial electron polarization is provided by the triplet mechanism.

2. THEORY

Fig. 1 depicts the three principal steps in an Overhauser mechanism of CIDNP for the case of a radical having a single hyperfine interaction with a nucleus of spin 1/2. The first step is production of electron-spin-polarized radicals from a photoexcited triplet precursor or by some other process, which is denoted by different quantum efficiencies, Q_+ and Q_-, for production of radicals in the e+ and e- states. Polarization is defined here as the deviation from the equilibrium distribution, which simplifies the treatment, because the relaxation processes can then be regarded as driving the difference in spin level population to zero. The

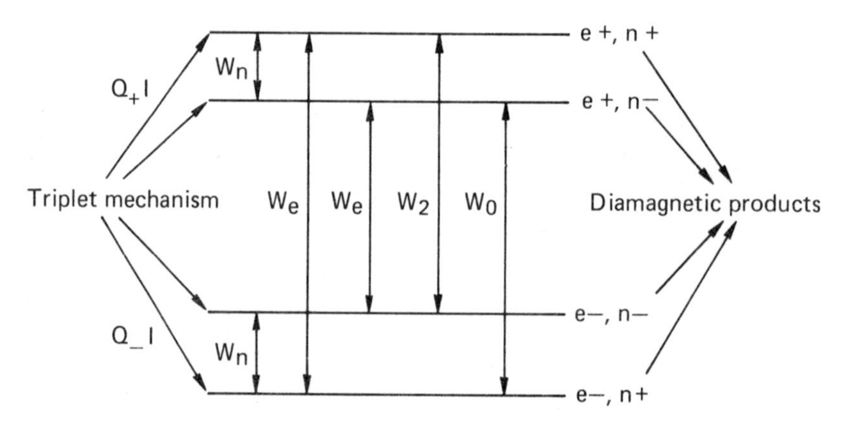

Fig. 1. Steps in the triplet Overhauser mechanism of CIDNP

second step is transfer of the electron spin polarization to the nuclear spins by the electron-nuclear cross-relaxation transitions, denoted as W_0 and W_2 in Fig. 1. Finally, the nuclear-spin-polarized radicals must react to transfer their polarization to diamagnetic products before the pure electron spin and pure nuclear spin transitions, W_e and W_n in Fig. 1, can remove all polarization from the radicals.

2.1 Triplet mechanism of CIDEP

The triplet mechanism of electron spin polarization is described in detail elsewhere in this book and the original references [7]. Thus, we limit present discussion to noting that the electron polarization originates in the spin-selective intersystem crossing of a photoexcited singlet molecule to a triplet and is transferred to a radical pair formed by a charge or proton transfer reaction of the triplet with a suitable substrate, provided that this reaction is rapid enough to compete with loss of polarization to the very efficient triplet spin relaxation mechanism (T_{1T} is typically 10^{-9} sec). The detailed theory gives [7,9]

$$\frac{2\delta Q}{Q} = \frac{2}{15} w_{ZFS} w_Z \tau_R^2 \left(\frac{1}{1+w_Z^2\tau_R^2} + \frac{4}{1+4w_Z^2\tau_R^2} \right) \frac{k_T}{k_T+W_T} \left(2P'_Z - P'_X - P'_Y \right), \quad (1)$$

where

$$2\delta Q = Q_+ - Q_- \quad : \quad Q = \frac{1}{2}(Q_+ + Q_-), \quad (2)$$

and the triplet spin relaxation rate is given by the equation

$$W_T = \frac{2}{15} w_{ZFS}^2 \tau_R \left(\frac{1}{1+w_Z^2\tau_R^2} + \frac{4}{1+4w_Z^2\tau_R^2} \right). \quad (3)$$

The quantity k_T is rate of the triplet reaction to form the radical pair, which quantity is proportional to the concentration of the substrate involved in the reaction. Other quantities in the foregoing equations are as follows: P'_Z, P'_X , and P'_Y are the populations of the various intermolecular spin states of the triplet immediately after its formation by intersystem crossing from the photoexcited singlet; w_{ZFS} is the zero-field-splitting parameter of the triplet; w_Z is the Zeeman splitting of the electron spin energy levels in the external magnetic field; and τ_R is the rotational correlation time of the triplet molecule.

For typical values of the parameters in these equations [7,8c] it can be shown that $(2\delta Q/Q)$ corresponds to an electron spin enhancement factor of roughly 1 to 100 times the equilibrium population difference given by the Boltzmann factor w_Z/kT, with 10 being a typical value. Perhaps more meaningful here, $(2\delta Q/Q)$ corresponds to nuclear spin enhancement factors of 100 to 10000 times the nuclear spin Boltzmann factor w_n/kT for w_n=60 MHz. Unfortunately, as will be seen, much of this potentially large enhancement is lost in the process of transfer of polarization from electrons to nuclei and thence to diamagnetic products.

2.2 Cross-relaxation mechanisms

The electron-nuclear cross-relaxation transitions arise from the magnetic hyperfine structure (hfs) interactions between the electron and nuclear spins. The nature of these transitions depends strongly on whether the isotropic or anisotropic hfs is the dominant interaction. Usually the anisotropic hfs dominates because this interaction is readily modulated by the rotational tumbling of the molecule, giving it the time dependent quality needed to stimulate spin transitions. For the case of an axially-symmetric anisotropic hfs the relaxation rates are [3]:

$$W_0=2W,\quad W_e'=3W,\quad W_2=12W;\quad W_n=\frac{3}{80}\frac{B_{ZZ}^2\tau_R}{\hbar^2}, \tag{4}$$

where

$$W=\frac{1}{80}\frac{B_{ZZ}^2\tau_R}{\hbar^2(1+w_Z^2\tau_R^2)}, \tag{5}$$

and B_{ZZ} is the component of the anisotropic hfs tensor along the symmetry axis. If the relaxation is determined by the isotropic hfs, which is possible in cases such as a methyl group, where the anisotropic splitting is small, and rotation of the methyl group modulates the isotropic hfs, the only relaxation process is W_0, whose rate is [3]

$$W_0=\frac{(\delta A)^2\tau_A}{2\hbar^2(1+w_Z^2\tau_A^2)}, \tag{6}$$

where δA is the mean-square variation in the isotropic hfs constant, A, and τ_A is the correlation time for the molecular motion which causes A to vary with time.

It has been estimated that $W \approx 10^4$ to 10^5 sec^{-1} for ^{19}F and ^{13}C nuclei, and an order of magnitude slower for protons which have smaller anisotropic hfs constants [10]. Estimates of W_0 from relaxation by modulation of an isotropic hfs interaction are harder to come by, but there are likely to be cases where the reorientation rate of CH_3 or CF_3 groups is such that W_0 is comparable to the rates found for the anisotropic case. It will be found that the sign of the Overhauser CIDNP depends on whether the anisotropic or isotropic hfs provides the electron-nuclear cross relaxation.

2.3 Radical reactions and efficiency of the Overhauser mechanism

Transfer of polarization from radicals to products may be achieved by one or more of the following classes of reactions:

(1) The first order fragmentation reaction

$$R\cdot{}^* \to R'\cdot{}^* + P^*, \tag{7a}$$

or the pseudo-first-order reaction

$$R\cdot{}^* + SH \to RH^* + S\cdot, \tag{7b}$$

where an asterisk denotes a nuclear-spin-polarized species.

(2) Radical-radical recombination reactions such as

$$R\cdot{}^* + R'\cdot \to R\text{-}R'^*. \tag{8}$$

Such reactions are always present, but in many cases they are so much slower than the various spin relaxation processes that they are relatively ineffective as a means of transferring polarization from the radicals to the products.

(3) Finally, there are the exchange reactions such as

$$R\cdot{}^* + RH \rightleftharpoons RH^* + R\cdot. \tag{9}$$

First order reaction. When the radicals react according to Eq. (7), the rate equations governing the populations of the various radical spin states are [8c]

$$\begin{aligned}
\frac{dN_{++}}{dt} &= Q_+IP-(W_n+W_e+W_2+k_1)N_{++}+W_nN_{+-}+W_eN_{-+}+W_2N_{--},\\
\frac{dN_{+-}}{dt} &= Q_+IP-(W_n+W_e+W_0+k_1)N_{+-}+W_nN_{++}+W_eN_{--}+W_0N_{-+},\\
\frac{dN_{--}}{dt} &= Q_-IP-(W_n+W_e+W_2+k_1)N_{--}+W_nN_{-+}+W_eN_{+-}+W_2N_{++},\\
\frac{dN_{-+}}{dt} &= Q_-IP-(W_n+W_e+W_0+k_1)N_{-+}+W_nN_{--}+W_eN_{++}+W_0N_{+-}.
\end{aligned} \tag{10}$$

Here k_1 is the rate constant of Eq.(7) which will depend on [SH] in a reaction of type [7b], and the rate of photolysis is given by the product of the light intensity I, the quantum yields Q_+ and Q_-, and the concentration of the reactant P, it being assumed that the probability of light absorption by the reactant is proportional to its concentration. If the sample is optically opaque, or if multiple reflections within the sample cell cause all entering light to be absorbed eventually, this last condition does not hold. The optically opaque case where the light is inhomogeneously absorbed in the sample is complicated but does not introduce any fundamentally new principles, so it will not be considered further in this introductory presentation. The case of complete absorption by multiple reflection is readily treated by replacing I in Eq.(10) and all subsequent equations by I/P [8c]. Solution of these equations for steady state conditions ($dN_{ij}/dt=0$) gives the following expression for the rate of production of nuclear-spin-polarized product

$$d(\delta P)/dt = -2IQP(2\delta Q/Q)\xi^{(1)}, \tag{11}$$

where δP is the difference between the amount of product in the n+ and n- states, and $\xi^{(1)}$ is an efficiency factor for transfer of the electron spin polarization $(2\delta Q/Q)$ to the nuclear spins. For the present case, it is given by the expression

$$\xi^{(1)} = \frac{(W_2-W_0)k_1}{(W_n+W_e+2W_2+k_1)(W_n+W_e+2W_0+k_1)-(W_n-W_e)^2}. \tag{12}$$

As defined by these equations, a negative δP corresponds to emissive CIDNP for the usual case of a nucleus with a positive gyromagnetic ratio.

Overhauser CIDNP in the anisotropic case is optimized for a total pure electron relaxation rate $W_e=W_e'=3W$, and $w_Z\tau_R<<1$, giving by Eqs.(4) and (5), $W_n=3W$. Usually, however, $W_e>W_e'$ because there are other sources of pure electron relaxation such as g factor anistropy of the radical [8c]. Also, one often has $W_n>3W$, thereby increasing the loss rate of the developing nuclear spin polarization. A plot of $\xi^{(1)}$ vs k_1/W for various values of W_e, assuming $W_n=3W$, given in Fig. 2, shows that the Overhauser mechanism is most efficient when the reaction rate is approximately equal to either $|(W_2-W_0)|$ or W_e, whichever is faster. For $k_1\gg|W_2-W_0|$, the product is produced before the polarization can be transferred from electrons to nuclei [although it might be important to consider the fate of the radicals which inherit the electron spin polarization of the original radicals via Eq.(7)], and for $k_1<<\max[|(W_2-W_0)|, W_e]$ all polarization is lost before it can be frozen into the diamagnetic

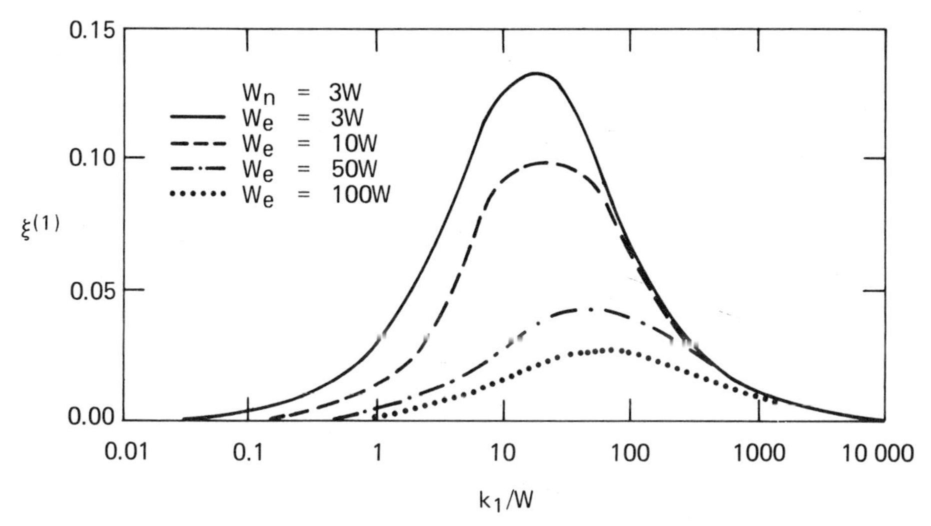

Fig. 2. Efficiency factor for transfer of polarization from electrons to nuclei in case of first-order radical reaction.

products(s) where the relaxation rates are much slower. From Fig. 2 it is seen that under optimum conditions only 13% of the electron polarization is transferred to the nuclei, and under typical conditions this efficiency factor will be at least an order of magnitude smaller. An initial electron polarization of the order of the thermal equilibrium population difference, as in the case of all electron levels equally populated, corresponds to a potential proton enhancement factor of 660 [3]. Taking $\xi^{(1)}$ into account, the maximum realizable enhancement is 80 and in practice will rarely exceed 10. This is quite small compared to the typical radical pair mechanism enhancement factors of 100 to 1000 [11] and shows that the radical pair mechanism will usually predominate. This conclusion could be different, however, in cases where the triplet mechanism gives much larger potential nuclear spin enhancement factors. In principle, cross relaxation by the isotropic hfs could yield larger Overhauser polarizations because the isotropic hfs gives only the polarizing W_0 transition, but in practice other contributions to W_e will generally give $W_e > W_0$, and the polarization by the isotropic mechanism will be the same order of magnitude as the anisotropic polarization.

Radical-radical reactions. If the product is formed by radical-radical reactions it can be shown that the efficiency factor for transfer of polarization to the product nuclei is much the same as the previous case insofar as the maximum and typical efficiencies are concerned [8c]. In this case, however, the efficiency does not fall off at high reaction rates, because radical recombination occurs only if the spins of the two radicals are antiparallel, and consequently the electron spin polarization remains in the same radical(s) and ultimately will be transferred to the product. Practically, however, this is of little importance, because at typical radical concentrations the second-order radical recombination rates are slow compared to typical relaxation rates.

Exchange reactions and cyclic reactions. The exchange reactions, cf. Eq.(9), can provide efficient transfer of nuclear polarization from radicals to products, and they cannot be too rapid because the reaction does not change the nature of the radical species or its electron polarization. They are reversible, and thus have the negative feature of reconverting an already polarized product to a radical where the polarization is lost via rapid radical relaxation processes; however, this affects both Overhauser and radical pair polarizations, and under some conditions gives the Overhauser mechanism an advantage.

An example is the cyclic photoreduction of 1,4-tetrafluoroquinone (FQ) by 1,4 tetrafluorohydroquinone (FQH_2). This reaction yields a net emissive CIDNP for FQ, which is difficult to explain by the radical pair mechanism because the pair consists of identical radicals [8c]. The reaction mechanism is

$$FQ_{(n\pm)} \xrightarrow{h\nu} {}^3FQ_{(e+,n\pm)} \xrightarrow{FQH_2} FQH\cdot_{(e+,n\pm)} + FQH\cdot \tag{13a}$$

$$FQ_{(n\pm)} \xrightarrow{h\nu} {}^3FQ_{(e-,n\pm)} \xrightarrow{FQH_2} FQH\cdot_{(e-,n\pm)} + FQH\cdot$$

$$FQH\cdot_{(e\pm,n\pm)} + FQ_{(n\mp)} \rightleftharpoons FQ_{(n\pm)} + FQH\cdot_{(e\pm,n\mp)} \quad (k_e) \tag{13b}$$

$$FQH\cdot_{(e\pm,n\pm)} + FQH\cdot_{(e\mp,n'\pm)} \rightarrow FQ_{(n\pm)} + FQH_{2(n'\pm)}\cdot \quad (k_2) \tag{13c}$$

Here, e± denotes the electron spin state of the radicals and n± denotes the nuclear spin states of the radicals and singlet molecules, and the quantities in parenthesis following the reactions are their rate constants. It is straightforward to write down the rate equations for the various species in different spin states, including the spin transitions. These inherently nonlinear equations can be linearized by neglecting second-order terms involving differences in spin level populations and δQ. The steady state solution for the polarization in the case of an efficient exchange reaction, and radical relaxation rates which are fast compared to the second-order radical recombination reactions, was shown to be [8c]

$$\delta P = -\sqrt{(k_2/2k_e^2 IQP)}\, IQP(2\delta Q/Q)\,\bar{\xi}^{(3)} \tag{14}$$

where

$$\bar{\xi}^{(3)} = \frac{\frac{1}{2}k_e P(W_2 - W_0)}{(W_n + W_e + 2W_2)(W_n + W_e + 2W_0) - (W_n - W_e)^2} \tag{15}$$

and P=[FQ]. It has been shown that under typical reaction conditions, with the triplet mechanism providing an initial electron polarization of the FQH radicals, this process can give nuclear spin enhancement factors of roughly 10 [8c].

Despite the efficiency of the exchange reactions, this polarization is not especially larger than can be achieved by the Overhauser mechanism with nonexchange type reactions, and the reason, as mentioned previously, is that the reversible exchange reactions also limit polarization by converting an already polarized product back to a radical. Most importantly, however, this loss mechanism also affects radical pair polarization, and the net result can be an advantage for the Overhauser mechanism in cyclic systems involving exchange reactions, albeit at the price of not being able to achieve very large polarizations by either mechanism. A model which demonstrates this is the cyclic photoreduction of a quinone (BQ) by a hydrogen donor (RH)

$$\left.\begin{array}{l} BQ_{(n\pm)} \xrightarrow{h\nu} {}^3BQ_{(n\pm)} \\ {}^3BQ_{(n\pm)} + RH \rightarrow {}^3\overline{BQH\cdot_{(n\pm)} + R\cdot} \\ {}^3\overline{BQH\cdot_{(n\pm)} + R\cdot} \begin{array}{l} \nearrow BQ_{(n\pm)} + RH \\ \searrow BQH\cdot_{(n\pm)} + R\cdot \end{array} \end{array}\right\}, \tag{16a}$$

$$BQH\cdot_{(n\pm)} + BQ_{(n\mp)} \rightleftharpoons BQ_{(n\pm)} + BQH\cdot_{(n\mp)} \qquad (k_e) \tag{16b}$$

$$BQH\cdot_{(n\pm)} + R\cdot \rightarrow BQ_{(n\pm)} + RH \qquad (k_2) \tag{16c}$$

Under the same conditions as applied to Eq.(14), this model was shown to give the steady state polarization [8c]

$$\delta P = -\tfrac{1}{2}\sqrt{(k_2/2k_e^2 IQP)}\; IQP(\delta Q_n/Q), \tag{17}$$

where δQ_n is the difference in quantum efficiencies for production of BQH radicals in the n+ and n- states resulting from the radical pair mechanism. To compare the two mechanisms we note that, as discussed previously, the triplet enhancement factor ($2\delta Q/Q$) can be comparable to or even an order of magnitude larger than the radical pair mechanism enhancement factor ($\delta Q_n/Q$), and the factor $\bar{\xi}^{(3)}$, unlike the factor $\xi^{(1)}$, can be quite large when k_eP is large. In this case comparison of Eqs.(14) and (17) shows that the triplet Overhauser mechanism will yield larger nuclear polarizations than the radical pair mechanism. Obviously, increasing k_eP by increasing P favors the Overhauser mechanism, and the photolysis of benzoquinone in $CDCl_3$ may be a case where the benzoquinone CIDNP changes from a radical pair process to an Overhauser process as the benzoquinone concentration is increased [8c]. Note, however, that a k_e value in excess of what is required to transfer nuclear spin polarization from radical to products before it is destroyed by the W_e and W_n relaxation processes, does not enhance the efficiency of the Overhauser mechanism, because the factor k_e in the expression for $\bar{\xi}^{(3)}$ is cancelled by the factor k_e^2 in the denominator of the square root in Eq.(14). Rather, a large k_e inhibits the radical pair mechanism.

3. PREDICTIONS OF THE THEORY

3.1 Magnetic field dependence

The magnetic field dependence of CIDNP by the triplet Overhauser mechanism results from the combined field dependences of the electron polarization mechanism, and the various cross relaxation processes. In general, the triplet enhancement factor ($2\delta Q/Q$) increases fairly rapidly with increasing fields at low fields (1 to 5 kG), increases less rapidly or levels off at intermediate fields (5 to 10 kG), and falls off at high fields. It behaves this way despite the fact that the maximum electron polarization is proportional to w_{ZFS}/w_Z in the usual case where $w_Z\tau_R>1$, cf. Eq.(1), because the electron polarization actually transferred to the radicals depends on the rate at which the triplet reacts to yield these radicals vs the triplet spin relaxation rate. According to Eq.(3), W_T decreases rapidly with increasing field in the region where $w_Z\tau_R>1$, and offsets the reduction in maximum polarization, until the point where $W_T<k_T$ is reached. The factors representing the efficiency of the transfer of electron polarization to the nuclei are largest at low fields, and decrease as one goes to intermediate and high fields, because of the decline of W_0 and W_2 at higher fields, while W_n and that part of W_e which is due to the g factor anisotropy do not decrease, cf. Eqs. (4) - (6). Thus, the nuclear spin polarization, which involves the product of these factors, has a broad maximum at intermediate fields and falls off at high and low fields. Such behavior, which has been observed in the cyclic photoreduction of 1,4-tetrafluoroquinone by 1,4-tetrafluorohydroquinone [8c], is a possible test for the Overhauser mechanism. Also, comparison of observed and computed field dependences can give estimates of the various parameters involved in the theory, such as the triplet reaction rate.

3.2 Effect of Oxygen

In all known cases where CIDNP is attributed to the Overhauser mechanism it is essential that the reaction mixture be carefully deoxygenated by freeze-pumping or helium bubbling [8], although it should be noted that some CIDNP-producing reactions where the radical pair mechanism is likely to predominate also require O_2 removal [12]. Dissolved oxygen can have a number of effects on the course of a radical reaction. Some of these effects, such as quenching a relatively long-lived triplet before it can react to yield a radical pair, or changing the course of a reaction by forming peroxy radicals, are likely to have comparable effects on CIDNP by both the radical pair and Overhauser mechanisms. In many cases, however, dissolved oxygen, although unable to affect the chemistry of the reaction, will adversely affect both radical pair polarization and Overhauser polarization by O_2-radical spin exchange [13]. This disrupts the coherent singlet-triplet mixing process required for radical pair polarization [14], and, in the case of the Overhauser mechanism, transfers electron polarization from the radicals to O_2 before it can be transferred to the nuclei. However, the Overhauser mechanism is more vulnerable because the electron nuclear cross relaxation step is relatively slow (typical required time: 10^{-4} to 10^{-6} sec), whereas the radical pair process requires only about 10^{-8} sec, this being the time required for complete singlet-triplet mixing in the typical radical pair wave function [14]. A typical O_2 concentration in an undegassed liquid reaction mixture could be around 10^{-4} to 10^{-5} mole/liter. Assuming that all CIDNP proscribing effects of O_2 are spin exchange processes which occur at the typical diffusion-controlled reaction rate of 10^{-11} cm^3 molecule^{-1} sec^{-1}, these O_2 concentrations correspond to half-lives of 10^{-5} to 10^{-6} sec for electron spin coherence or polarization lifetime in any process susceptible to O_2 interference. This is too slow to interfere with a radical pair step, but will remove the electron polarization in a radical before it can be transferred to the nuclei. It is possible, therefore, that controlled oxygenation of CIDNP producing reaction systems can test for contributions from the triplet Overhauser mechanism, and give information about the rates of certain processes involved in the reaction.

ACKNOWLEDGEMENT

I wish to dedicate my contributions to this book, along with that part of these summaries which represents my contributions to the field of Chemically Induced Magnetic Polarization, to the memory of my father, Frank L. Adrian, who died just before work on these manuscripts was completed. Although not a scientist himself, I feel that he could have been a much more accomplished one than I can ever hope to be, given the opportunities which his self-sacrificing devotion to his family provided for me.

REFERENCES

1. J. Bargon and H. Fischer, Z. Natürforsch 22a, 1556 (1967).
2. R.G. Lawler, J. Am. Chem. Soc. 89, 5518 (1967).
3. (a) K.H. Hauser and D. Stehlik, Adv. Magn. Reson. 3, 79 (1968).
   (b) A. Carrington and A.D. McLachlan, Introduction to Magnetic Resonance (Harper & Row, New York, 1967) pp. 229-233.
4. (a) G.L. Closs, J. Am. Chem. Soc. 91, 4552 (1969).
   (b) G.L. Closs and A.D. Trifunac, J. Am. Chem. Soc. 92, 2183 (1970).
5. R. Kaptein and L.J. Oosterhoff, Chem. Phys. Lett. 4, 195 (1969); ibid. p. 214.
6. (a) P.W. Atkins, I.C. Buchanan, R.C. Gurd, K.A. McLauchlan and A.F. Simpson, Chem. Commun., 513 (1970).
   (b) S.K. Wong, D. A. Hutchinson and J.K.S. Wan, J. Am. Chem. Soc. 95, 622 (1973).
   (c) A. J. Dobbs and K.A. McLauchlan, Chem. Phys. Lett. 30, 257 (1975).
   (d) B.B. Adeleke, K.Y. Choo and J.K.S. Wan, J. Chem. Phys. 62, 3822 (1975).
7. (a) S.K. Wong, D. A. Hutchinson and J.K.S. Wan, J. Chem. Phys. 58, 985 (1973).
   (b) P. W. Atkins and G.T. Evans, Mol. Phys. 27, 1633 (1974).
   (c) F.J. Adrian, J. Chem. Phys. 61, 4875 (1975).
8. (a) J. Bargon and K. G. Seifert, Ber. Bunsenges. Phys. Chem. 78, 1180 (1974).
   (b) H.M. Vyas and J.K.S. Wan, Chem. Phys. Lett. 34, 470 (1975).
   (c) F.J. Adrian, H.M. Vyas and J.K.S. Wan, J. Chem. Phys. 65, 1454 (1976).

9. This derivation proceeds as described for Eq.(19) of reference 8c, except that here we omit the step setting $k_T/(k_T+W_T)=[1/(k_T^{-1}+W_T^{-1})]W_T^{-1}$ and using Eq.(3) for W_T^{-1}.
10. F.J. Adrian, Chem. Phys. Lett. 26, 437 (1974).
11. G.L. Closs and L.E. Closs, J. Am. Chem. Soc. 91, 4549 (1969); ibid. p. 4550.
12. A.A. Lamola, M.L. Manion, H.D. Roth and G. Tollin, Proc. Nat. Acad. Sci. USA 72, 3265 (1975).
13. E.M. Purcell and G.B. Field, Astrophys. J. 124, 542 (1956).
14. F.J. Adrian, J. Chem. Phys. 53, 3374 (1970); J. Chem. Phys. 54, 3912 (1971).

CHAPTER XXII

THE EFFECTS OF MAGNETIC FIELDS ON CHEMICAL REACTIONS

P.W. Atkins

Physical Chemistry Laboratory, South Parks Road, Oxford, England.

ABSTRACT. The theories of the ways in which a magnetic field can affect a chemical reaction are outlined. In Section (1) the range of observations and the general approach is outlined. In Section (2) the problem is cut down to its simplest form, and the 1-dimensional reaction is built into the model. Coherent and incoherent processes are distinguished. In Section (3) the 3-dimensional model system is treated and explicit expressions derived for the field effects.

1. INTRODUCTION

A variety of magnetic field effects on chemical reactions have been reported (see [1] for a review) but few of them can be taken to be firmly reliable. The difficulty arises from the reproducibility of the experiments. Nevertheless there are good reasons for expecting magnetic field effects in radical reactions [2-5] and some fully substantiated work in related systems, namely electrochemiluminescence (e.c.l.) [6] and radical recombination in radiolysis experiments [7]. In this Chapter we outline why chemical reactions should respond to magnetic fields, and then outline the approach to the e.c.l. and radiolysis experiments.

The radical pair approach to spin polarization provides a ready way of understanding why reactions should be field sensitive. As has been fully established in other Chapters, when a radical pair separates, its overall spin state evolves because the individual component spins are subject to different perturbations.

L. T. Muus et al. (eds.), Chemically Induced Magnetic Polarization, 383-392.

The probability of a cage product being formed depends on the retention (or acquisition) of overall singlet character. If a magnetic field can adjust the rate of spin evolution, the probability of formation of cage product is modified. Therefore, a magnetic field can be expected to modify the proportions of cage and non-cage products. The calculations involved are exactly the same as in RP-CIDNP, the only difference being that the quantities of interest, the yields, do not make reference to the nuclear spin states of the products: it is not polarization state but overall population that is required.

A formal calculation can be made in a variety of ways. A basic approach would be to solve the stochastic Liouville equation for the spin-position density matrix of the mobile, evolving radical pair. This can be approached analytically if special forms of the distance dependence of the exchange interaction and the recombination probability are assumed, or it can be solved numerically, taking note of the role of the exchange interaction in governing the mobility, and more realistic exchange and reaction terms used. Both approaches are simple extensions of the techniques described elsewhere in this volume. For the purposes of presenting the underlying ideas we adopt the simplest approach possible, and base the calculation on an elementary combination of radical diffusion with spin evolution, and build up a model of the process in stages.

2. DIFFUSION, EVOLUTION, and REACTION.

As a first step consider a <u>linear</u> system. All the mathematical techniques are displayed by this model, and moving on to a realistic three-dimensional model is a matter of algebra rather than of concepts. Furthermore, in this first approach, assume that the singlet and one of the triplet states are degenerate and that some perturbation has matrix elements $\hbar V$ between them. Then the probability of being in a singlet state is equal to $\cos^2 Vt$ if the initial state is singlet, and $\sin^2 Vt$ if the initial state is triplet. We consider only the former; extension to the latter case is trivial. Note that this approach corresponds to the assumption of the <u>coherent</u> evolution of the spins, and therefore corresponds to the type of experiment discussed by Brocklehurst [7]. We shall see that it is a comparatively simple matter to modify the equations to allow for <u>incoherent</u> evolution, where the two radicals are allowed to undergo independent spin relaxation.

The probability of being separated by x at a time t if initially the radical pair is separated by x_0 (corresponding to instantaneous formation with a separation x_0) can be denoted $g(x|t|x_0)$. There is the possibility, however, that the radicals of the pair are scavenged during their lifetime. If the pseudo-first-order scavenging rate coefficient is k_S, the probability g also declines exponentially with time, and we should use

$$g^*(x|t|x_o) = g(x|t|x_o)\exp(-k_s t) \tag{2.1}$$

Let the (conditional) probability of recombination occurring if the radicals are separated by x and the pair is an overall singlet, be $K(x-x^*)$, and denote the time-dependent singlet probability as $p_S(t)$. Then the total cage recombination probability P_{cage}, after all reaction has ceased, is

$$P_{cage} = \int_o^\infty dx \int_o^\infty dt K(x-x^*) g^*(x|t|x_o) p_S(t) \ . \tag{2.2}$$

For simplicity we allow cage reaction to occur only at x^*, and so set $K(x-x^*) = K\delta(x-x^*)$. It follows that

$$P_{cage} = K\int_o^\infty dt g^*(x^*|t|x_o) p_S(t). \tag{2.3}$$

This expression can be simplified by noting that the Laplace transform of a function f(t) can be expressed as

$$\phi(s) = \mathcal{L}(s,t)f(t) \equiv \int_o^\infty dt f(t) e^{-st} \tag{2.4}$$

Then, since $p_S(t) = \cos^2 Vt = \frac{1}{2}(1 + \cos 2Vt)$, the cage recombination probability is

$$\begin{aligned} P_{cage}^{coh} &= \tfrac{1}{2}K\{\mathcal{L}(k_s,t) + \mathrm{re}\,\mathcal{L}(k_s+2iV,t)\}g(x^*|t|x_o) \\ &= \tfrac{1}{2}K\{\gamma(x^*|k_s|x_o) + \mathrm{re}\gamma(x^*|k_s+2iV|x_o) \end{aligned} \tag{2.5}$$

and the problem reduces to finding the Laplace transform γ of g.

In the case of incoherent spin evolution the time-dependence of the singlet probability can be modelled by writing

$$p_S(t) = \tfrac{1}{4} + \tfrac{3}{4}\, e^{-Wt} \tag{2.6}$$

where W = 1/T is some composite spin relaxation rate. (This expression models the relaxation from an initial singlet towards equal probabilities of singlet and each triplet state.) It follows from eqn (2.3) that

$$\begin{aligned} P_{cage}^{incoh} &= \tfrac{1}{4}K\{\mathcal{L}(k_s,t) + 3\mathcal{L}(k_s+W,t)\}g(x^*|t|x_o) \\ &= \tfrac{1}{4}K\{\gamma(x^*|k_s|x_o) + 3\gamma(x^*|k_s+W|x_o)\}. \end{aligned} \tag{2.7}$$

The determination of g is clearly central to the calculation, and we continue with the one-dimensional model (which, as it turns out, underlies the three-dimensional model). g is a solution of the 1-dimensional diffusion equation

$$D\partial^2 g(x|t|x_o)/\partial x^2 = \partial g(x|t|x_o)/\partial t \tag{2.8}$$

subject to the initial condition $g(x|0|x_o) = \delta(x-x_o)$ and to the two boundary conditions that g is finite for indefinitely large x, $g(\infty|t|x_o) < \infty$, and that the particles do not approach more closely than the distance x^* at which they react. The second condition corresponds to the existence of a reflecting wall at x^* which confines all material to the domain $x > x^*$, and can be expressed mathematically by requiring $\{\partial g(x|t|x_o)/\partial x\}_{x=x^*} = 0$.

The solution of eqn (2.8) subject to these constraints is a standard problem, and runs as follows [8]. Multiplication by e^{-st}, integration over $0 \leqslant t \leqslant \infty$ and imposition of the initial condition and use of eqn (2.4) gives

$$D\partial^2\gamma(x|s|x_o)/\partial x^2 = -\delta(x-x_o) + s\gamma(x|s|x_o). \tag{2.9}$$

Since a representation of the δ-function is

$$\delta(x-x_o) = (1/2\pi)\int_{-\infty}^{\infty} dk\ e^{ik(x_o-x)} \tag{2.10}$$

it follows that the solution of the equation for γ is of the form

$$\gamma(x|s|x_o) = Ae^{-px} + A'e^{px} + \frac{1}{2\pi}\int_{-\infty}^{\infty} dk \left(\frac{e^{ik(x_o-x)}}{k^2 + p^2}\right) \tag{2.11}$$

with $p^2 = s/D$. The first boundary condition implies that $A' = 0$. The integral can be evaluated by contour integration around poles at $k = \pm ip$:

$$\int_{-\infty}^{\infty} dk \left(\frac{e^{ik\alpha}}{k^2+p^2}\right) = (\pi/p)\exp(-p|\alpha|)\ . \tag{2.12}$$

The constant A is determined from the second boundary condition. Since $x^* < x_o$ (a physical rather than a mathematical constraint), and since $g'(x^*|t|x_o) = 0$ implies $\gamma'(x^*|s|x_o) = 0$, A is determined from

$$\gamma'(x^*|s|x_o) = -\ pAe^{-px^*} + \tfrac{1}{2}e^{-px_o+px^*} = 0$$

so that

$$A = (1/2p)\exp\{-p(x_o-2x^*)\}.$$

It follows that

$$\gamma(x|s|x_o) = (1/2p)\{e^{-p(x_o-x^*)-p(x-x^*)} + e^{-p|x-x_o|}\}. \quad (2.13)$$

This is virtually the end of the calculation. There is no need to take the inverse transform, but for completeness we note that use of $\mathcal{L}^{-1}(t,s)(e^{-p\alpha}/p) = (1/\pi tD)^{\frac{1}{2}}\exp(-\alpha^2/4Dt)$ generates the normal 1-dimensional solution of the diffusion equation. Here we concentrate on the transform γ, and for convenience set $\gamma^* = 0$ (for this 1-dimensional problem that choice could have been made at the outset, but we need the greater generality for the 3-dimensional problem). Since $\{\gamma(x^*|s|x_o)\}_{x^*=0} = e^{-px_o}/p$ it follows that

$$P^{coh}_{cage} = \tfrac{1}{2}K\left\{\frac{\exp[-x_o\surd(k_s/D)]}{\surd(k_s/D)} + re\,\frac{\exp[-x_o\surd(k_s+2iV)/D]}{\surd[(k_s+2iV)/D]}\right\} \quad (2.14)$$

On writing $\kappa^2 = k_s^2 + 4V^2$ and $\tan\theta = 2V/k_s$, this turns into

$$P^{coh}_{cage} = \tfrac{1}{2}K\left\{\frac{\exp[-x_o\surd(k_s/D)]}{\surd(k_s/D)} + \frac{\exp[-x_o(\cos\frac{1}{2}\theta)\surd(\kappa/D)]\cos[\frac{1}{2}\theta+x_o(\sin\frac{1}{2}\theta)\surd(\kappa/D)]}{\surd(\kappa/D)}\right\} \quad (2.15)$$

When $V = 0$, $\theta = 0$ and $\kappa = k_s$. Then

$$P^{coh}_{cage}(0) = \{K/\surd(k_s/D)\}\exp\{-x_o\surd(k_s/D)\}\,. \quad (2.16)$$

It follows that

$$P^{coh}_{cage}(V)/P^{coh}_{cage}(0) = 1 + \phi(V) \quad (2.17)$$

with

$$\phi(V) = \tfrac{1}{2}(k_s/\kappa)^{\frac{1}{2}}\exp\{x_o\surd(k_s/D)[1-C\surd(\kappa/k_s)]\}\cos\{\tfrac{1}{2}\theta+x_oS\surd(\kappa/D)\}-\tfrac{1}{2} \quad (2.18)$$

with $C = \cos\frac{1}{2}\theta$ and $S = \sin\frac{1}{2}\theta$.

Consider the limit of this equation when the perturbation is so weak that $4V^2 \ll k_s^2$. Then $\kappa \sim k_s(1+2V^2/k_s^2)$, $\theta \sim 2V/k_s$, and

$$\phi(V) \sim -(3V^2/4k_s^2)\{1 + x_o\surd(k_s/D) + \tfrac{1}{3}x_o^2(k_s/D)\}. \quad (2.19)$$

If $x_o \sim 0$ we obtain simply $\phi(V) \sim -3V^2/4k_s^2$, corresponding to a reduction of cage product as the perturbation V is increased.

In the case of incoherent relaxation, we combine eqn (2.13) with eqn (2.7) and obtain

$$P_{cage}^{incoh} = \tfrac{1}{4}K\left\{\frac{e^{-\surd(k_s/D)x_o}}{\surd(k_s/D)} + \frac{3e^{-\surd[(k_s+W)/D]x_o}}{\surd[(k_s+W)/D]}\right\} \qquad (2.20)$$

This may be expressed in the form

$$P_{cage}^{incoh}(W)/P_{cage}^{incoh}(0) = 1 + \psi(W) \qquad (2.21)$$

by writing

$$\psi(W) = \tfrac{3}{4}\left(\frac{k_s}{k_s+W}\right)^{\frac{1}{2}} \exp\{-\left(\frac{k_s}{D}\right)^{\frac{1}{2}}\left[\left(\frac{k_s+W}{k_s}\right)^{\frac{1}{2}} -1\right]x_o\} - \tfrac{3}{4}\,. \qquad (2.22)$$

If the relaxation is slow in the sense $W \ll k_s$, this becomes

$$\psi(W) \sim -3W/8k_s \qquad (2.23)$$

showing how the cage product declines in importance as W increases.

Both V and W may depend on the field. For example, $S-T_o$ mixing is driven by a g-value difference, and so $\hbar V$ may be identified with $\tfrac{1}{2}(g_A-g_B)\mu_B B$. Spin relaxation is also field dependent so long as $\omega^2\tau^2$ is not large. Hence both ψ and ϕ are field dependent, and the proportion of cage product is field dependent at about the level of a few percent (e.g. $V \sim 10^8\ s^{-1}$, $k_s \sim 10^9\ s^{-1}$, $W \sim 10^6\ s^{-1}$).

3. THE 3-DIMENSIONAL PROBLEM

The calculation in the last section was unrealistic in a variety of ways, but it contained all the features necessary for solving more complicated versions. We still confine ourselves to the problem of piecing together a model of the process rather than setting out a complicated stochastic Liouville problem with sources and sinks. The calculation just done will be generalized as follows.

In the first place we allow reaction to occur anywhere on a spherical surface of radius r^*. The recombination rate can then be expressed as $K(\vec{r}) = K\delta(r-r^*)/r^{*2}$. The cage product probability is then

$$P_{cage} = \int_0^\infty dt \int d\vec{r} K(\vec{r}) g^*(\vec{r}|t|\vec{r}_o) p_S(t)$$

$$= K \int_0^\infty dt p_S(t) e^{-k_s t} \int d\Omega g(\vec{r}^*|t|\vec{r}_o). \tag{3.1}$$

From now on we need only the radial component of g because of the integration over the angles Ω; we write $g(r|t|r_o) = \int d\Omega g$, and s_o are on the way to converting the problem into 1-dimensional diffusion along r. The initial condition is radical pair formation with separation r_o and isotropic distribution: $g(r|t|r_o) = \delta(r-r_o)/r^2$ expresses this requirement together with $\int d\vec{r} g(\vec{r}|\overset{o}{t}|\vec{r}_o) = 1$. The boundary conditions are finiteness at great distances and the existence of a spherical reflecting barrier at r^*. The latter is expressed by the requirement that $\{\partial g(r|t|r_o)/\partial r\}_{r^*} = 0$.

The diffusion equation in 3-dimensions reads

$$D\nabla^2 g(r|t|r_o) = \partial g(r|t|r_o)/\partial t \tag{3.2}$$

and integration over the angles reduces it to

$$D(1/r^2)(\partial/\partial r)r^2(\partial/\partial r)g(r|t|r_o) = \partial g(r|t|r_o)/\partial t \tag{3.3}$$

because only the spherically symmetric part of g(r) survives after the integration, and its angular derivatives are zero. Make the substitution $g = h/r$, and the equation reduces to

$$D\partial^2 h(r|t|r_o)/\partial r^2 = \partial h(r|t|r_o)/\partial t \tag{3.4}$$

and the reflecting barrier condition is expressed by $h'(r^*|t|r_o) = h(r^*|t|r_o)/r^*$. Eqn (3.4) is exactly the same as the one-dimensional problem treated before, and so writing its transform as η gives

$$\eta(r|s|r_o) = Ae^{-pr} + (1/pr_o)\exp\{-p|r-r_o|\} \tag{3.5}$$

(the r_o^{-1} factor comes from the modified initial condition). The boundary condition is different because we have $\eta' = \eta/r^*$ at r^* in place of $\gamma'(r^*) = 0$. Nevertheless it is just as simple to obtain, and we find

$$A = \left\{\frac{pr^* - 1}{pr^* + 1}\right\} \left\{\frac{e^{-p(r_o - 2r^*)}}{pr_o}\right\} \tag{3.6}$$

for $r^* < r_o$. Consequently,

$$\gamma(r|s|r_o) = \left\{\frac{pr^*-1}{pr^*+1}\right\}\left\{\frac{e^{-p(r_o+r-2r^*)}}{pr_o r}\right\} + \frac{1}{prr_o}\, e^{-p|r-r_o|} \tag{3.7}$$

and so, for $r^* < r_o$,

$$\gamma(r^*|s|r_o) = \frac{2e^{-p(r_o-r^*)}}{r_o(pr^*+1)} \tag{3.8}$$

with $p^2 = s/D$.

In the case of independent spin relaxation we use eqn (2.7) in the form

$$P_{cage}^{incoh} = \tfrac{1}{4}K\{\gamma(r^*|k_s|r_o) + 3\gamma(r^*|k_s+W|r_o)\} \tag{3.9}$$

and obtain

$$P_{cage}^{incoh} = (K/2r_o)\left\{\frac{e^{-(r_o-r^*)\surd(k_s/D)}}{r^*\surd(k_s/D)+1} + \frac{3e^{-(r_o-r^*)\surd[(k_s+W)/D]}}{r^*\surd[(k_s+W)/D]+1}\right\} \tag{3.10}$$

By writing $\alpha = \surd(k_s/D)$ and $\beta = \surd(1+W/k_s)$ this simplifies to

$$P_{cage}^{incoh}(W)/P_{cage}^{incoh}(0) = 1 + \Phi(W) \tag{3.11}$$

$$\Phi(W) = \tfrac{3}{4}\left\{\left(\frac{\alpha r^*+1}{\alpha\beta r^*+1}\right) e^{-\alpha(\beta-1)(r_o-r^*)} - 1\right\} \tag{3.12}$$

If $r^* = 0$ so that reaction occurs on contact,

$$\Phi_o(W) = \tfrac{3}{4}\{e^{-\alpha(\beta-1)r_o} - 1\} \tag{3.13}$$

This shows that the further apart the radicals are when the point is formed, the closer $\Phi(W)$ is to $-\tfrac{3}{4}$, and the closer the proportion of cage product is to random singlet-triplet encounters. If $W = 0$, $\beta-1 = 0$ and $\Phi = 0$. When the relaxation is slow, in the sense $W \ll k_s$,

$$\Phi(W) \sim \tfrac{3}{4}\{e^{-\frac{1}{2}Wr_o/\surd(k_sD)} - 1\} \sim -\,3Wr_o/8\surd(k_sD) \tag{3.14}$$

Since W is field dependent, at two fields B_1 and B_2 we have

$$\Phi(W_1)/\Phi(W_2) \sim W_1/W_2 = T(B_2)/T(B_1) \tag{3.15}$$

showing how the cage product yield depends on the relaxation times at the two fields. Note that in this limit ($W \ll k_s$ and $Wr_o/2\surd(k_s D) \ll 1$) the ratio of Φs is independent of details of the reaction, but this fails when the latter limit fails.

So much for incoherent relaxation. When we turn to coherent processes, we generalize the former expression for $p_S(t)$ to the case in which there are two levels separated by an energy $\hbar\omega$ and joined by a constant perturbation $\hbar V$:

$$p_S(t) = \frac{\omega^2 + 4V^2\cos^2\{\frac{1}{2}(\omega^2 + 4V^2)^{\frac{1}{2}}t\}}{\omega^2 + 4V^2}$$

$$= [1 - (2V^2/\nu^2)] + [2V^2/\nu^2]\cos \nu t \qquad (3.16)$$

where $\nu^2 = \omega^2 + 4V^2$. This expression can be dealt with immediately because it modifies only the arguments of the Laplace transforms of g:

$$P^{coh}_{cage} = [1 - (2V^2/\nu^2)]K\gamma(r^*|k_s|r_o) + [2V^2/\nu^2]K\mathrm{re}\gamma(r^*|k_s+i\nu|r_o) \qquad (3.17)$$

A great deal of labour can be saved by setting $r^* = 0$ at this stage (this restriction can easily be removed if required). When $r^* = 0$ we have $\gamma(0|s|r_o) = 2e^{-pr_o}/r_o$ and so

$$P^{coh}_{cage} = (2K/r_o)\{[1-(2V^2/\nu^2)]e^{-r_o\surd(k_s/D)} + (2V^2/\nu^2)\mathrm{re} \times e^{-(r_o/\surd D)\surd(k_s+i\nu)}\}$$

$$= (2K/r_o)\{[1-2V^2/\nu^2]e^{-r_o\surd(k_s/D)} + (2V^2/\nu^2)e^{-r_oC\surd(\kappa/D)} \times \cos[r_oS\surd(\kappa/D)]\} \qquad (3.18)$$

where $\kappa^2 = k_s^2 + \nu^2$, $\tan\theta = \nu/k_s$, $C = \cos\frac{1}{2}\theta$, and $S = \sin\frac{1}{2}\theta$.

The last expression is difficult to analyse because both V and ν (through ω in the case of S-$T_\pm$ mixing) can be field dependent. First suppose that there is S-$T_\pm$ mixing under the influence of a field independent perturbation (e.g., hyperfine interactions) and that in the absence of a field S, T_o, $T_\pm$ are generate. Then the amount of mixing <u>decreases</u> as the field increases because the perturbation is unable to populate the $T_\pm$ states so effectively when they are no longer degenerate with the S state. Therefore P_{cage} <u>increases</u> with the strength of the applied field. When the field is so large that the $T_\pm$ states are well away from the S level only S-T_o mixing remains. This may be driven by the $g_A - g_B$ effect already described, and the mixing increases with applied field. Therefore the proportion of cage product is expected to drop again when this region of field strengths is attained. Both effects (including the possibility of having a constant exchange interaction splitting S and T_o) are contained in eqn (3.18).

To go any further needs a more sophisticated analysis, and reliance on a numerical analysis of the effects of exchange interactions, sinks, sources, and relaxation. The kind of analysis needed in the case of coherent effects of several hyperfine interactions has been described in detail by Brocklehurst [7]. The adaptation to the effect of magnetic fields on colliding excited triplets has been treated in detail elsewhere [6], and application of techniques similar to those described here, with numerical examples, have been given by Buchachenko [1] and Sagdeev *et al*.[5]. But what really is required before the theories are developed any further, is much more experimental work.

REFERENCES

1. P.W. Atkins and T.P. Lambert, *Ann. Repts.*, *A*, 67 (1975).
2. A.L. Buchachenko, *Chemical polarization of electrons and nuclei*, Nauka, Moscow 1974.
3. R. Kaptein, Thesis, Leyden 1971.
4. R. Lawler and G.T. Evans, *Ind. chem. Belges*, *36*, 1087, (1971).
5. R. Sagdeev, Yu. N. Molin, K.M. Salikhov, T.V. Leshina, M.A. Kamha, and S.M. Shein, *Org. Mag. Resonance*, *5*, 603, (1973).
6. P.W. Atkins and G.T. Evans, *Molec. Phys.*, *29*, 921 (1975).
7. B. Brocklehurst, *Chem. Phys. Lett.*, *29*, 635 (1974).
8. J. Crank, *The mathematics of diffusion*, Clarendon Press, Oxford, 1975.

CHAPTER XXIII

CIDNP FROM TRIPLET PAIR RECOMBINATION

Joachim Bargon

IBM Research Laboratory
San Jose, California 95193

ABSTRACT: The phase of CIDNP resonances and the spin selectivity which governs the recombination of radical ion pairs is shown to depend upon the relation between the change of free enthalpy (ΔG) associated with the electron transfer from the radical anion to the radical cation and the triplet energies (E_T) of the products. If triplet recombination is energetically allowed ($\Delta G > E_T$), it is typically more efficient as a consequence of the energy gap law.

1. INTRODUCTION

The radical pair theory of CIDNP (1,2) is based upon the assumption that the chemical fate of radical pairs is electron spin dependent. Thus the probability of product formation from a radical pair via either recombination or disproportionation is typically set to be proportional to the *singlet* character of the total wave function describing the two unpaired electrons. In the reactions studied so far with the CIDNP technique, triplet pairs did not lead to product formation.

In reactions of radical ions it has been frequently observed that the annihilation between a cation radical ($D^{\cdot +}$) and an anion radical ($A^{\cdot -}$) is accompanied by the emission of light. The latter requires that the energy liberated in the step:

$$A^{\cdot -} + D^{\cdot +} \longrightarrow A^{*} + D \quad \text{or} \quad D^{*} + A \qquad (1)$$

is greater than is necessary to produce electronically excited states (A^* or D^*) of A, D, or molecular complexes thereof. The excited states formed can be singlet or triplet, depending upon

L. T. Muus et al. (eds.), Chemically Induced Magnetic Polarization, 393-398.

the energetics of reaction 1, and upon the energies of the excited singlet and triplet states of A and of D, respectively. Weller et al. (3) have shown that it is the change of enthalpy (ΔG) which is the useful parameter to characterize the energetics of reaction 1. ΔG can be estimated from electrochemical data of the acceptor molecule (A) and of the donor molecule (D). Thus for example ΔG follows from the electrochemical cyclic voltammetric peak potentials (4) for the $A/A^{\cdot -}$ and $D^{\cdot +}/D$ couples via the relation:

$$|\Delta G| \text{ (in eV)} = E_p(D^{\cdot +}/D) - E_p(A/A^{\cdot -}) - \text{const.} \qquad (2)$$

The constant depends upon the solvent, and for acetonitrile it has been approximated (3) to be:

$$\text{const.} = 0.06 \text{ eV} \qquad (3)$$

Recently the fast occurrence of triplets ^3A or ^3D during the photoinduced electron transfer between various donors and acceptors in polar solvents has been demonstrated to be the consequence of the geminate recombination of *triplet* radical ion pairs (5,6). We have been interested in the consequences of allowed triplet pair recombination for the CIDNP resulting from the radical ion pair annihilation. Therefore, we have investigated various photoinduced electron transfer reaction in polar protic and aprotic solvents with the CIDNP technique, thereby adjusting the energetics of reaction 1 relative to the triplet energies of the acceptors ($E_T(A)$) using appropriate substituents.

2. GENERATION OF RADICAL ION PAIRS

2.1 Excitation of the Charge Transfer Band of EDA Complexes

We have previously observed (7) that tertiary aliphatic amines and acetonitrile in their ground states form weak electron donor acceptor (EDA) complexes, which give rise to a weak charge transfer (CT) band centered around 3000 A ($\varepsilon \sim 10$). Here acetonitrile acts as both the acceptor and the polar solvent. Upon excitation of the CT band of a 10^{-3} molar solution of triethylamine (TEA) in CD_3CN inside a modified all-quartz insert of a 100 MHz NMR spectrometer with the filtered UV light 2900 A $< \lambda <$ 3100 A of a 1 kW high pressure mercury xenon light source, we observed the methylene protons of TEA in emission. The methyl protons of TEA displayed no CIDNP. Identical results were obtained with benzonitrile as the acceptor and as the solvent. From the CIDNP pattern and from the known (8) EPR parameters of the TEA radical cation ($a_{CH_3} = 0$, $a_{CH_2} = +2{,}6$ mT, $g = 2.0041$) we conclude that excitation of the CT band promotes an electron from the highest occupied molecular orbital (HOMO) of TEA to the lowest unoccupied molecular orbital

(LUMO) of the acceptor *without* a change of the multiplicity. Accordingly, in the polar solvents used here, this electron transfer reaction yields radical ion pairs with electronic singlet character. For A = benzonitrile with $E_T(A) = 3.33$ eV (9) and $\Delta G = 3.2$ eV (10), the energetics of reaction 1 excludes the formation of triplet products ($\Delta G < E_T(A), E_T(D)$). Assuming that $g(A^{\cdot -}) < g(TEA^{\cdot +})$ and applying the Kaptein Rule (11), the emission of the methylene protons of TEA is consistent with these assumptions. For A = CD_3CN as well singlet pairs are being generated and only singlet pairs may recombine.

Since acetonitrile is a poor acceptor (12), addition of a stronger acceptor is expected to lead to an electron transfer:

$$(CD_3CN)_n^{\cdot -} + A \longrightarrow n \cdot CD_3CN + A^{\cdot -} \qquad (4)$$

Table 1 lists the CIDNP phase observed in the methylene protons of TEA during photolysis in the presence of various substituted benzenes as acceptors in CD_3CN as the solvent together with data for the energetics of reaction 1 and the triplet energies of A:

TABLE 1: CIDNP DURING PHOTOLYSIS OF $(CH_3\overset{*}{C}H_2)_3N + X\text{-}C_6H_4\text{-}Y$ in CD_3CN

Substituents		CIDNP	Multiplicity	$E_T(A)$ in eV		ΔG in eV
H-	-H	E	Singlet	3.66	>	
H-	-CN	E	"	3.33	>	3.25
NC-	-CN	E	"	3.16	>	2.46
CH_3O-	-CN	A	Triplet	3.26	<	3.46
$H_2C{=}CH$-	-H	A	"	2.68	<	3.35

From the data in Table 1 we conclude that an electron transfer according to reaction 4 does occur. Furthermore the inversion of the CIDNP phase at the point where the energetics of reaction 1 becomes favorable for formation of ^3A indicates that for:

$$\left.\begin{array}{l}\Delta G < E_T(A)\\ \Delta G > E_T(A)\end{array}\right\} \text{ the CIDNP in TEA is governed by } \left\{\begin{array}{l}\text{singlet pairs}\\ \text{triplet pairs}\end{array}\right. \qquad (5)$$

2.2 Excitation of the Acceptor

In order to probe whether relation 5 holds also if the acceptor molecule is directly excited, we investigated the system TEA plus substituted naphthalenes in CD_3CN or dimethylformamide as solvents. The system A=naphthalene and D=TEA has originally been investigated by Taylor. His unpublished results have been accidentally misquoted (13). In agreement with Roth (14) we observed the methylene protons of TEA in enhanced absorption. For this system $\Delta G > E_T(A)$ holds (9,10). The system A = 1-cyanonaphthalene, however, is characterized by $\Delta G < E_T(A)$, and consistently here the methylene protons of TEA display *emission*.

3. DISCUSSION

To explain the observed change of the CIDNP phase depending upon the correlation between ΔG and $E_T(A)$, we use the energy diagram shown in Figure 1. We assume that $E_T(D) > E_T(A)$, and that D absorbs at shorter wavelengths than A.

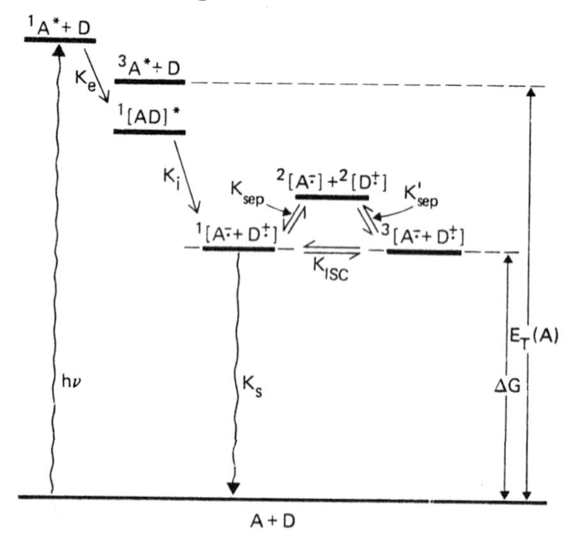

FIGURE 1: Energy diagram for a radical ion pair with $\Delta G < E_T(A)$

Figure 1 illustrates the case $\Delta G < E_T(A)$. Here only singlet pairs can undergo geminate recombination (k_s). Accordingly, in this case the spin selectivity is identical to that encountered in reactions of most neutral free radicals, and hence the Kaptein Rule may be applied in the conventional way.

Figure 2 represents the case $\Delta G > E_T(A)$. Now triplet pairs may combine as well (k_T) to yield $^3A + D$ in addition to the product A + D, which results from singlet pair recombination (k_s). Furthermore the radical ions may separate (k_{sep}) to become individually solvated.

For $k_s = k_T$ and $k_{sep} = k_{sep}'$, no CIDNP would be expected. Therefore, the observation of CIDNP even for $k_T \neq 0$ requires that $k_T > k_s$. Furthermore, the nuclear spins in the individually solvated radical ions must become randomized by efficient relaxation processes, since they carry the complementary CIDNP to that of D.

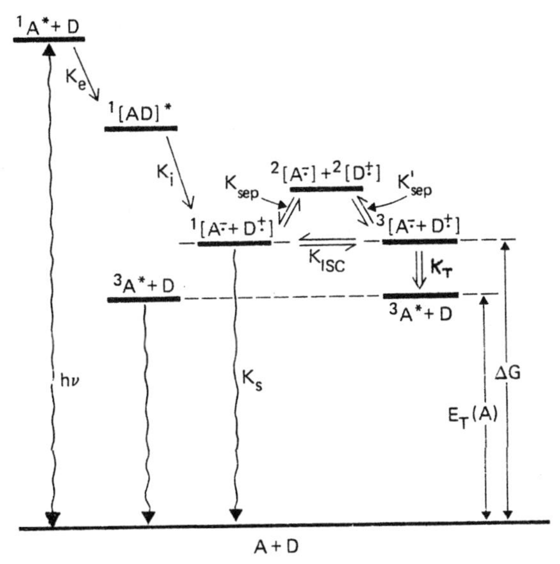

FIGURE 2; Energy diagram for a radical ion pair with $G > E_T(A)$

According to our preliminary findings, the CIDNP enhancement reaches a maximum for $\Delta G = E_T(A)$. We explain these observations as follows: The CIDNP phase and intensities depend upon the relative magnitude of k_s and of k_T. The ratio of these rate constants can be approximated using the *energy gap law* of spectroscopy (15). The energy gap law states: The smaller the gap, the bigger the rate.

Whether the energy gap law can be universally applied has yet to be investigated. Further details of the systems studied here and of investigations carried out in protic solvents will be published elsewhere. Here we have shown that the energetics of the radical ion annihilation has an important consequence for the phase and the intensities of CIDNP: If triplet pair recombination is energetically feasible, it is also more efficient ($k_T > k_s$). Whether a similar relation holds for pairs of neutral radicals, for example two acetyl radicals, for which triplet recombination is also energetically feasible (16), is not yet known. Surprisingly biacetyl, the recombination product of two acetyl radicals, shows no multiplet effect in the ^{13}C-CIDNP spectrum (17).

Introducing substituents into radical ions, as has been done in this study, also influences the g-value of the radical ions. This is especially the case for the donor substituents OH or $-OCH_3$, and for the acceptor substituents $-CO-O-CH_3$, which both increase the g-values of radical ions substantially. This may change Δg, a parameter governed by the Kaptein Rule, and hence the phase of CIDNP resonances, as will be demonstrated elsewhere (18).

REFERENCES

1. a) G. L. Closs, J. Am. Chem. Soc. 91, 4552, (1969)
   b) G. L. Closs and A. D. Trifunac, J. Am. Chem. Soc. 92,2183, (1970)
2. R. Kaptein and L. J. Oosterhoff, Chem. Phys. Lett. 4, 195, 214, (1969)
3. a) A. Weller and K. Zachariasse, J. Chem. Phys. 46, 4984, (1967)
   b) D. Rehm and A. Weller, Z. Phys. Chem. NF 69, 183, (1970)
   c) D. Rehm and A. Weller, Ber. Bunsenges. Phys. Chem. 73, 834, (1969)
4. L. R. Faulkner, H. Tachikawa, and A. J. Bard, J. Am. Chem. Soc. 94, 691, (1972)
5. K. Schulten, H. Staerk, A. Weller, H.-J. Werner, and B. Nickel, Zeitschr. Physk. Chem. NF 101, 371, (1976)
6. M. E. Michel-Beyerle, R. Haberkorn, W. Bube, E. Steffens, H. Schröder, H.J. Neusser, E. W. Schlag, and H. Seidlitz, Chem. Phys. 17, 139, (1976)
7. J. Bargon in "Magnetic Resonance and Related Phenomena", Proceedings of the XIXth Congress Ampere, Heidelberg 1976, p. 145
8. H. D. Roth, Molec. Photochem. 5, 91, (1973)
9. a) E. T. Harrigan, T. C. Wong, and N. Hirota, Chem. Phys. Lett. 14, 579, (1972)
   b) J. B. Birks, "Photophysics of Aromatic Molecules", Wiley Interscience, New York (1970)
   c) D. Haarer, unpublished results
10. a) A. Mann, and K. K. Barnes, "Electrochemical Reactions in Nonaqueous Systems", M. Dekker, Inc., New York (1970)
   b) K. A. Zachariasse, Thesis, Amsterdam, (1972)
11. R. Kaptein, Chem. Comm. 732, (1971)
12. K. D. Jordan, and J. J. Wendoloski, Chem. Phys. 21, 145, (1977)
13. G. N. Taylor, quoted in reference 8
14. H. D. Roth, private communication
15. J. A. Barltrop and J. D. Coyle, "Excited States in Organic Chemistry", J. Wiley and Sons, New York (1975), p. 92
16. J. G. Calvert and J. N. Pitts, "Photochemistry", J. Wiley and Sons, New York (1966)
17. H. Benn and H. Dreeskamp, Zeitschr. Physk. Chem. NF 101, 11, (1976)
18. J. Bargon, to be published

CHAPTER XXIV

LIGHT-INDUCED MAGNETIC POLARIZATION IN PHOTOSYNTHESIS

A.J. Hoff and H. Rademaker

Department of Biophysics, Huygens Laboratory,
State University of Leiden, Leiden, The Netherlands

In the primary event of photosynthesis, a light quant impinging on a light-collecting array of pigment molecules in membranes of photosynthetic material (higher plants, some bacteria) is transported by resonant energy transfer to the so-called "reaction-center", where it gives rise to charge separation: $DA \xrightarrow{h\nu} D^*A \rightarrow D^+A^-$, where D is the primary electron donor and A the first electron acceptor. Normally, the donor is subsequently reduced by a secondary donor, and the acceptor loses its electron to secondary acceptors, and so on. After numerous cycles an electron is transported along the photosynthetic electron transport chain from a reductant with low (less negative) redox potential to $NADP^+$ which is reduced to NADPH, an essential catalyst in metabolic reactions. In plants, two photosystems, PS1 and PS2 act in tandem, in photosynthetic bacteria only one photosystem is present. The details of these processes will not concern us here (see for recent reviews refs. 1 and 2); we will focus our attention on the radical pair D^+A^-.

Depending on the system and the temperature, the pair has a lifetime ranging from 10 ns to several ms. This time is sufficiently long to permit, in principle, the induction of magnetic polarization. Very recently, electron polarization (CIDEP) and effects of a magnetic field on the yield of reactants have indeed been observed. In this section we will briefly review these observations. First we will report on CIDEP in plant material, then on similar observations in reaction centers of photosynthetic bacteria, and finally on the magnetic field dependence of the triplet yield, also in photosynthetic bacteria.

1. CIDEP IN PLANT PHOTOSYNTHESIS

In 1975 Blankenship, McGuire and Sauer (3) observed a light-induced emissive EPR signal in chloroplasts (organelles in the

L. T. Muus et al. (eds.), Chemically Induced Magnetic Polarization, 399-404.

plant cell which carry the photosynthetic apparatus). They noted that after a 1 μs light flash a strong emissive EPR spike was visible. A plot of the amplitude of the spike as a function of the magnetic field showed a somewhat asymmetric, emissive ESR line, with a g-value slightly above that of P^+700. At that time the authors concluded that the CIDEP was caused by the triplet mechanism and displayed the polarized spectrum of the primary acceptor of PS1. Later work revealed that the form of the emissive spectrum was strongly dependent on orientation induced by flow of chloroplasts through the flat EPR cell, the emissive line being the mirror image of P^+700 in the absence of flow, and having various peaks for high flow rates (4). The orientation effects could be accounted for by a model (5) which employs the radical pair mechanism in the form of the multiple reencounter theory of Pedersen and Freed (6). The P^+700 is assumed to posses negligible g- and hf-anisotropy, the acceptor A is considered to consist of several sites (7) with highly anisotropic g-factors and different exchange interactions with P^+. The electron is thought to hop from one site to another. The authors now conclude that the emissive spectrum is a polarized P^+700 signal and its precursor is an excited P700 singlet state.

McIntosh and Bolton (8) observed a similar signal in the alga Scenedesmus obliquus (fig. 1). Using fully deuterated algae they found that the polarized EPR spectrum consisted of two well-separated lines, with g-values of 2.0042 and 2.0025 (± 0.0003), which were both in emission. They attributed the first signal to the primary acceptor of photosystem 2 (probably a plastoquinone) and the latter signal to P^+680, the primary donor of photosystem 2. If this assignment is correct, both radicals are in emission and it follows that the triplet state of the primary donor is the precursor of charge separation. This would require extremely fast intersystem crossing (in less than 6 ps). It would be desirable to obtain independent experimental support for this unusual process.

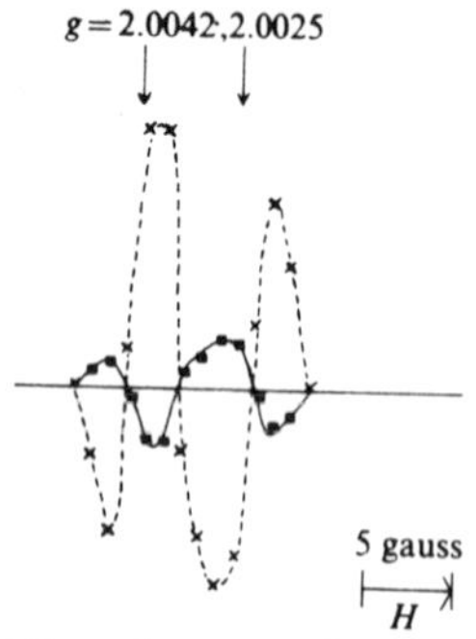

Fig. 1 Transient emissive (x) EPR signals in algae.

2. CIDEP IN BACTERIAL PHOTOSYNTHESIS

Recently we found in our laboratory that at temperatures between 4.1 °K and 150 °K reaction centers from Rhodopseudomonas sphaeroides wild type which were treated with the detergent SDS (sodium dodecyl sulphate) show strong enhanced absorption and emission 25-400 μs after a saturating xenon - or dye laser flash (fig. 2) (9). These reaction centers are full photoactive and show identical behaviour in all respects compared to chromatophores except for the back reaction $P^+X^- \rightarrow PX$ at high temperatures (above 200 °K) which is a factor of two slower. Flash and passage artifacts were excluded. The polarization phenomena were not observed in non-SDS treated reaction centers of the Rps. sphaeroides mutant R-26, nor in whole cells or chromatophores.

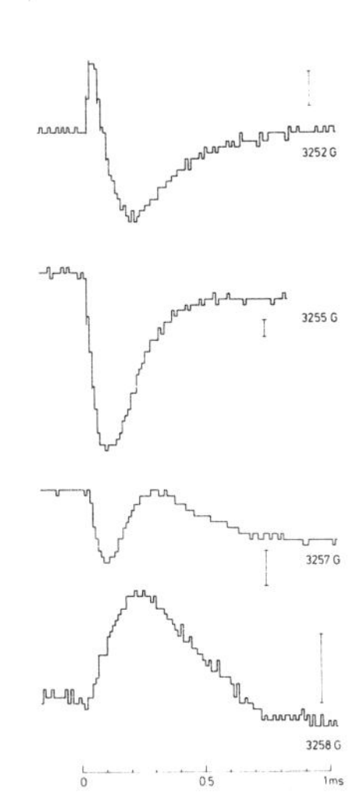

Fig.2 Transient EPR signals in bacterial RC.

The bacterial primary reaction can be represented by $PIX \xrightarrow[<6ps]{h\nu} P^+I^-X \xrightarrow[100-200\ ps]{} P^+I\ X^-$, where P^+ is the oxidized primary donor (a bacteriochlorophyll (Bchl) dimer on which the electron spin is fully delocalized), I^- is monomeric bacteriopheophytin (Bph), and X is, in normal reaction centers, an iron-ubiquinone (Fe-UQ) complex. In the reduced state X^-, the electron spin sits almost solely on the UQ, where it experiences an exchange interaction with the $S = 2$ Fe^{2+} atom. As a consequence X^- has a strongly anisotropic g-factor (principal values 2.05, 1.83, 1.68) and it can only be observed by EPR at very low temperatures (below about 15 ^oK). In our SDS-treated reaction centers, however, the electron on UQ^- is not magnetically coupled to an Fe atom; its EPR spectrum has isotropic $g = 2.0046$. Apparently, the observation of CIDEP in bacterial reaction centers is linked to the absence of exchange coupling to the Fe^{2+} ($S = 2$) atom. This is not unreasonable, since electron spin polarization, unlike CIDNP, depends according to the radical pair mechanism (RPM) on aninterplay of exchange interaction between members of the RP, the time span of this interaction and the lifetime of the RP. From kinetic curves, such as displayed in fig. 2, time resolved spectra were constructed which could fairly well be reproduced by assuming the RPM operative and taking the usual RP spin hamiltonian , but not by assuming the triplet mechanism of electron polarization to be responsible for polarization. In fact, it can be directly seen from the kinetic curves of fig. 2 that one does not observe two radicals which are both emissively polarized.

Since CIDEP depends on the value of the exchange parameter J and J in turn depends strongly on distance and geometry of the reactants, the CIDEP phenomenon can serve as a tool to probe the detailed structure of the reaction center. Future work will certainly focus on this aspect.

3. MAGNETIC FIELD EFFECTS ON THE YIELD OF THE TRIPLET STATE IN REACTION CENTERS OF PHOTOSYNTHETIC BACTERIA

When the acceptor X in the primary reaction in the bacterial system is reduced chemically by dithionite prior to illumination, not the familar P^+ EPR spectrum is observed but (at low temperatures) a triplet spectrum emerges, which is strongly spin polarized (cf. Closs, this volume). By fast optical spectroscopy (10) it has been determined that at room temperature the following reactions take place: $PIX^- \rightarrow P^*IX^- \xrightarrow[<6ps]{} P^+I^-X^- \rightarrow P^TIX^- \xrightarrow[6\mu s]{} PIX^-$. The system $(P^+I^-)X^-$ consists of at least one radical pair, P^+I^- which decays into the triplet state P^T (15 % at room temperature) and into the singlet state P (or $P^* \rightarrow P$). In a magnetic field the yield of P^T

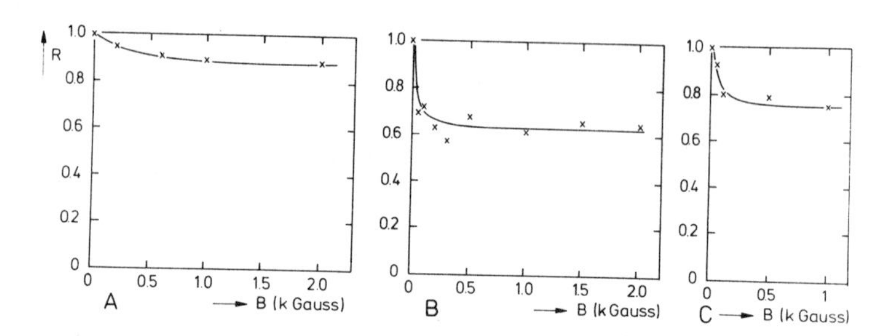

Fig.3 Magneto-dependence of triplet yield
A. Chromatophores B. RC wild type C. RC R-26

should depend on the strength of the field, being lower at high field for a singlet state P* precursor (13). Such an effect has indeed been observed, first by Blankenship et al. (11) and later also by Hoff et al. (12). Originally the effect was observed in chromatophores of R. sphaeroides. At high field it amounted to a lowering of the yield by 20-30%, the field at which half the effect was attained, $B_{\frac{1}{2}}$, being about 500 Gauss (50 mT) (11). Later work in our laboratory on chromatophores and reaction centers with, and without intact Fe-UQ acceptor complex (12), showed that the amplitude of the effect was rather variable, ranging from 10% (chromatophores) to 35% (SDS-treated reaction centers) (fig. 3). The $B_{\frac{1}{2}}$ values also showed large variations on sample type, being about 500 G for chromatophores and 50-100 G for reaction centers, both with and without intact Fe-UQ. Using the kinetic scheme of fig. 4, a simple theory was given (12) based on the radical pair formalism and assuming that the pair P^+I^- does not undergo diffusion, but has a constant lifetime τ (i.e. the original Kaptein theory). The result for the ratio of the yield in high field to that without field, R, then is:

$$R = \phi_T^H/\phi_T^o = \frac{P_T}{1-P_T}\left[3 - \phi_T^o\left(3 - \frac{P_T}{1-P_T}\right)\right]^{-1} \qquad (1)$$

where $P_T = 2\,a_n^2\tau^2(1 + 4\omega^2\tau^2)^{-1}$, and a_n and ω have their usual meaning (see Kaptein, this volume). a_n was calculated assuming $\Delta g = 0.0009$ and summing the largest hyperfine interactions over 4096 configurations. The result is $a_n = 4.5 \times 10^3\ H + 0.6 \times 10^8$ rad s^{-1}, ie. the hyperfine term is dominant for the fields considered. The simple theory cannot account for the magnitude of the effect in high field (it predicts R = 0.2 for $\phi^o = 0.15$, whereas the measured values range between R=0.85-0.65), nor does it give a satis-

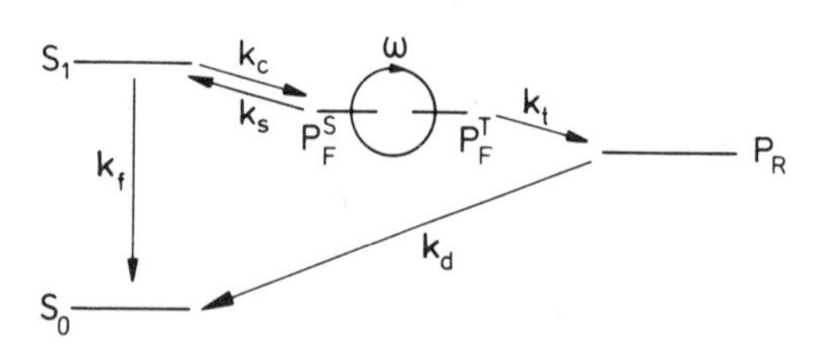

Fig.4 Kinetic scheme of primary photoreactions for the state PIX^-.

factory explanation of the differences in the value of $B\frac{1}{2}$. In fact, one would have expected for $J \sim 0$, that $B\frac{1}{2}$ would roughly correspond to the hyperfine fields (some 10 G).

A more sophisticated calculation can be made by actually solving the differential equations describing the kinetic scheme of fig. 4. A similar analysis has independently been given by Blankenship, Schaafsma and Parson (to be published). From the scheme of fig. 4 it can be easily deduced that the triplet yield $Y_T = \phi^T(1-p_c\phi_S)^{-1}$ in which p_c is the probability for charge separation $p_c = k_c(k_c + k_f)^{-1}$. ϕ_T is the probability for triplet formation for a RC in the state P_F and $\phi_S = 1-\phi_T$ is the probability for singlet formation. The lifetime is increased by the same factor $\tau \sim \tau_i(1-p_c\phi_S)^{-1}$. τ_i is the lifetime of P_F for one turnover. $\tau_i \sim k_S^{-1}$ if $k_S >> \omega$. To calculate the triplet yield Y_T and the lifetime τ we only have to calculate ϕ_T and τ_i from the following scheme

$$\xleftarrow{k_S} \; P_F^S \overset{\omega}{\rightleftharpoons} P_F^T \; \xrightarrow{k_T}$$

The decay kinetics of P_F can be described by

$$\frac{dF(t)}{dt} = -\left[k_S f_S(t) + k_T f_T(t)\right] F(t) \tag{2}$$

in which $f_S(t)$ is the fraction of singlet and $f_T(t)$ the fraction of triplet in P_F. For one reaction center $f_T = \sin^2\omega t = \frac{1}{2}(1-\cos 2\omega t)$, $f_S = (1-f_T)$, and for an ensemble of RC's with a normal distribution of ω's, centered around $\omega = 0$, with variance σ^2 (or $\omega_{rms} = \sigma$): $f_T = \frac{1}{2}(1-\exp(-2\sigma^2t^2))$. Assuming $F(0) = 1$, integration of eq. 2 yields: $F(t) = \exp{-g(t)}$ with

$$g(t) = \frac{k_S + k_T}{2} t + \frac{k_S - k_T}{2} \frac{\sqrt{\pi}}{2} \frac{\operatorname{erf}(t\sigma\sqrt{2})}{\sigma\sqrt{2}}$$

The triplet yield ϕ_T is

$$\phi_T = \int_0^\infty k_T f_T(t)\, F(t)dt = \frac{k_T}{2}\int_0^\infty (1-\exp{-2\sigma^2t^2})\, \exp{-g(t)}dt \tag{3}$$

and the lifetime $\tau_i \sim \int_0^\infty F(t)\,dt = \int_0^\infty \exp -\; g(t)\,dt$ (4)

In Table I some results are given of numerical calculations of τ_i and ϕ_T from eq. 3 and 4 and the resulting values of τ and Y_T. The parameters are chosen, such that $\tau \sim 10$ ns and $Y_T \sim 0.10$.

Table I

k_S	k_T	σ	ϕ_T	τ_i	p_c	Y_T	τ
2	7	0.06	0.006	0.50	0.95	0.10	9.1
3	14	0.06	0.004	0.33	0.97	0.10	10.2
5	50	0.06	0.003	0.20	0.98	0.12	9.2
10	160	0.06	0.001	0.10	0.99	0.10	9.9

τ_i and τ in ns; k_S, k_T, σ in ns^{-1}.

In fig. 5 the kinetics of P_F for one turnover is plotted for a single reaction center (a) and for an ensemble (b). If the back reaction $P_F^S \xrightarrow{k_S} P$ is directly to the ground state, these are the exact kinetics of P_F. If the back reaction to the ground state is <u>via</u>

P^* the kinetics can be calculated from $P_F(t) = k_c \int_0^\infty P^*(t)F(t-\tau)dt$
P^* is the solution of the equation:

$$\frac{dP^*}{dt} = -(k_f + k_c)P^* + k_S k_c \int_0^t P^*(\tau) f_S(t-\tau)F(t-\tau)d\tau$$

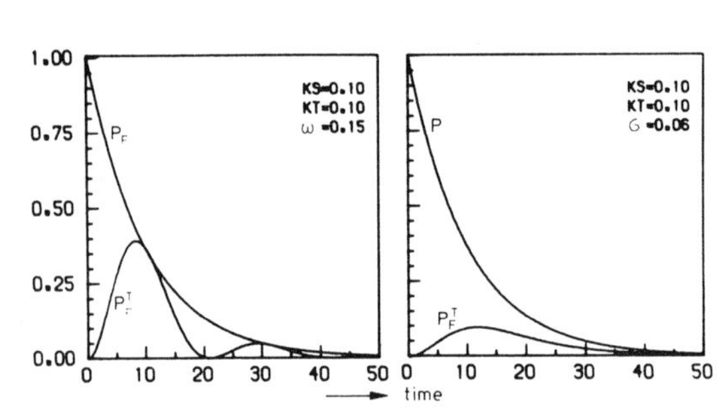

Fig.5 Calculation of decay of P_F and generation of P_F^T. Time in ns.

To deduce values of R from the above theory, one needs to calculate the kinetics of P_F^T formation for B = 0. Assuming that the hyperfine fields lift the degeneracy of the singlet and triplet levels and that they are the prime cause of S-T mixing, the function F(t) remains the same, and the ratio R becomes R =(1/2)/3/4) = 2/3, if all three triplet levels are equally available for mixing. This value agrees better with our experimental results than the one from eq. 1. Work is in progress to calculate S-T mxing in low fields for given hyperfine constants A_i and exchange parameter J, taking all three triplet levels into account.

ACKNOWLEDGEMENT

One of us (A.J.H.)thanks staff,lecturers and participants of the CIMP course for many enjoyable discussions.

REFERENCES

1. Feher, G. and Okamura, M.Y. (1977) in Bacterial Photosynthesis (Clayton, R.K. and Sistrom, W.R. eds.) Plenum Press, N.Y. in the press.
2. Amesz, J. (1977) Progress in Botany 39, in the press.
3. Blankenship, R., McGuire, A. and Sauer, K. (1975) Proc. Natl. Acad. Sci. U.S. 72, 4943.
4. Sauer, K.H., Blankenship, R.H., Dismuke, G.C. and McGuire, A. (1977) Biophys. J. 17, abstr. F-AM-F2.
5. Dismuke, C., Friesner, R. and Sauer, K. (1977) Biophys. J. 17, abstr. F-AM-F3.
6. Pedersen, J.B. and Freed, J. (1974) J. Chem. Phys. 61, 1517.
7. Evans, M.C.W., Shira, C.K., Bolton, J.R. and Cammack, R. (1975) Nature 256, 668.
8. McIntosh, A.R. and Bolton, J.R. (1976) Nature 263, 443.
9. Hoff, A.J., Gast, P. and Romijn, J.C. (1977) FEBS Lett. 73, 185.
10. Parson, W.W. and Monger, T.G. (1976) Brookhaven Symp. Biol. 28, in the press.
11. Blankenship, R.E., Schaafsma, T.J. and Parson, W.W. (1977) Biochim. Biophys. Acta, in the press.
12. Hoff, A.J., Rademaker, H., van Grondelle, R. and Duysens, L.N.M. (1977) Biochim. Biophys. Acta 460, 547.
13. Schulten, K., Staerk, H., Weller, A., Werner, H.-J., Nickel, B. (1976) Z. Phys. Chem. NF101, 371-390.

SUBJECT INDEX